Viscosity-Modifying

Chemical Admixtures for Concrete

From Formulation to Field: Applications, Troubleshooting & Case Studies

A Comprehensive Resource for Formulators, Ready-Mix Concrete Producers, and Consultants

About the Book

This book provides a comprehensive, industry-oriented treatment of viscosity-modifying admixtures for concrete, spanning polymer chemistry, rheological mechanisms, formulation practice, digitalization, and field deployment. It is written to bridge the persistent gap between fundamental research and reliable industrial application, with particular emphasis on variability management, robustness, and technology transfer under real production constraints.

The content integrates polymer science with cement chemistry and fresh concrete rheology, explaining how molecular architecture, adsorption behavior, and water structuring translate into measurable performance in modern concrete systems. Special attention is given to complex and emerging binder technologies, including low-clinker cements, high-SCM formulations, recycled aggregates, limestone-rich systems, and carbon-mineralized binders, where viscosity control is both critical and challenging. Throughout the book, VMAs are treated not as isolated additives but as system-level enablers interacting with dispersants, fillers, environmental conditions, and processing history.

A distinguishing feature of the book is its strong focus on industrial execution and Industry 4.0 integration. Digital twins, sensor-based rheology monitoring, predictive analytics, and standardized digital workflows are presented as practical tools for transforming VMAs from static formulation components into adaptive rheology control elements. Case studies trace the pathway from laboratory formulation through pilot validation to full-scale field application, highlighting common failure modes and proven mitigation strategies.

The book is intended for admixture formulators, concrete producers, researchers, process engineers, and technical managers involved in high-performance and low-carbon concrete technologies. It also serves as a reference for those responsible for scale-up, quality systems, and regulatory compliance, offering structured frameworks for documentation, standardization, and knowledge retention. By combining scientific rigor with manufacturing realism, the book aims to support the confident deployment of next-generation VMAs in an increasingly automated, sustainability-driven concrete industry.

American Industrial Technologies LLC.

About the Series

The "*Concrete Chemical Admixtures*" series is a comprehensive, multi-volume technical reference set designed for professionals across the concrete industry, including engineers, researchers, contractors, and materials scientists, who rely on advanced admixture knowledge to meet modern construction demands. Each volume focuses on a specific category of chemical admixtures, offering in-depth insights into their formulation principles, field applications, compatibility considerations, and troubleshooting strategies.

This series bridges the critical gap between laboratory-based research and practical implementation, ensuring that readers can effectively translate scientific understanding into real-world performance improvements. Whether working on high-rise buildings, infrastructure projects, or sustainable development initiatives, professionals will find actionable guidance tailored to the evolving needs of the global concrete industry.

Books in the Series

1. Concrete Chemical Admixtures
2. Type A: Water-Reducing Chemical Admixtures
3. Type B: Set-Retarding Chemical Admixtures
4. Type C: Accelerating Chemical Admixtures
5. Type D: Water Reducing and Retarding Chemical Admixtures
6. Type E: Water-Reducing and Accelerating Chemical Admixtures
7. Type F: High-Range Water-Reducing Chemical Admixtures (HRWR)
8. Type G: High-Range Water-Reducing and Retarding Chemical Admixtures
9. Air-Entraining Chemical Admixtures
10. Shrinkage-Reducing Chemical Admixtures
11. Corrosion-Inhibiting Chemical Admixtures
12. **Viscosity-Modifying Chemical Admixtures (VMA) for Concrete (This book)**
13. Non-Chloride Accelerating Chemical Admixtures
14. Pozzolanic Admixtures (Dry Blends)
15. Self-Healing Concrete Chemical Admixtures
16. Testing and Evaluation of Concrete Chemical Admixtures
17. Sustainability and Environmental Aspects of Concrete Chemical Admixtures

Viscosity-Modifying

Chemical Admixtures for Concrete

From Formulation to Field: Applications, Troubleshooting & Case Studies

A Comprehensive Resource for Formulators, Ready-Mix Concrete Producers, and Consultants

1st Edition

Volume 12

Copyright

Viscosity-Modifying Chemical Admixtures for Concrete - From Formulation to Field: Applications, Troubleshooting & Case Studies
Author: Mostafa A. Ebrahim, PhD
Published by: American Industrial Technologies LLC.

Disclaimer

The field of concrete technology and chemical admixtures is continually advancing. As new research, industry standards, and practical experiences emerge, formulations, application techniques, and troubleshooting methods may evolve.

The content of this book reflects the author's knowledge and perspectives at the time of writing and does not necessarily represent the views of the publisher or editors. Neither the author nor the publisher guarantees the accuracy, completeness, or suitability of the information provided for any specific purpose.

The reader assumes all responsibility for the proper application, testing, and verification of any procedures, formulations, or recommendations outlined in this book. The author and publisher disclaim any liability for damages, including but not limited to structural failure, material incompatibility, or financial loss, arising from the use or misuse of the information contained herein.

Professionals and practitioners are advised to consult current industry standards, local regulations, and qualified experts before implementing any techniques or formulations in real-world applications.

Trademark and Brand Disclaimer

All trademarks, service marks, trade names, trade dress, logos, and other brand identifiers featured in this publication are the property of their respective owners. Their inclusion is solely for identification and illustrative purposes and does not imply any endorsement or affiliation with the book's author or publisher. This publication does not intend to infringe upon the rights of any trademark holder.

Intended Use Statement

This book is designed as a technical reference and educational resource for professionals involved in concrete materials science, admixture formulation, and construction practices. It is not intended to replace formal engineering judgment, laboratory validation, or regulatory compliance. Users should always consult applicable ASTM, EN, IS, or other recognized standards before applying any admixture-related technique in the field.

Complimentary Technical Advice

As a thank-you to registered readers, I offer free technical support via email to help with questions or challenges related to concrete admixture formulation and application. Whether you need clarification, product guidance, or best practice recommendations, I’m available to assist.

This service is limited to 2 hours per reader and must be requested within one month of purchasing a new copy from an authorized source. Please use the email address associated with your purchase to ensure prompt assistance.

Please note: This advisory service is provided voluntarily and does not constitute a contractual obligation. It reflects my commitment to supporting professionals in the field and promoting knowledge sharing across the industry.

I look forward to engaging with readers and contributing to their success through this opportunity.

Dedication

To my beloved father, whose steadfast guidance and unwavering work ethic have been a constant source of inspiration. Thank you for teaching me the true value of perseverance and the relentless pursuit of excellence.

To my caring mother, whose unconditional love and nurturing spirit have profoundly shaped me into the person I am today. Your endless encouragement has empowered me to boldly chase my dreams and passions.

To my wonderful wife, Monica, whose unwavering support and love have been my anchor. Your belief in me and our shared dreams inspire me every day.

And to my darling daughters, Fatima and Mona, whose boundless enthusiasm and insatiable curiosity remind me daily of the profound importance of wonder, discovery, and building a better future. This work is a testament to the world I hope to create for you.

This book is a humble offering to the four pillars of my life, may it serve as a small expression of my deepest, most heartfelt love and gratitude.

Mostafa Ebrahim

Acknowledgments

The knowledge presented in this book is built upon years of hands-on experience working with concrete technologists, admixture formulators, and construction professionals worldwide. What began as practical field observations and technical notes has now developed into this comprehensive guide.

I am deeply grateful to the many industry experts who generously shared their insights through productive discussions and rigorous peer reviews. Their constructive feedback on the manuscript significantly enhanced both the technical depth and practical relevance of this work.

Special thanks go to my family for their unwavering support throughout this project. To my wife Monica and daughters Mona and Fatima, your patience and encouragement made this endeavor possible.

I also wish to acknowledge the valuable contributions from colleagues and reviewers who provided thoughtful suggestions on chapter outlines and manuscript drafts. Their expertise helped shape this book into a more useful resource for the concrete industry.

This work stands as a collective achievement of the global concrete community, and I am honored to contribute to our shared knowledge.

Dr. Mostafa Ebrahim
Additives Formulator and Consultant
American Industrial Technologies LLC.

Forward

The global concrete industry is currently navigating its most significant transformation since the invention of Portland cement. Driven by the urgent mandate for decarbonization, we are witnessing a rapid shift toward low-clinker binders, high-volume supplementary cementitious materials (SCMs), and the integration of recycled constituents. While these innovations are essential for sustainability, they introduce unprecedented rheological volatility. In this new landscape, viscosity-modifying admixtures (VMAs) have evolved from simple "bleed-control" additives into the strategic backbone of modern concrete stability.

Viscosity-Modifying Chemical Admixtures for Concrete arrives at a critical juncture. For too long, a persistent gap has existed between the sophisticated polymer chemistry developed in the laboratory and the rugged, variable reality of the batching plant and construction site. This book effectively bridges that divide. It offers a comprehensive treatment that begins with the molecular architecture of natural and synthetic polymers and follows their journey through formulation, manufacturing, and field deployment.

What distinguishes this work is its forward-looking embrace of Industry 4.0. The author does not merely treat VMAs as static chemical components; instead, they are presented as adaptive tools within a digital ecosystem. By integrating concepts such as digital twins, sensor-based rheology monitoring, and predictive analytics, the manuscript provides a roadmap for transforming "trial-and-error" concrete production into a precise, data-driven industrial process.

The inclusion of twenty-one detailed formulation examples, covering everything from 3D-printed concrete to underwater tremie applications, provides the reader with a practical "field manual" for troubleshooting and optimization. This level of industrial granularity, combined with rigorous scientific analysis of adsorption kinetics and thixotropic rebuild, makes this volume an indispensable resource for formulators, ready-mix producers, and consultants alike.

As we strive to build a more resilient and sustainable infrastructure, the ability to control the "flow" of concrete under increasingly complex conditions will remain a defining challenge. This book provides the scientific clarity and technical leadership needed to meet that challenge head-on.

Mostafa A. Ebrahim

About the Author

Dr. Mostafa Ebrahim is a recognized expert in polymer additives and chemical admixtures, with over 7 years of experience as a materials scientist, concrete additives formulator, and industry consultant. He has developed advanced admixture solutions for infrastructure, high-rise, marine, and sustainable construction projects, earning recognition as a leading authority in the field.

With deep expertise in chemical additives, Dr. Ebrahim specializes in optimizing material compatibility and performance. His work bridges scientific understanding with real-world application, making him a trusted advisor to construction firms and materials suppliers worldwide.

A dedicated educator, he has led numerous training programs and workshops, helping professionals implement cutting-edge admixture technologies with confidence. This book reflects his extensive research and hands-on experience, offering readers both technical depth and practical insights into the role of admixtures in modern, sustainable concrete systems.

Join Dr. Ebrahim on this in-depth exploration of admixture science, application, and troubleshooting, to unlock new opportunities for innovation in concrete technology.

American Industrial Technologies LLC.

Preface

The construction industry is currently undergoing a period of profound transformation, driven by the dual imperatives of sustainability and digitalization. As we move toward a "Net Zero" future, the concrete we produce is becoming increasingly complex. The widespread adoption of low-clinker binders, high-volume supplementary cementitious materials (SCMs), and recycled aggregates has made rheological stability, the ability of concrete to remain homogeneous and workable, more difficult to achieve than ever before.

In this challenging environment, Viscosity-Modifying Admixtures (VMAs) have transitioned from niche "problem-solvers" used to control bleeding into essential, system-level enablers. They are the "invisible hand" that allows for the successful deployment of Self-Consolidating Concrete (SCC), 3D-printed concrete, and ultra-high-performance mixes. Yet, despite their importance, a persistent gap remains between the high-level polymer science conducted in laboratories and the practical, often unpredictable, reality of the concrete batching plant.

This book, Viscosity-Modifying Chemical Admixtures for Concrete, was written to bridge that gap. My goal is to provide a comprehensive resource that speaks to both the chemist formulating the next generation of polymers and the consultant troubleshooting a mix design in the field.

The manuscript is structured to take the reader on a journey from "Formulation to Field." We begin by exploring the fundamental molecular architectures, such as cellulose ethers, biopolymers, and synthetic polyacrylamides, and how their chemical structures dictate their performance in cementitious systems. From there, we move into the "industrial kitchen," detailing the manufacturing processes, quality control protocols, and the critical importance of digital integration.

A defining feature of this work is its focus on Industry 4.0. We explore how digital twins, real-time sensor data, and predictive analytics are transforming how admixtures are utilized, moving the industry away from reactive troubleshooting toward proactive, data-driven stability.

Ultimately, this book is a testament to the idea that the future of construction lies at the intersection of chemistry and technology. It is my hope that the insights, case studies, and twenty-one formulation examples provided herein will serve as a valuable roadmap for professionals striving to build a more resilient, efficient, and sustainable built environment.

Mostafa A. Ebrahim

Disclosure: Acknowledging Collaborative Technology

This book was developed using advanced writing technology as a supportive tool. While all key concepts, technical analyses, and original insights remain the author's own work, artificial intelligence assisted by suggesting relevant research, verifying facts, improving grammar, and enhancing readability through structural refinements.

This technological collaboration was designed to complement human expertise, not replace it. The author maintains complete oversight and responsibility for all technical accuracy, professional judgments, and the work's overall integrity. Every recommendation and conclusion has been carefully reviewed to meet industry standards for concrete admixture technology.

This project represents an innovative approach in technical publishing, where AI augmentation helps subject-matter experts produce more precise, thorough, and accessible content. As you engage with this material, we hope you'll appreciate how thoughtfully applied technology can enhance - while never supplanting - expert knowledge and rigorous scholarship.

Declaration of Competing Interest

I hereby declare that there are no known competing financial interests, professional affiliations, or personal relationships that could be perceived as influencing this publication, content, or recommendations presented in this work. No financial support or compensation was received from any organization that might have an interest in the outcomes of this publication.

1. Fundamentals of Viscosity-Modifying Admixtures for Concrete

Viscosity-modifying admixtures form the rheological backbone of modern concrete technology, enabling precise control over stability, flow, and robustness in increasingly complex mix designs. Understanding their fundamental principles is essential to appreciate how polymer chemistry, rheology, and microstructural mechanisms translate into practical performance benefits at plant and jobsite scale. This introductory framework establishes the technical basis for VMA function, their evolution alongside modern admixture systems, and their strategic role in performance-based, digitally managed concrete production environments.

1.1 Definition, Functional Scope, and Value Proposition

Technical Definition of Viscosity-Modifying Admixtures

Viscosity-modifying admixtures are water-soluble polymeric systems intentionally designed to increase the plastic viscosity and structural cohesion of fresh concrete without causing excessive yield stress or permanent particle flocculation. They operate primarily within the aqueous phase of the cement paste, modifying flow resistance at low shear rates while maintaining shear-thinning behavior under mixing, pumping, and placement conditions.

From a formulation standpoint, VMAs are distinct from mineral viscosity enhancers and air-based stabilization mechanisms. Their performance is governed by polymer molecular architecture, hydration kinetics, and ionic sensitivity rather than physical filler effects, allowing precise rheology control at low dosage levels.

Functional Scope in Fresh Concrete Performance

The functional scope of VMAs encompasses stabilization of fresh concrete against bleeding, segregation, and aggregate settlement across a wide range of mix designs. VMAs enable consistent rheological performance in systems characterized by high workability, low fines content, or aggressive particle dispersion induced by modern superplasticizers.

In industrial practice, VMAs extend mix robustness by compensating for variability in cement chemistry, aggregate grading, moisture fluctuations, and water quality. This stabilization function is critical in ready-mix environments where raw material variability is unavoidable and real-time corrective actions are limited.

Role in Decoupling Flowability from Stability

A defining value of VMAs lies in their ability to decouple flowability from stability, a requirement in self-consolidating, pumpable, and extrudable concretes. By selectively increasing plastic viscosity while preserving low yield stress, VMAs allow high slump flow without compromising homogeneity.

This decoupling mechanism enables formulators to design mixes that satisfy both placement efficiency and structural integrity requirements. It also reduces reliance on excess cement or fines as stabilization tools, directly supporting cost optimization and carbon reduction strategies.

Industrial Value Proposition and Economic Impact

From an economic perspective, VMAs contribute to reduced rejected loads, fewer placement defects, and lower rework rates. These benefits translate into measurable reductions in operational cost, fuel consumption, and embodied CO_2 per delivered cubic meter of concrete.

For admixture suppliers and ready-mix producers, VMAs enhance formulation resilience, facilitate cross-plant standardization, and shorten commissioning cycles for new raw materials. Their contribution extends beyond fresh concrete aesthetics to overall production reliability and profitability.

Strategic Positioning in Performance-Based Specifications

In performance-based concrete specifications, VMAs function as enablers of compliance rather than prescriptive additives. Their inclusion allows producers to meet stability, flow, and durability requirements under broader material variability envelopes.

When integrated into digitally managed production systems, VMAs support adaptive dosing strategies linked to mixer energy, slump evolution, and temperature data. This positions VMAs as key components in Industry 4.0-aligned concrete production, bridging formulation science with real-time process control.

1.2 Historical Evolution of VMA Technology

Early Bleed Control and Anti-Washout Applications

The earliest viscosity-modifying systems were introduced to address excessive bleeding and cement washout in high water content concretes and underwater placement scenarios. Natural polysaccharides, particularly cellulose ethers and starch-based derivatives, were adopted due to their strong water retention capacity and availability. These materials provided basic stability but exhibited high dosage sensitivity, limited tolerance to shear variation, and inconsistent performance across cement types.

In industrial practice, first-generation VMAs were applied in a corrective manner, often added reactively at the plant or jobsite. Their use was constrained by slow hydration kinetics and susceptibility to microbial degradation during storage, which limited shelf life and formulation robustness.

Transition to Polymer-Engineered VMA Systems

The widespread adoption of high-range water-reducing admixtures fundamentally altered fresh concrete rheology, creating ultra-dispersed systems prone to segregation. This shift necessitated independent viscosity control, driving the development of synthetic and semi-synthetic polymer VMAs with controlled molecular weight distribution and predictable hydration behavior.

Polyacrylamide- and polyacrylate-based systems enabled improved stability under variable shear conditions and elevated temperatures. These systems marked a transition from purely water-retention mechanisms to rheology-engineered solutions capable of balancing plastic viscosity and thixotropic rebuild in highly flowable concretes.

Co-Evolution with Polycarboxylate Ether Technology

The emergence of polycarboxylate ether superplasticizers accelerated innovation in VMA design. Excessive particle dispersion and reduced yield stress introduced by advanced PCEs exposed limitations of legacy viscosity modifiers, particularly in self-consolidating concrete and long-haul ready-mix applications.

Next-generation VMAs were engineered to minimize competitive adsorption while maintaining compatibility with dispersing polymers. Associative polymers and functionalized backbones were developed to deliver reversible rheological structuring without suppressing flow. This co-evolution established VMAs as integral components of modern admixture systems rather than standalone additives.

Adaptation to Low-Clinker and SCM-Rich Binder Systems

As sustainability objectives intensified, binder systems incorporating fly ash, slag, limestone, and calcined clays became more prevalent. These systems introduced higher variability in early-age rheology due to differences in particle morphology, surface chemistry, and dissolution kinetics.

VMA technology evolved to address these challenges through improved ionic tolerance and controlled hydration rates. Modern VMAs are formulated to stabilize paste rheology across wide sulfate and alkali ranges, supporting consistent performance in low-clinker and blended cement systems increasingly adopted for carbon reduction.

Digitalization and Industry 4.0-Driven Evolution

Recent advancements in VMA technology reflect the integration of digital manufacturing and process control in ready-mix operations. VMAs are now designed for compatibility with adaptive dosing strategies informed by real-time plant data such as mixer power, temperature, and slump flow evolution.

This digital alignment shifts VMA usage from fixed empirical dosages toward data-driven optimization. The evolution toward digitally enabled rheology control positions VMAs as key functional elements within Industry 4.0 concrete production frameworks, supporting predictive quality control and reduced material waste.

1.3 Fundamental Rheological Concepts Relevant to VMAs

Yield Stress and Initiation of Flow

Yield stress defines the minimum stress required to initiate flow in fresh concrete and governs the onset of deformation under gravity, vibration, or pumping. While VMAs are not primarily designed to increase yield stress, their presence can indirectly influence it through modification of interparticle interactions and paste cohesion. Effective VMA systems maintain low yield stress to preserve workability while preventing premature phase separation under static conditions.

From an industrial formulation perspective, excessive yield stress increase is undesirable as it compromises pumpability and finishing. Modern VMAs are therefore engineered to selectively target viscosity control rather than rigid structural buildup, ensuring flow initiation remains within practical operational limits.

Plastic Viscosity and Flow Resistance Under Shear

Plastic viscosity represents the resistance to flow once yield stress has been exceeded and is the primary rheological parameter influenced by VMAs. By increasing resistance to internal flow at low to moderate shear rates, VMAs suppress sedimentation of aggregates and limit bleeding without eliminating shear-thinning behavior.

The magnitude of viscosity increase depends on polymer molecular weight, solvation state, and concentration relative to the aqueous phase volume. In high-performance concretes, controlled plastic viscosity is essential to balance pump pressure requirements with segregation resistance, particularly during long-distance transport and placement.

Thixotropy and Structural Rebuild Behavior

Thixotropy describes the time-dependent recovery of internal structure after shear cessation. VMAs contribute to reversible structural rebuild through transient polymer networks that reform when shear is reduced. This behavior is critical for maintaining shape stability after casting while allowing easy flow during placement.

In practical terms, thixotropic rebuild supported by VMAs reduces formwork pressure, improves surface finish, and enhances buildability in layered placement systems. The kinetics of rebuild must be tuned to match placement methods and ambient conditions to avoid either excessive stiffening or insufficient stability.

Shear-Dependent Viscosity and Shear-Thinning Behavior

Fresh concrete exhibits non-Newtonian shear-thinning behavior, where apparent viscosity decreases with increasing shear rate. VMAs must preserve this characteristic to avoid excessive energy demand during mixing and pumping. Well-designed VMAs increase viscosity primarily at low shear rates while exerting minimal influence at high shear conditions.

This selective rheological response is achieved through polymer architectures that align or disentangle under shear. From a manufacturing standpoint, maintaining shear-thinning behavior ensures compatibility with existing plant equipment and avoids the need for process modifications.

Time-Dependent Rheology and Structural Evolution

Concrete rheology evolves continuously due to cement hydration, dissolution of ionic species, and temperature effects. VMAs interact with this evolving system by delaying or moderating viscosity increase during early hydration stages while stabilizing rheology over transport and placement windows.

Understanding time-dependent rheology is essential for defining dosage windows and sequencing with other admixtures. Improper alignment between VMA hydration and cement reaction kinetics can result in delayed thickening or unexpected slump loss, particularly in warm weather conditions.

Relevance of Rheological Models to VMA Optimization

Rheological models such as the Bingham and Herschel–Bulkley frameworks are commonly used to describe fresh concrete behavior. VMAs primarily shift plastic viscosity parameters while preserving model validity across operational shear ranges.

In advanced applications, digital rheology monitoring and data-driven modeling allow real-time estimation of viscosity and thixotropy from indirect measurements such as mixer power consumption. These approaches support Industry 4.0-enabled optimization of VMA dosing, improving consistency and reducing material waste.

1.4 Mechanisms of Action at the Microstructural Level

Water Structuring and Pore Solution Viscosity Modification

At the microstructural scale, viscosity-modifying admixtures primarily act by increasing the apparent viscosity of the pore solution through polymer hydration and chain expansion. Hydrophilic functional groups along the polymer backbone bind and immobilize free water molecules, reducing their mobility under gravitational and capillary forces. This mechanism directly suppresses bleeding and limits upward water migration in freshly placed concrete.

The efficiency of water structuring depends on polymer solubility, hydration rate, and ionic tolerance. High alkali or sulfate concentrations can alter polymer conformation, influencing the degree of viscosity enhancement and necessitating formulation adjustment for cement-specific chemistries.

Steric Stabilization of Cement and Fine Particles

VMAs provide steric stabilization by creating a hydrated polymer layer around cement grains and fine particles, increasing effective particle spacing without inducing electrostatic repulsion. This steric barrier limits close particle approach and reduces the likelihood of density-driven segregation, particularly in highly dispersed systems containing modern superplasticizers.

Unlike flocculating agents, steric stabilization induced by VMAs is weak and reversible, allowing particles to rearrange under shear. This reversibility preserves flowability during mixing and placement while maintaining stability at rest, a critical balance in self-consolidating and pumpable concretes.

Transient Polymer Network Formation

Certain VMA chemistries form transient, three-dimensional polymer networks within the aqueous phase through chain entanglement or associative interactions.

These networks increase resistance to low-shear deformation without generating rigid structures that impede flow initiation.

Under shear, these networks partially disentangle, reducing apparent viscosity. Upon cessation of shear, rapid re-entanglement occurs, contributing to thixotropic rebuild. The density and strength of these transient networks are governed by molecular weight distribution, functional group density, and dosage relative to the paste volume.

Interaction with Cement Hydration Products

VMAs interact indirectly with early hydration products by modifying the transport of ions and water within the paste. By slowing capillary flow and limiting local concentration gradients, VMAs can moderate early-age rheological evolution without significantly retarding hydration reactions.

In systems rich in supplementary cementitious materials, VMAs help stabilize the evolving microstructure during the induction period when particle morphology and surface chemistry differ from ordinary Portland cement. Proper alignment between polymer hydration and cement reaction kinetics is essential to avoid delayed thickening or premature viscosity buildup.

Aggregate Suspension and Density Compensation Effects

In concrete containing coarse aggregates, VMAs contribute to suspension stability by increasing the viscous drag acting on particles. This mechanism counteracts density differences between aggregates and paste, reducing settlement and segregation during transport and placement.

The effectiveness of aggregate suspension is influenced by paste volume, grading curve, and particle shape. VMAs enable broader tolerance to aggregate variability, particularly when recycled or manufactured sands introduce irregular particle morphology and higher fines fluctuation.

Microstructural Mechanisms and Industrial Implications

Understanding microstructural action mechanisms is critical for translating laboratory performance to plant-scale consistency. Over-reliance on water immobilization can lead to excessive viscosity under low temperatures, while insufficient steric stabilization may fail to prevent segregation in highly dispersed mixes.

For industrial formulation and application, microstructural insight supports rational selection of VMA chemistry, dosage optimization, and sequencing with other admixtures. When combined with digital monitoring of rheological indicators, these mechanisms form the basis for predictive control strategies aligned with Industry 4.0 concrete production systems.

1.5 Role of VMAs in Modern Concrete Mix Design

Enabling High-Workability and Stable Mix Designs

In modern concrete mix design, viscosity-modifying admixtures function as enabling components that allow high workability to be achieved without compromising stability. VMAs make it possible to design mixes with elevated slump flow or pumpability while maintaining resistance to bleeding, segregation, and aggregate settlement. This role is particularly critical in self-consolidating, pumped, and highly flowable concretes where traditional stability mechanisms based on fines content are insufficient or economically unfavorable.

From an industrial perspective, VMAs reduce dependency on narrow aggregate grading envelopes and excessive paste volumes. This flexibility allows mix designers to accommodate local raw material variability while maintaining consistent fresh-state performance across production batches and plant locations.

Supporting Cement Reduction and Binder Optimization

VMAs play a strategic role in binder optimization by enabling reductions in total cement content without sacrificing rheological stability. By stabilizing the paste phase, VMAs compensate for lower fines levels and reduced early-age cohesion typically associated with cement reduction strategies.

This functionality directly supports sustainability objectives by lowering embodied CO_2 and energy consumption per cubic meter of concrete. In performance-based mix design frameworks, VMAs enable higher substitution levels of supplementary cementitious materials while preserving placement reliability and surface quality.

Managing Variability in SCM-Rich and Low-Clinker Systems

Modern binder systems increasingly incorporate fly ash, slag, limestone, and calcined clays, each introducing distinct rheological behaviors. VMAs provide a buffering mechanism that dampens variability arising from differences in particle morphology, surface chemistry, and dissolution kinetics.

In SCM-rich systems, VMAs stabilize early-age rheology during the induction period, improving consistency during transport and placement. Their use reduces the need for frequent mix redesign when raw material sources or proportions change, enhancing operational robustness.

Enhancing Robustness in Ready-Mix Production

Ready-mix production environments are characterized by continuous variability in aggregate moisture, grading, and water quality. VMAs expand the acceptable operating window of mix designs by mitigating the impact of these fluctuations on fresh concrete performance.

When integrated into standardized mix designs, VMAs reduce sensitivity to operational deviations such as haul time variation and temperature changes. This robustness translates into fewer corrective adjustments at the plant or jobsite and improved delivery reliability.

Integration with Digital Mix Design and Industry 4.0 Systems

In digitally enabled concrete production, VMAs support adaptive mix design strategies linked to real-time process data. Parameters such as mixer power consumption, slump flow evolution, and temperature profiles can be used to adjust VMA dosage dynamically within predefined limits.

This integration aligns VMA usage with Industry 4.0 principles, enabling closed-loop control of rheology and predictive quality assurance. VMAs thus serve as key actuators in data-driven mix optimization, bridging formulation science with automated plant operation.

1.6 Functional Positioning of VMAs Relative to Other Admixtures

Complementary Role to Superplasticizers

Viscosity-modifying admixtures are functionally complementary to high-range water-reducing admixtures, particularly polycarboxylate ether systems. While superplasticizers primarily reduce yield stress through particle dispersion, VMAs selectively increase plastic viscosity to restore stability in highly dispersed mixes. This complementary interaction allows independent tuning of flowability and segregation resistance, a requirement in modern high-performance concretes.

From a formulation perspective, competitive adsorption between VMAs and dispersing polymers must be controlled through molecular design and sequencing. Improper timing or overdosing can suppress dispersion efficiency, highlighting the need for coordinated admixture selection rather than isolated optimization.

Interaction with Air-Entraining Admixtures

Air-entraining admixtures introduce microscopic air voids to enhance freeze–thaw durability, relying on stable foam formation within the paste. VMAs can influence air content and bubble stability by modifying surface tension and increasing aqueous phase viscosity.

In practice, certain VMA chemistries may destabilize entrained air or alter bubble size distribution. Industrial formulations therefore require careful balancing of VMA dosage and air-entrainer chemistry to maintain target air content while achieving adequate rheological stability.

Relationship with Set Modifying Admixtures

Set retarders and accelerators primarily influence cement hydration kinetics rather than fresh-state rheology. However, changes in early-age hydration directly affect rheological evolution, indirectly interacting with VMA performance.

VMAs must be compatible with set modifiers across temperature ranges to avoid delayed thickening or premature stiffening. Seasonal formulation adjustments often involve coordinated changes to both VMA and set-modifying admixture dosages to maintain consistent placement windows.

Differentiation from Mineral and Filler-Based Stabilization

Mineral fillers and fine materials increase viscosity through physical packing and surface area effects. While they can partially substitute for VMAs, they lack the tunability and low-dosage efficiency of polymer-based systems.

VMAs offer superior control over rheology without significantly increasing paste volume or water demand. This distinction is critical for optimizing cost, durability, and sustainability, particularly in mixes targeting low binder content or high recycled material usage.

Functional Boundaries and Misapplication Risks

VMAs are not designed to compensate for fundamentally poor aggregate grading, excessive water addition, or inadequate mix design. Misapplication often results in overdosing, excessive viscosity, and workability loss, which can be incorrectly attributed to cement or aggregate quality.

Clear functional positioning of VMAs relative to other admixtures enables rational formulation strategies and effective troubleshooting. Understanding these boundaries is essential for preventing unintended interactions and ensuring reliable performance across diverse production conditions.

Table 1.6: Admixture Interaction Overview

Admixture Type	Primary Function	Interaction with VMAs	Key Control Parameter
Superplasticizers	Yield stress reduction	Complementary, adsorption-sensitive	Addition sequence
Air Entrainers	Freeze–thaw durability	Foam stability impact	Dosage balance
Set Modifiers	Hydration control	Indirect rheology effect	Temperature alignment
Mineral Fillers	Physical viscosity increase	Partial substitution	Paste volume

1.7 Industrial Relevance and Technology Transfer Perspective

VMAs as Industrial Robustness Enablers

In industrial concrete production, viscosity-modifying admixtures function primarily as robustness enhancers rather than performance luxury additives. Their key industrial relevance lies in expanding the acceptable variability window of raw materials, process conditions, and logistics without compromising fresh-state stability. VMAs mitigate the impact of fluctuations in cement chemistry, aggregate grading, moisture content, and haul time, all of which are inherent to ready-mix operations.

From a production reliability standpoint, VMAs reduce the frequency of corrective actions, rejected loads, and jobsite complaints. This positions VMAs as risk-management tools that stabilize output quality under non-ideal and rapidly changing operational conditions.

Bridging Laboratory Formulation and Plant-Scale Reality

A recurring challenge in admixture deployment is the performance gap between laboratory-scale mix optimization and full-scale plant execution. VMAs play a critical role in narrowing this gap by providing rheological buffering capacity that absorbs scale-dependent variability such as mixer energy differences, batching tolerances, and temperature gradients.

Effective technology transfer requires that VMA formulations be validated under realistic shear histories and residence times. Laboratory rheology data must therefore be translated into plant-relevant indicators such as mixer power consumption and slump flow retention, ensuring that formulation intent survives industrial scale-up.

Standardization and Cross-Plant Consistency

For multi-plant ready-mix operations, VMAs enable standardization of mix designs across geographically dispersed facilities with differing raw material sources. By stabilizing paste rheology, VMAs reduce the need for plant-specific mix redesigns and simplify quality management frameworks.

This standardization facilitates centralized formulation governance and supports consistent customer-facing performance. In practice, VMAs allow corporate-level mix design strategies to be deployed locally with minimal adjustment, improving operational efficiency and knowledge transfer within organizations.

Integration into Digitally Managed Production Systems

In Industry 4.0-aligned concrete production, VMAs act as controllable actuators within digitally managed rheology control loops. When linked to manufacturing execution systems, VMA dosing can be adjusted based on real-time signals such as mixer torque, temperature, and slump flow evolution.

This digital integration transforms VMA usage from fixed empirical dosing to adaptive, data-driven optimization. The result is improved consistency, reduced material waste, and predictive quality control, aligning rheology management with broader digital transformation initiatives in construction materials manufacturing.

Training, Knowledge Transfer, and Misuse Prevention

Misapplication of VMAs remains a common root cause of fresh concrete performance issues, often due to inadequate understanding of their functional boundaries. Structured training based on mechanistic understanding is essential to prevent overdosing, false segregation diagnostics, and incorrect attribution of failures to cement or aggregate quality.

A technology transfer-oriented approach emphasizes shared understanding among formulation chemists, plant operators, and field personnel. Clear communication of VMA function, limitations, and interaction mechanisms is critical for sustainable adoption and long-term performance reliability.

Strategic Importance in Sustainable Concrete Production

As the industry transitions toward low-clinker, SCM-rich, and recycled-material concrete systems, VMAs become increasingly strategic. Their ability to stabilize rheology under chemically and physically diverse conditions enables aggressive sustainability targets without sacrificing constructability.

From a lifecycle perspective, VMAs contribute indirectly to carbon reduction by enabling cement optimization, reducing waste, and minimizing rework. Their industrial relevance therefore extends beyond immediate performance to long-term sustainability and resilience of concrete production systems.

2. Market Drivers, Performance Expectations, and Regulatory Context

The adoption and evolution of viscosity-modifying admixtures are shaped by a convergence of market pressures, performance demands, and regulatory frameworks rather than by polymer innovation alone. Increasing construction complexity, tighter quality tolerances, sustainability mandates, and the globalization of concrete supply chains have elevated rheology control from an operational adjustment to a strategic requirement. Within this landscape, VMAs are expected to deliver predictable performance across diverse materials, comply with regionally differentiated standards, and enable low-carbon concrete technologies, all while integrating seamlessly into digitally controlled, Industry 4.0 oriented production environments.

2.1 Evolution of VMA Usage in Modern Concrete Technologies

The industrial use of viscosity-modifying admixtures has progressed in direct response to increasing complexity in concrete mix design and placement technologies. Early concrete systems relied primarily on aggregate grading and water content control to manage segregation and bleeding. VMAs were initially introduced as secondary additives, typically based on simple water-retentive polymers, to correct deficiencies in marginal mixes rather than as core formulation components. Their function was largely passive, acting through bulk viscosity increase with limited predictability across varying raw material sources.

As high-range water reducers, particularly polycarboxylate ether superplasticizers, became dominant in concrete formulation, traditional stability mechanisms proved insufficient. High dispersion efficiency reduced yield stress to levels where particle settlement, laitance formation, and dynamic segregation became critical risks. VMAs evolved accordingly into active rheology control agents capable of independently adjusting plastic viscosity and yield stress. Polymer chemistry advanced toward associative and partially adsorbing macromolecules that interact with cement particles, hydration products, and dissolved ions, enabling reproducible rheological profiles even under aggressive water reduction.

The emergence of self-consolidating concrete marked a decisive inflection point. VMAs transitioned from optional additives to mandatory formulation tools, defining the stability window between flow under self-weight and resistance to segregation. This role further expanded with ultra-high performance concrete, where extreme particle packing density and ultrafine powders required precise control of interstitial fluid viscosity. In parallel, additive manufacturing and 3D-printed concrete imposed new rheological demands, notably thixotropic structuration and rapid rebuild of yield stress after shear. VMAs are now engineered to deliver time-dependent rheology rather than static viscosity enhancement.

In modern production environments, VMA usage is increasingly integrated into digitalized batching and quality control systems. Real-time monitoring of slump flow, torque, and energy input allows adaptive dosing strategies that compensate for raw material variability, temperature fluctuations, and recycled content. This evolution aligns VMA technology with Industry 4.0 principles, where rheology becomes a controllable process variable rather than an empirical outcome. VMAs have thus shifted from corrective bleed control agents to foundational enablers of advanced, low-carbon, and digitally managed concrete technologies.

2.1.1 Transition From Bleed Control Agents to Multifunctional Rheology Stabilizers

Early viscosity-modifying admixtures were introduced to address visible defects in fresh concrete, primarily bleeding and particle segregation in mixes with low fines content or poorly graded aggregates. These first-generation products were largely based on water-retentive polymers such as cellulose ethers and natural gums. Their mode of action relied on increasing pore solution viscosity and delaying water migration, with limited interaction at the cement particle level. While effective in reducing surface bleed, their influence on overall rheological behavior was coarse and often accompanied by undesirable side effects such as delayed setting or reduced early strength.

The widespread adoption of high-efficiency polycarboxylate ether superplasticizers fundamentally altered fresh concrete rheology by drastically lowering yield stress and enhancing dispersion. Under these conditions, simple viscosity increase was no longer sufficient to stabilize the system. This limitation drove the development of second-generation VMAs designed as multifunctional rheology stabilizers. These materials are typically synthetic polymers such as hydrophobically modified polyacrylamides, partially hydrolyzed polyacrylonitriles, and associative copolymers with controlled molecular weight and side-chain architecture. Their functionality extends beyond bleed suppression to independent modulation of plastic viscosity, yield stress, and thixotropic rebuilding.

At the microstructural level, modern VMAs operate through a combination of mechanisms including weak adsorption on cement and SCM particle surfaces, formation of transient polymer networks in the pore solution, and hydrophobic associations that respond dynamically to shear. This multiscale interaction allows VMAs to stabilize particle suspensions without excessive water retention, preserving early hydration kinetics and mechanical performance. In practice, such polymers enable precise rheological tuning across different shear regimes encountered during mixing, pumping, placement, and finishing.

From an industrial formulation perspective, multifunctional VMAs are now selected based on compatibility with other admixtures, robustness against raw material variability, and predictable response to temperature and mixing energy. Their integration into digitally controlled batching systems supports adaptive dosing strategies consistent with Industry 4.0 manufacturing concepts. As a result, VMAs have transitioned from corrective additives addressing isolated defects to essential formulation tools that define the rheological envelope of advanced concrete technologies.

2.1.2 Role of VMAs in SCC, 3D-printed concrete, UHPC, and low-cement systems

The functional role of viscosity-modifying admixtures becomes most evident in concrete technologies where conventional rheological margins are no longer adequate. In self-consolidating concrete, VMAs are essential to establish a stable balance between high flowability and resistance to static and dynamic segregation. By selectively increasing plastic viscosity while maintaining low yield stress, VMAs suppress aggregate settlement and bleeding during flow under self-weight. Industrial formulations rely on this mechanism to achieve reproducible slump flow, passing ability, and surface finish across variable aggregate sources and ambient conditions.

In 3D-printed concrete, VMAs enable a fundamentally different rheological response. The material must flow under high shear during extrusion yet rapidly rebuild internal structure once deposited. VMAs contribute to controlled thixotropy by forming reversible polymer networks that collapse under shear and reform during rest. This behavior governs buildability, interlayer bonding, and dimensional accuracy of printed elements. Polymer architecture, particularly hydrophobic associations and molecular weight distribution, is tuned to synchronize structural build-up with printing speed and layer height, allowing industrial-scale additive manufacturing with consistent output.

Ultra-high performance concrete presents extreme formulation challenges due to very low water-to-binder ratios, ultrafine powders, and dense particle packing. VMAs stabilize the interstitial fluid, mitigating microsegregation and ensuring homogeneous distribution of cement, silica fume, and fillers. Their role extends to improving fiber dispersion in fiber-reinforced UHPC by increasing matrix cohesion without impairing flow. This rheological control supports uniform mechanical performance and reduces sensitivity to mixing energy and sequence in production environments.

Low-cement and low-clinker systems incorporating supplementary cementitious materials and recycled constituents exhibit elevated surface area and variable particle chemistry, which often destabilize fresh concrete rheology. VMAs compensate for these effects by increasing suspension stability and reducing water demand, enabling cement reduction while preserving workability and strength development. When combined with digital batching control and inline rheology monitoring, VMA dosing can be dynamically adjusted to accommodate variability in SCM reactivity and recycled aggregate quality, aligning low-cement concrete production with Industry 4.0 driven sustainability strategies.

2.2 Global Standards, Specifications, and Acceptance Criteria

The regulatory acceptance of viscosity-modifying admixtures is governed by international standards that emphasize demonstrated performance in concrete rather than prescriptive polymer chemistry. This approach reflects the diversity of VMA molecular architectures and their formulation-dependent interaction with cement, SCMs, and other admixtures. Acceptance criteria typically focus on fresh concrete stability, strength neutrality, and durability compatibility, ensuring VMAs function as rheology control tools without compromising structural performance.

Global standards increasingly coexist with project-specific and performance-based specifications. This hybrid regulatory environment requires admixture suppliers and concrete producers to demonstrate compliance through standardized testing while maintaining flexibility to adapt formulations to local raw materials, climatic conditions, and construction methods. Digital documentation, batch traceability, and statistical quality control aligned with Industry 4.0 principles are now commonly integrated into acceptance procedures, particularly for large infrastructure and high-performance concrete projects.

2.2.1 ASTM C494, EN 934-2, JIS specifications

ASTM C494/C494M provides the primary regulatory framework for chemical admixtures in North America and many export-driven markets. VMAs are typically classified under Type S, defined as specific performance admixtures that deliver a declared function beyond water reduction or set control. Acceptance under ASTM C494 requires demonstration that the admixture does not adversely affect setting time, compressive strength at specified ages, or durability-related properties when compared with a control mix. Although rheology-specific metrics are not explicitly mandated, segregation resistance and workability stability are assessed indirectly through slump or flow retention and visual stability observations.

EN 934-2 adopts a more function-oriented classification system. VMAs are categorized based on their ability to modify consistency and improve cohesion. The standard specifies test methods to evaluate water retention, flow behavior, and stability in mortar and concrete. Compliance requires performance within defined tolerances relative to reference mixes, reinforcing the principle that VMAs must enhance rheology without negative interaction with cement hydration or air entrainment systems.

Japanese Industrial Standard JIS A 6204 emphasizes reproducibility and precision in testing protocols. The standard requires controlled evaluation of fresh concrete properties using standardized materials and mixing sequences. VMAs are assessed based on their contribution to consistency control and resistance to segregation under defined conditions. The Japanese framework reflects a strong focus on quality consistency, which aligns with high-performance and seismic-resistant concrete applications prevalent in the region.

Table 2.2.1: Comparative Overview of Major VMA-Related Standards

Standard	Primary Scope	Key Acceptance Criteria	Market Relevance
ASTM C494/C494M	Chemical admixtures	Set time, strength neutrality, workability	North America, global
EN 934-2	Functional admixture performance	Consistency, stability, water retention	Europe
JIS A 6204	Chemical admixture quality	Reproducibility, rheology control	Japan

2.2.2 Regional deviations and performance-based specifications

Despite harmonized international standards, regional deviations are common due to differences in climate, aggregate geology, construction practices, and regulatory philosophy. In hot-climate regions such as the Middle East and parts of Africa, specifications often include extended workability retention and enhanced segregation resistance under elevated temperatures. These requirements frequently exceed baseline ASTM or EN criteria, necessitating VMAs with high thermal stability and predictable interaction hooking with PCE superplasticizers.

In North America, transportation agencies and infrastructure owners increasingly rely on performance-based specifications. Departments of Transportation specify target rheological windows, stability indices, and pumping behavior rather than admixture type or dosage. VMAs must demonstrate consistent performance across variable cement sources and recycled materials, with acceptance supported by statistical process control data and historical production records.

Asian markets, particularly Japan and South Korea, apply stringent quality control frameworks emphasizing uniformity and repeatability. Performance-based specifications often incorporate rheological measurements, such as flow loss over time, to ensure reliability in high-strength and seismic-resistant concrete. These regional practices favor VMAs with narrow performance variability and robust tolerance to raw material fluctuations.

The shift toward performance-based acceptance aligns with Industry 4.0 manufacturing models. Real-time monitoring of fresh concrete properties, automated dosage adjustment, and digital compliance reporting are increasingly used to satisfy regulatory and contractual requirements. VMAs therefore function not only as rheology modifiers but also as enablers of standardized, data-driven concrete production across diverse regulatory environments.

2.3 Sustainability and Carbon Reduction Drivers

Sustainability imperatives have become a primary driver shaping the formulation and application strategy of viscosity-modifying admixtures in modern concrete. Regulatory pressure to reduce embodied carbon, combined with rising cement costs and increasing variability in raw materials, has shifted VMAs from performance enhancers to enablers of low-carbon concrete technologies. Their contribution lies in stabilizing fresh concrete rheology under reduced binder content and increased material heterogeneity, thereby expanding the feasible operating window for sustainable mix designs.

From an industrial perspective, VMAs support sustainability by reducing corrective actions during production, minimizing rejected batches, and improving consistency when alternative materials are introduced. Integration of rheology control with digital batching systems enables data-driven optimization of cement content, water demand, and admixture dosage, consistent with Industry 4.0 manufacturing and environmental reporting frameworks.

2.3.1 Cement reduction via rheology stabilization

Cement reduction strategies in modern concrete are fundamentally constrained by fresh-state stability rather than hardened-state performance. Lowering cement content decreases paste volume and weakens the rheological buffer that traditionally prevents segregation, bleeding, and non-uniform flow. Viscosity-modifying admixtures address this limitation by stabilizing the suspension through controlled increases in plastic viscosity and yield stress, allowing cohesive behavior to be maintained at reduced binder levels.

At the microstructural scale, VMAs enhance the effectiveness of particle packing by increasing interparticle friction and forming transient polymer networks within the pore solution. This mechanism reduces the need for excess cement to compensate for fines deficiency or poor grading. In practice, VMAs enable the use of optimized aggregate skeletons with lower paste demand, supporting systematic cement reduction without sacrificing workability or placement reliability.

Industrial implementations in self-consolidating and performance concretes demonstrate cement savings typically ranging from 10 to 20 percent when VMAs are properly matched with superplasticizer chemistry and aggregate grading. These reductions directly translate into lower CO_2 emissions and energy consumption, given that cement clinker production remains the dominant contributor to concrete's carbon footprint. The benefit becomes particularly significant in large-volume infrastructure projects, where small percentage reductions in cement content yield substantial absolute emission savings.

In digitally enabled production environments, rheology stabilization through VMAs supports closed-loop optimization of mix designs. Real-time monitoring of flow, torque, and stability allows incremental cement reduction while VMA dosage compensates for resulting rheological changes. This data-driven approach aligns cement reduction programs with Industry 4.0 principles, ensuring consistency, traceability, and measurable carbon intensity reduction across production cycles.

2.3.2 Compatibility with SCMs, recycled aggregates, and wash water reuse

The incorporation of supplementary cementitious materials, recycled aggregates, and process water reuse is central to contemporary low-carbon concrete strategies, yet each introduces significant rheological variability. Increased specific surface area, heterogeneous particle morphology, and altered ionic composition collectively destabilize fresh concrete behavior. Viscosity-modifying admixtures play a stabilizing role by compensating for these disruptions through controlled enhancement of suspension cohesion and pore solution viscosity.

SCMs such as fly ash, ground granulated blast furnace slag, limestone fillers, and calcined clays modify adsorption equilibria and water demand due to their fine particle size and surface chemistry. VMAs designed with associative or weakly adsorbing polymer structures stabilize the paste without competing excessively with polycarboxylate ether superplasticizers for surface sites. This compatibility preserves dispersion efficiency while reducing segregation risk, particularly in low-clinker and ternary blended systems where fines content is high and particle interactions are complex.

Recycled aggregates introduce angularity, residual mortar, and elevated water absorption, which disrupt flow uniformity and promote localized stiffening. VMAs mitigate these effects by increasing yield stress margins and maintaining homogeneous particle suspension during mixing, transport, and placement. This stabilization improves reproducibility of fresh concrete properties and reduces the need for corrective water or cement additions, which would otherwise undermine sustainability targets.

Wash water reuse introduces suspended fines and dissolved ions that significantly influence admixture performance. Elevated alkalinity and ionic strength can reduce superplasticizer efficiency and accelerate rheological drift. VMAs with tolerance to variable ionic environments stabilize rheology by buffering changes in pore solution viscosity and fine particle concentration. When integrated with sensor-based monitoring of solids content and conductivity, VMA dosage can be adjusted dynamically, enabling closed-loop water management consistent with Industry 4.0 driven resource efficiency.

3. Chemical Families and Molecular Design of VMAs

The performance of viscosity-modifying admixtures is fundamentally dictated by polymer chemistry and molecular architecture. Unlike dispersants, VMAs derive functionality from their ability to structure the aqueous phase, interact weakly with solid surfaces, and respond predictably to shear. Industrial selection and formulation therefore depend on understanding how polymer origin, molecular weight, substitution pattern, and backbone flexibility translate into macroscopic rheological control across diverse concrete systems.

3.1 Natural Polymer-Based VMAs

Natural polymer-based VMAs represent the earliest and most widely adopted class, valued for their availability, water solubility, and predictable thickening behavior. Their performance is primarily governed by hydrophilic chain expansion in the pore solution and limited surface interactions with cement particles.

3.1.1 Cellulose ethers, starch derivatives, guar-based systems

Cellulose Ether VMAs

Cellulose ethers remain the industrial benchmark among natural polymer-based VMAs due to their consistent rheology control, chemical robustness in alkaline pore solutions, and scalable manufacturing routes. Commercial grades are typically derived from wood pulp or cotton linters and chemically modified to form methyl cellulose, hydroxyethyl cellulose, hydroxypropyl methyl cellulose, or mixed substitution variants. The degree of substitution and molar substitution determine solubility kinetics, hydration rate, and the extent of chain expansion in cement pore water. High molecular weight fractions promote yield stress enhancement through water structuring and micro-gel formation, while maintaining shear-thinning behavior under mixing and pumping conditions.

From a formulation standpoint, cellulose ethers act primarily through aqueous phase thickening rather than particle adsorption, minimizing interference with cement hydration when properly selected. Excessive substitution or overly high molecular weight can, however, delay setting by restricting ionic mobility near clinker surfaces. Industrial practice increasingly applies narrow molecular weight distribution grades and controlled substitution patterns to balance viscosity retention with acceptable early strength development. Digital rheometry combined with inline torque monitoring enables Industry 4.0 driven optimization of cellulose ether dosage across variable cement chemistries.

Starch Derivative VMAs

Starch-based VMAs are produced via etherification or esterification of amylose and amylopectin backbones, introducing hydroxyalkyl or carboxymethyl functionalities that enhance water solubility and alkaline stability. Compared to cellulose ethers, starch derivatives typically exhibit lower intrinsic viscosity at equivalent molecular weight, resulting in softer thickening profiles and reduced yield stress buildup. This makes them attractive for self-consolidating concrete and pumped mortars where segregation resistance must be achieved without excessive flow resistance.

Their branched molecular architecture promotes rapid hydration and early viscosity development, although long-term rheological stability can be sensitive to enzymatic degradation or microbial activity if not properly stabilized. Industrial formulations often combine starch derivatives with secondary stabilizers or preservatives to ensure shelf life and performance consistency. Advanced production plants increasingly integrate real-time viscosity tracking and predictive modeling to adjust starch modification levels in response to raw material variability.

Guar and Modified Guar Systems

Guar-based VMAs originate from galactomannan polysaccharides extracted from guar beans and are characterized by extremely high molecular weight and strong hydration-driven thickening efficiency. Even at low dosages, guar polymers impart significant viscosity enhancement and segregation control, making them effective in low binder or high water content mixes. Chemical modification through hydroxypropylation or carboxymethylation is commonly employed to improve solubility rate and tolerance to high pH environments.

Despite their efficiency, guar-based systems exhibit sensitivity to shear degradation and ionic strength, which can lead to viscosity loss under aggressive mixing or in high-alkali cement systems. Industrial users often apply enzyme-deactivated or partially depolymerized grades to improve robustness. Industry 4.0 frameworks increasingly utilize rheology sensors and machine learning algorithms to predict guar performance under variable mixing energy and temperature profiles, reducing the risk of over-thickening or batch-to-batch inconsistency.

Table 3.1.1: Comparative Molecular and Functional Characteristics

Polymer Family	Typical Molecular Weight	Primary Thickening Mechanism	Key Advantages	Principal Limitations
Cellulose ethers	100,000 to >1,000,000 g/mol	Water structuring and micro-gel formation	Stable rheology, broad cement compatibility	Potential setting delay at high dosage
Starch derivatives	50,000 to 500,000 g/mol	Hydration-driven chain expansion	Soft viscosity profile, good pumpability	Biological stability sensitivity
Guar-based systems	>1,000,000 g/mol	Rapid hydration and entanglement	High efficiency at low dosage	Shear and alkali sensitivity

Industrial and Technology Transfer Perspective

Natural polymer-based VMAs continue to play a critical role in cost-sensitive and sustainability-driven concrete formulations. Their molecular design flexibility allows tailoring of rheological profiles without reliance on synthetic polymers, supporting bio-based additive strategies. Successful technology transfer requires close control of raw material sourcing, substitution chemistry, and molecular weight distribution, supported by digital quality control platforms and predictive rheology models. Integration of formulation databases with plant-level process analytics enables consistent VMA performance across regions and cement sources.

3.1.2 Molecular weight distribution and substitution effects

Role of Molecular Weight Distribution in Rheology Control

Molecular weight distribution exerts a primary influence on the rheological efficiency and robustness of natural polymer-based VMAs. Broad distributions introduce a combination of long-chain fractions responsible for yield stress development and shorter chains that contribute to solution viscosity without excessive network formation. Narrow molecular weight distributions, by contrast, deliver more predictable thickening behavior and reduced sensitivity to mixing energy, which is advantageous for ready-mix and precast operations requiring tight process control. Industrial experience shows that bimodal distributions are often deliberately engineered to stabilize segregation resistance while preserving pumpability under high shear.

At the microstructural level, higher molecular weight fractions enhance chain entanglement and water immobilization within the cement pore solution. This mechanism increases plastic viscosity and static stability but also raises the risk of shear degradation during aggressive mixing. Industry 4.0 enabled rheology monitoring systems increasingly correlate inline power consumption and torque signals with molecular weight fingerprints, allowing early detection of polymer degradation or off-spec batches before concrete placement.

Substitution Type and Degree as Functional Design Parameters

Chemical substitution determines polymer solubility, hydration kinetics, and alkaline tolerance. In cellulose ethers, methyl, hydroxyethyl, and hydroxypropyl substituents regulate the balance between hydrophilicity and backbone flexibility. Higher substitution levels promote rapid dissolution and stronger water retention but can suppress ionic diffusion in the pore solution, influencing early hydration reactions. Lower substitution grades hydrate more slowly and tend to produce delayed viscosity build-up, which can be beneficial for extended workability applications.

For starch and guar derivatives, substitution disrupts native hydrogen bonding within the polysaccharide backbone, reducing crystallinity and enabling controlled chain expansion in alkaline environments. Carboxymethyl substitution introduces weak anionic character, enhancing compatibility with certain cement chemistries while increasing sensitivity to calcium ion concentration. Industrial formulations typically optimize substitution degree to remain below thresholds where competitive adsorption or excessive retardation becomes significant.

Interaction Between Molecular Weight and Substitution

Molecular weight and substitution effects are interdependent rather than additive. High molecular weight polymers with low substitution may hydrate incompletely, leading to heterogeneous viscosity development and poor reproducibility. Conversely, highly substituted low molecular weight polymers may dissolve rapidly but fail to generate sufficient network strength to suppress bleeding and segregation. Successful VMA design therefore targets a defined molecular weight window combined with substitution patterns that ensure controlled hydration and stable chain conformation under shear.

Advanced manufacturing increasingly applies statistical process control and spectroscopic fingerprinting to align molecular weight distribution with substitution chemistry. Machine learning models trained on rheological datasets are now being deployed to predict concrete performance based on polymer structural descriptors, supporting rapid formulation adjustments across different cement sources.

Practical Implications for Industrial Formulation

From a technology transfer perspective, molecular weight distribution and substitution control represent critical quality attributes rather than secondary specifications. Variability in either parameter can outweigh dosage adjustments in determining field performance. Producers supplying multiple regional markets often tailor polymer architecture to local cement alkali levels, fineness, and sulfate balance, using digital twins of mixing and placement processes to validate robustness prior to commercialization.

In high-performance and self-consolidating concrete, tighter molecular weight distributions combined with moderate substitution levels are favored to achieve stable flow profiles. In contrast, mass concrete and mortar applications may tolerate broader distributions where cost efficiency and segregation control dominate performance criteria.

Typical Structure–Property Relationships

Table 3.1.2: Typical Structure–Property Relationships

Structural Parameter	Increase Effect	Positive Outcome	Potential Risk
Average molecular weight	Higher chain entanglement	Improved segregation resistance	Shear degradation, pump pressure rise
Molecular weight distribution width	Broader distribution	Balanced yield stress and viscosity	Batch-to-batch variability
Degree of substitution	Faster hydration, higher solubility	Rapid viscosity development	Hydration delay at high levels
Substitution type	Tailored polymer–ion interaction	Cement compatibility tuning	Calcium sensitivity or retardation

3.2 Synthetic and Semi-Synthetic Polymer VMAs

Synthetic and semi-synthetic VMAs have become the preferred rheology modifiers for high-performance, self-consolidating, and specialty concretes where consistency across variable raw materials is critical. Their controlled polymerization routes allow precise adjustment of molecular weight, charge density, and chain architecture, enabling reproducible viscosity control independent of cement source, filler mineralogy, or mixing energy. Unlike natural polymers, these systems exhibit higher resistance to alkaline hydrolysis and microbial degradation, supporting long storage stability and predictable field performance. Industrial formulation increasingly relies on digitally defined polymer specifications linked to rheological performance databases, aligning with Industry 4.0 driven admixture production and deployment.Polyacrylamides, polyacrylates, associative polymers

3.2.1 Polyacrylamides, Polyacrylates, and Associative Polymers

Polyacrylamide-Based VMAs

Polyacrylamide-based VMAs represent one of the most widely engineered synthetic systems for viscosity control in cementitious formulations. These polymers are typically produced via free-radical polymerization to high molecular weights, enabling effective aqueous phase structuring through chain entanglement rather than surface adsorption. Partial hydrolysis introduces controlled anionic functionality, improving solubility and dispersion in alkaline pore solutions while maintaining compatibility with calcium-rich environments. Excessive hydrolysis increases charge density, which can compromise viscosity stability due to ionic shielding and competitive interactions with dispersing admixtures.

From an industrial processing perspective, polyacrylamides exhibit strong shear-thinning behavior, supporting pumpability and placement without sacrificing static segregation resistance. Their resistance to biological degradation and predictable hydration kinetics favor liquid admixture formats with long shelf life. Advanced admixture plants increasingly correlate polymer molecular fingerprints with inline rheological response to detect mechanical degradation during handling and dosing.

Polyacrylate-Based VMAs

Polyacrylates used as VMAs differ structurally and functionally from dispersant-type polycarboxylates. They are designed with higher molecular weights and lower effective charge densities, allowing them to contribute to viscosity enhancement without inducing excessive particle dispersion. Thickening arises from a combination of water retention, steric hindrance, and weak polymer–particle interactions, yielding smoother flow curves compared to polyacrylamide systems.

Industrial formulation advantages include high tolerance to cement variability and improved compatibility with modern superplasticizers when properly balanced. Polymer architecture can be tuned through copolymer composition and branching, enabling customization for self-consolidating concrete, mortars, and sprayed applications. Industry 4.0 practices increasingly integrate polymer synthesis data with field rheology feedback to optimize polyacrylate-based VMAs across regional cement sources.

Associative Polymer VMAs

Associative polymers constitute a specialized class of VMAs characterized by hydrophobic groups grafted onto hydrophilic backbones such as polyacrylamides or polyethers. These hydrophobic moieties form transient physical junctions in aqueous solution, creating reversible networks that impart pronounced thixotropy and rapid viscosity recovery after shear. Such behavior is particularly advantageous in self-consolidating and shotcrete systems where flow under shear must be followed by rapid structural rebuild.

Performance sensitivity to hydrophobe content, distribution, and spacing is significantly higher than in non-associative systems. Over-association can lead to excessive yield stress and workability loss, while under-association limits segregation control. Industrial suppliers increasingly rely on rheological profiling and predictive analytics to align associative polymer architecture with specific mixing energy and placement conditions.

Comparative Performance Characteristics

Synthetic and semi-synthetic viscosity-modifying admixtures exhibit clearly differentiated performance profiles that align with specific concrete processing and application requirements. Polyacrylamide-based systems thicken predominantly through molecular chain entanglement and structured water domains, generating pronounced shear-thinning behavior that translates into high efficiency at low addition rates and effective resistance to segregation. Their performance, however, can deteriorate under highly alkaline conditions if excessive hydrolysis alters molecular architecture. Polyacrylates act mainly through water retention combined with steric stabilization, resulting in more moderate shear-thinning behavior and excellent compatibility with polycarboxylate ether superplasticizers, although their thickening contribution is comparatively limited at very low dosages. Associative polymers rely on reversible hydrophobic interactions to form transient networks, imparting strong thixotropy and rapid structural rebuild after shear cessation, a property highly valued in applications requiring shape stability, while simultaneously demanding tight dosage control due to their relatively narrow operational window.

Table 3.2.1: Comparative Performance Characteristics

Polymer Class	Primary Thickening Mechanism	Shear Response	Key Industrial Advantage	Principal Limitation
Polyacrylamides	Chain entanglement and water structuring	Strong shear thinning	High efficiency, robust segregation control	Sensitivity to over-hydrolysis
Polyacrylates	Water retention and steric effects	Moderate shear thinning	Compatibility with dispersants	Lower thickening at low dosage
Associative polymers	Hydrophobic network formation	Pronounced thixotropy	Rapid viscosity recovery	Narrow formulation window

Industrial and Technology Transfer Considerations

The selection between polyacrylamide, polyacrylate, and associative polymer VMAs is governed by performance envelope, process robustness, and formulation compatibility rather than cost alone. Effective technology transfer requires translating polymer structural parameters into actionable dosage and mixing guidelines, supported by digital rheology monitoring and feedback loops from field applications. Integration of polymer design, admixture production, and concrete performance data is increasingly essential to maintain consistency under Industry 4.0 manufacturing paradigms.

3.2.2 Tailoring Hydrophilicity and Backbone Flexibility

Hydrophilicity as a Control Lever for Aqueous Phase Structuring

Hydrophilicity governs polymer solvation, chain expansion, and the kinetics of viscosity development in cement pore solutions. In synthetic and semi-synthetic VMAs, hydrophilicity is precisely adjusted through monomer selection, copolymer ratio, and functional group density along the backbone. Highly hydrophilic polymers hydrate rapidly and generate immediate viscosity build-up, which is advantageous for segregation control in low-fines or high water-to-cement ratio systems. Excessive hydrophilicity, however, increases sensitivity to calcium ion shielding and can reduce rheological stability in highly alkaline environments.

Industrial formulation strategies typically target intermediate hydrophilicity windows where chain expansion is sufficient to structure the aqueous phase without inducing excessive electrostatic interactions. Digital solubility modeling and pore-solution simulations are increasingly used to match polymer hydrophilicity profiles with cement alkali levels and sulfate balance, reducing trial-based optimization.

Backbone Flexibility and Shear-Responsive Behavior

Backbone flexibility determines how polymer chains deform, align, and disentangle under shear. Flexible backbones such as polyacrylamides exhibit pronounced shear-thinning behavior, enabling low pumping pressures while maintaining static viscosity after placement. More rigid backbones or highly substituted chains increase resistance to deformation, contributing to higher yield stress but also elevating sensitivity to mechanical degradation during mixing.

Controlled incorporation of comonomers, branching, or short side chains allows modulation of backbone flexibility without sacrificing molecular integrity. Industrial producers increasingly quantify flexibility indices using rheological fingerprinting combined with molecular modeling, enabling predictive adjustment of polymer architecture for specific mixing energy and temperature regimes.

Balancing Hydrophilicity and Flexibility in Polymer Design

Hydrophilicity and backbone flexibility act synergistically in determining VMA performance. Polymers with high hydrophilicity but insufficient flexibility may dissolve rapidly yet fail to sustain viscosity under shear. Conversely, highly flexible but weakly hydrophilic systems may hydrate slowly, delaying rheological control during early mixing. Successful designs therefore align hydration rate, chain mobility, and network formation to achieve stable viscosity profiles across processing stages.

Industry 4.0 driven formulation platforms increasingly integrate polymer synthesis parameters with real-time rheology data, enabling closed-loop optimization of hydrophilicity and flexibility for different concrete classes. This approach improves robustness against raw material variability and reduces formulation drift during scale-up.

Structure–Property Design Map

Engineering synthetic viscosity-modifying admixtures through controlled adjustment of hydrophilicity and polymer backbone architecture enables performance to be tailored with a high degree of specificity, moving beyond the concept of universal thickening agents. Elevated hydrophilicity promotes rapid solvation and early viscosity build-up, which is advantageous for immediate stability control, yet it concurrently increases sensitivity to ionic strength and cement chemistry. Enhanced backbone flexibility facilitates molecular deformation under shear, improving pumpability and flow under dynamic conditions, while potentially accelerating viscosity decay during prolonged or high-energy mixing. The deliberate use of comonomers, along with controlled branching, allows precise regulation of hydration kinetics, entanglement density, and structural recovery, yielding rheological profiles aligned with defined mixing, transport, and placement scenarios. Such molecular-level optimization inevitably increases formulation and processing complexity, necessitating advanced process control, digital monitoring, and well-defined industrial guidelines to mitigate risks related to mixing sensitivity, ionic interference, and scale-up during technology transfer.

Table 3.2.2: Structure–Property Design Map

Design Variable	Increase Effect	Functional Outcome	Industrial Risk
Hydrophilicity	Faster solvation	Rapid viscosity development	Ionic sensitivity
Backbone flexibility	Enhanced chain deformation	Improved pumpability	Viscosity loss under shear
Comonomer content	Tuned hydration rate	Balanced rheology	Process complexity
Branching level	Controlled entanglement	Structural rebuild	Mixing sensitivity

Industrial Relevance and Technology Transfer

Tailoring hydrophilicity and backbone flexibility transforms synthetic VMAs from generic thickeners into engineered rheology modifiers. Effective technology transfer requires translating molecular design choices into clear formulation guidelines and processing windows, supported by digital monitoring of mixing energy and rheological response. Integration of laboratory polymer design with plant-scale analytics ensures consistent performance across production sites and aligns VMA development with advanced manufacturing frameworks.

3.3 Hybrid and Next-Generation Architectures

The progressive densification of concrete formulations, coupled with higher fines content, recycled constituents, and tighter performance tolerances, has accelerated the development of hybrid VMA architectures. These systems integrate multiple molecular functions within a single polymer framework, enabling simultaneous control of aqueous phase structuring, shear response, and compatibility with modern dispersants. Hybrid designs reduce formulation complexity by minimizing additive stacking while delivering robust rheology across variable cement chemistries. Industrial development increasingly relies on digitally assisted polymer design and performance mapping to translate laboratory concepts into scalable admixture solutions.

3.3.1 Comb polymers and functionalized backbones

Comb Polymer Architecture in VMA Design

Comb polymers represent a strategically engineered class of hybrid VMAs in which a primary backbone is decorated with pendant side chains that introduce secondary control functions. In viscosity-modifying applications, the backbone typically provides bulk rheology modification through chain entanglement and aqueous phase structuring, while the side chains regulate solubility, steric effects, and shear response. Unlike superplasticizer comb polymers, which emphasize particle dispersion through dense grafting, VMA comb polymers employ lower graft densities and shorter side chains to preserve viscosity enhancement without inducing excessive deflocculation.

The spatial arrangement of side chains dictates polymer conformation in cement pore solutions. Optimized architectures promote extended yet flexible chain geometries that maintain static viscosity and segregation resistance while allowing rapid alignment and disentanglement under shear. Industrial synthesis increasingly relies on controlled radical polymerization to achieve consistent graft distribution, supported by inline molecular weight and graft density monitoring.

Functionalized Backbones for Multi-Mechanism Performance

Functionalized backbones incorporate targeted chemical groups along the main chain to introduce additional mechanisms beyond simple thickening. Weakly anionic functionalities improve compatibility with calcium-rich pore solutions while avoiding strong adsorption that could interfere with dispersants. Hydrophobic moieties enable associative interactions, forming reversible junctions that impart thixotropy and rapid structural rebuild after shear cessation. In advanced systems, functional groups capable of moderating local ionic environments are incorporated to stabilize rheology under fluctuating alkali and sulfate levels.

From an industrial formulation perspective, functionalized backbones reduce reliance on additive stacking by consolidating multiple performance attributes into a single polymer. This approach improves robustness in blended cement systems and concretes containing recycled fines, where pore solution chemistry and surface reactivity vary significantly. Digital modeling of polymer conformation and interaction profiles increasingly supports early-stage selection of functionalized backbone designs.

Structure–Property Relationships in Comb and Functionalized Systems

The performance envelope of comb and functionalized VMAs is governed by the balance between backbone length, graft density, and functional group distribution. Excessive side-chain density or strong functionality can shift the polymer toward dispersant-like behavior, reducing viscosity efficiency. Insufficient functionalization, by contrast, limits adaptability to modern concrete systems. Successful designs occupy a narrow architectural window that delivers viscosity control, shear responsiveness, and admixture compatibility simultaneously.

Industry 4.0 driven development frameworks increasingly integrate polymer synthesis parameters, rheological fingerprints, and field performance data into closed-loop optimization platforms. This enables rapid refinement of comb and functionalized architectures with reduced experimental iteration.

Key Design Parameters and Functional Impact

The performance of viscosity-modifying admixtures is governed by a set of interdependent architectural parameters that must be balanced to achieve robust industrial behavior. Increasing backbone molecular weight strengthens chain entanglement within the aqueous phase, thereby enhancing segregation resistance and stability, yet simultaneously elevates susceptibility to mechanical degradation under high shear mixing and pumping conditions. Control of side-chain length influences polymer solubility and hydration stability, supporting consistent viscosity development across variable cement chemistries, while excessive extension of side chains can promote over-dispersion and diminish effective thickening. Graft density serves as a sensitive lever for tailoring shear response and flow modulation, enabling precise rheological adjustment, but it operates within a narrow formulation window that increases the risk of performance drift during scale-up. Selection of functional group chemistry further enables targeted interactions with cement phases and supplementary cementitious materials, improving compatibility and robustness, although it introduces additional chemical complexity that must be carefully managed during formulation development, quality control, and manufacturing transfer.

Table 3.3.1: Key Design Parameters and Functional Impact

Architectural Parameter	Primary Function	Performance Benefit	Design Limitation
Backbone molecular weight	Chain entanglement	Segregation resistance	Shear sensitivity
Side-chain length	Solubility control	Stable hydration	Over-dispersion risk
Graft density	Shear response tuning	Controlled flow	Narrow design window
Functional group type	Targeted interactions	Cement compatibility	Interaction complexity

Industrial and Technology Transfer Considerations

Comb polymers and functionalized backbones exemplify the transition from empirical VMA selection to molecularly engineered rheology control. Effective technology transfer requires clear translation of architectural parameters into formulation guidelines, supported by digital analytics, inline quality monitoring, and feedback from field applications. These systems enable high-performance rheology control at reduced dosage, aligning with advanced manufacturing and sustainability objectives.

3.3.2 Design Trends for Low-Dosage, High-Efficiency Systems

Molecular Efficiency as a Primary Design Metric

Current VMA development increasingly prioritizes molecular efficiency, defined by the magnitude and stability of rheological modification achieved per unit mass of polymer. Low-dosage systems rely on precise alignment of polymer architecture with dominant structuring mechanisms in the cement pore solution, reducing reliance on bulk concentration effects. Hybrid polymers combining aqueous phase structuring, weak particle interaction, and reversible association enable significant viscosity enhancement at addition levels well below those required for conventional VMAs.

Industrial evaluation increasingly replaces single-point viscosity measurements with multi-parameter efficiency indices incorporating yield stress, plastic viscosity, and structural build-up rate. These datasets support algorithm-driven ranking of candidate polymers, accelerating selection and reducing formulation iteration during scale-up.

Architectural Optimization for Maximum Impact

Low-dosage efficiency is achieved through deliberate control of molecular weight distribution, functional group placement, and chain conformation. High molecular weight backbones deliver strong structuring, while localized functional domains amplify interaction efficiency without increasing overall polymer mass. Comb and functionalized architectures are increasingly favored, as they enable decoupling of solubility control from thickening performance.

Polymer designs now routinely target narrow architectural windows where small dosage changes translate into predictable rheological shifts. This precision reduces overdosing risk and improves reproducibility across batching plants. Industry 4.0 manufacturing platforms integrate synthesis data, rheological fingerprints, and application feedback to continuously refine these architectural targets.

Compatibility-Driven Design in Complex Admixture Systems

Low-dosage VMAs must operate effectively within dense admixture packages that include high-range water reducers, accelerators, and supplementary cementitious materials. Modern designs emphasize low effective charge density and steric shielding to avoid competitive adsorption and performance loss. Functional groups are selected to maintain compatibility with evolving binder chemistries, including limestone-rich and blended cements.

Industrial formulation workflows increasingly employ digital twins of admixture interaction networks to pre-assess compatibility, minimizing costly field corrections. These tools enable selection of high-efficiency VMAs that maintain performance across diverse concrete compositions.

Sustainability and Process Robustness Considerations

Reduced dosage directly supports sustainability objectives by lowering polymer consumption, transportation impact, and formulation complexity. High-efficiency systems also reduce sensitivity to raw material variability, improving process robustness in ready-mix operations. Advanced polymerization techniques minimize off-spec production, aligning economic and environmental performance.

Integration of real-time synthesis control, inline rheology monitoring, and field data analytics creates a closed-loop development environment. This Industry 4.0 approach ensures that low-dosage, high-efficiency VMAs deliver consistent performance from polymer reactor to concrete placement.

Design Levers for Low-Dosage VMA Systems

The development of low-dosage viscosity-modifying admixtures hinges on strategic molecular design levers that enhance efficiency without excessive polymer use. Hybrid polymer architectures integrate multiple functional mechanisms into a single molecule, yielding high rheological impact but increasing synthesis complexity. Functional localization focuses interactive groups into specific zones, reducing required dosage within a narrow, precise tolerance window. Controlling molecular weight distribution ensures predictable performance and dosage precision, albeit at higher production costs. Additionally, minimizing polymer charge density improves admixture compatibility and stabilizes performance in variable ionic environments, though it may diminish beneficial ion-sensitive structuring. Successfully deploying these advanced systems requires careful balancing of their significant efficiency gains against inherent implementation risks.

Table 3.3.2: Design Levers for Low-Dosage VMA Systems

Design Lever	Functional Strategy	Efficiency Gain	Implementation Risk
Hybrid architecture	Multiple mechanisms per molecule	High rheological impact	Synthesis complexity
Functional localization	Targeted interaction zones	Reduced dosage	Narrow tolerance window
Controlled MWD	Predictable response	Dosage precision	Production cost
Low charge density	Admixture compatibility	Stable performance	Reduced ion sensitivity

Industrial and Technology Transfer Perspective

Design trends toward low-dosage, high-efficiency VMAs reflect a broader shift toward molecularly precise admixture engineering. Successful technology transfer depends on converting architectural design principles into practical formulation guidelines supported by digital quality systems and field feedback loops. As concrete systems continue to evolve, these trends provide a scalable pathway to maintain performance while reducing additive load and operational variability.

4. Rheological Principles Governing VMA Performance

Viscosity-modifying admixtures operate by selectively altering the rheological response of cementitious suspensions under static and dynamic conditions. Their primary function is to decouple stability from workability, allowing modern concrete systems to maintain segregation resistance without excessive yield stress penalties during placement and pumping. From an industrial formulation standpoint, VMAs are designed to operate within the Bingham or Herschel–Bulkley frameworks that govern fresh concrete flow behavior, with tailored sensitivity to shear rate, solid volume fraction, and ionic environment.

The performance envelope of VMAs becomes increasingly critical in low water-to-binder ratio systems, self-consolidating concrete, and mixtures incorporating supplementary cementitious materials or recycled fines. Industry 4.0 integration enables real-time rheological fingerprinting using inline rheometers and AI-driven feedback loops, allowing VMA dosage and chemistry to be dynamically optimized during batching and mixing to maintain target rheological windows across variable raw material quality.

4.1 Yield Stress, Plastic Viscosity, and Thixotropy Control

Yield stress represents the minimum stress required to initiate flow in cementitious systems and is directly linked to static stability, formwork pressure, and resistance to segregation. Viscosity-modifying admixtures influence yield stress primarily through weak, reversible structuring of the solid–liquid network rather than strong particle dispersion. Industrially optimized VMAs aim to stabilize the paste matrix at rest while avoiding excessive yield stress escalation that would compromise pumping efficiency and surface finish quality in high-performance concrete formulations.

Plastic viscosity governs the resistance to flow once yielding has occurred and is the dominant parameter controlling pumping energy, flow uniformity, and mold filling behavior. VMAs are specifically engineered to increase plastic viscosity through polymer chain entanglement and water phase thickening, allowing fine particles to remain suspended under dynamic conditions. From a formulation standpoint, this selective viscosity enhancement enables low water-to-binder ratio mixes to maintain robust rheological stability even when incorporating fillers, recycled fines, or high volumes of supplementary cementitious materials.

Thixotropy describes the time-dependent breakdown and rebuild of internal structure under shear and rest cycles. VMAs contribute to controlled thixotropic behavior by forming transient physical networks that collapse under shear and rapidly re-form when shear is removed. This property is critical in applications such as self-consolidating concrete, slipform paving, and 3D concrete printing, where rapid structural recovery ensures shape retention and dimensional accuracy. The kinetics of thixotropic rebuilding are strongly influenced by polymer molecular weight, charge density, and adsorption dynamics at the cement particle surface.

In modern production environments, Industry 4.0 tools enable quantitative control of yield stress, plastic viscosity, and thixotropy through real-time rheological monitoring and digital process control. Inline rheometers, mixer power signature analysis, and AI-assisted data models allow continuous adjustment of VMA dosage in response to raw material variability and environmental conditions. Such data-driven control strategies significantly improve batch-to-batch consistency, reduce overdesign, and accelerate technology transfer from laboratory formulations to full-scale concrete manufacturing.

Table 4.1: Key Rheological Effects of VMAs on Fresh Concrete

Rheological Parameter	Primary VMA Influence Mechanism	Industrial Relevance
Yield stress	Weak reversible network formation	Static stability, segregation control
Plastic viscosity	Water phase structuring and polymer entanglement	Pumpability, flow uniformity
Thixotropy	Time-dependent structural rebuild	Shape retention, buildability

4.1.1 Interaction with Cement Particle Flocculation

Fresh cement paste is characterized by a dynamic flocculated structure arising from van der Waals attraction, electrostatic interactions, and ionic bridging mediated by calcium and sulfate species. These flocs form a space-filling network that contributes significantly to static yield stress and early structural buildup. Viscosity-modifying admixtures interact with this flocculated system without inducing full dispersion, differentiating their function from polycarboxylate-based superplasticizers used for flow enhancement.

At the molecular level, VMAs adsorb partially onto cement particle surfaces or remain in the pore solution, depending on their functional group chemistry and molecular architecture. Cellulose ethers, polysaccharide derivatives, and synthetic water-soluble polymers introduce steric constraints that limit excessive floc growth while preserving reversible particle associations. This controlled modulation reduces the brittleness of flocculated clusters under shear, enabling smoother breakdown during pumping and improved homogeneity during placement.

From an industrial formulation perspective, the interaction between VMAs and cement flocculation is highly sensitive to clinker phase composition, fineness, and sulfate availability. Variations in C_3A content or alkali levels alter polymer adsorption kinetics and floc strength, often requiring cement-specific VMA optimization. Advanced manufacturing plants increasingly rely on inline torque monitoring and AI-assisted pattern recognition to detect abnormal flocculation behavior in real time, allowing proactive dosage correction and minimizing fresh concrete rejection.

During rest periods, VMAs promote rapid yet controlled re-flocculation by facilitating weak physical bridging through hydrated polymer chains. This behavior supports thixotropic rebuilding without inducing irreversible agglomeration that would impair later flow. Such flocculation management is particularly critical in self-consolidating concrete and additive manufacturing, where repeated shear-rest cycles occur. Digital twins of fresh concrete rheology, integrated into MES platforms, are now used to simulate flocculation dynamics across mixing, transport, and placement stages, accelerating robust technology transfer from laboratory-scale studies to industrial deployment.

4.1.2 Water structuring and shear-dependent behavior

Water structuring is a primary mechanism through which viscosity-modifying admixtures regulate the rheological response of cementitious systems. VMAs introduce hydrophilic polymer chains capable of forming extensive hydrogen bonding networks with free and weakly bound water in the pore solution. This interaction reduces water mobility at low shear rates, effectively increasing apparent viscosity and enhancing suspension stability without significantly increasing interparticle attractive forces.

Under increasing shear, the structured water domains undergo progressive alignment and partial disentanglement of polymer chains. This shear-induced reorganization results in pronounced shear-thinning behavior, allowing the material to flow readily during mixing, pumping, and placement. From an industrial processing standpoint, this reversible viscosity reduction is essential for maintaining energy-efficient pumping while preserving resistance to bleeding and segregation at rest.

The extent and reversibility of shear-dependent behavior are strongly influenced by polymer molecular weight distribution, substitution degree, and backbone flexibility. High-molecular-weight, lightly substituted polymers tend to generate stronger water structuring and higher low-shear viscosity, while more substituted or branched architectures favor faster alignment under shear. Industrial formulation strategies therefore tailor VMA chemistry to the target shear profile associated with specific placement technologies, including tremie casting, slipforming, and 3D concrete printing.

Industry 4.0 implementation enables quantitative mapping of shear-dependent viscosity profiles using inline rheometers and mixer power analytics. These datasets are increasingly coupled with machine learning models that correlate water structuring behavior to field performance indicators such as formwork pressure evolution and surface finish quality. Such data-driven control frameworks facilitate rapid formulation optimization and reliable technology transfer across plants, cement sources, and environmental conditions.

4.2 Bleeding, Segregation, and Stability Mechanisms

Bleeding in cementitious systems originates from the upward migration of free water driven by density differences between solid particles and the pore solution, combined with insufficient network strength to retain water under gravity. In modern low water-to-binder ratio concretes, bleeding is less visible but remains technologically relevant due to its impact on surface defects, interlayer bonding, and early-age durability. Viscosity-modifying admixtures mitigate bleeding primarily by immobilizing free water through polymer-induced structuring, thereby reducing permeability pathways within the fresh paste matrix.

Segregation represents the spatial separation of constituents, typically coarse aggregates settling downward while paste and fines migrate upward. This phenomenon becomes critical in self-consolidating concrete, pumped concrete, and mixtures containing recycled aggregates or fillers with broad particle size distributions. VMAs counteract segregation by increasing the viscosity of the continuous phase and enhancing the suspension capacity of the paste, ensuring uniform stress transfer to aggregates during flow. Unlike yield-stress-driven stabilization alone, viscosity-driven mechanisms remain effective under dynamic shear, which is essential during transport and placement.

Stability is the integrated outcome of controlled bleeding and segregation resistance under both static and dynamic conditions. VMAs contribute to stability by forming weak, reversible polymer–water networks that adapt to shear history and rest periods. These networks do not rely on strong particle flocculation, which could compromise workability, but instead provide a continuous rheological backbone that supports fines and aggregates uniformly. From an industrial manufacturing standpoint, this balance is crucial for achieving robustness against raw material variability, particularly when incorporating supplementary cementitious materials, manufactured sands, or reclaimed fines.

Industry 4.0 frameworks increasingly quantify bleeding and segregation risk through digital rheological signatures rather than empirical visual tests alone. Inline density profiling, mixer torque analysis, and AI-based prediction models enable proactive adjustment of VMA dosage and chemistry in real time. Such data-driven stabilization strategies significantly reduce quality deviations, minimize on-site corrective actions, and accelerate reliable technology transfer across plants, climates, and application segments.

Table 4.2: Comparative Overview of Bleeding, Segregation, and VMA-Based Stability Control

Aspect	Bleeding	Segregation	Stability via VMAs
Primary driving force	Water density gradient and gravity	Particle density and size disparity	Rheological network formation
Critical phase affected	Free and weakly bound water	Aggregate and fine particle distribution	Entire continuous phase
Sensitivity to shear	Low shear dominant	High shear critical	Effective across shear regimes
Typical consequences	Surface laitance, weak top layer	Non-uniform strength, poor finish	Uniform placement and durability
VMA control mechanism	Water immobilization and retention	Increased paste viscosity and suspension capacity	Reversible polymer–water structuring
Relevance to SCC and 3D printing	Moderate	High	Very high
Industry 4.0 monitoring tools	Bleed water sensors, surface analytics	Inline density and particle tracking	Rheological digital twins and AI control

4.2.1 Capillary Water Immobilization

Capillary water immobilization is a central stability mechanism through which viscosity-modifying admixtures suppress bleeding and enhance uniformity in fresh cementitious systems. In the absence of rheology control, capillary water migrates upward through interconnected pore channels formed between settling solid particles. VMAs disrupt this mobility by increasing the effective viscosity of the pore solution and introducing polymer–water interactions that reduce capillary flow continuity, thereby lowering the hydraulic conductivity of the fresh paste.

At the physicochemical level, hydrophilic VMA polymers establish extensive hydrogen bonding with capillary water, converting free water into a semi-bound state with reduced translational freedom. This transformation does not rely on chemical fixation but on reversible physical interactions that remain sensitive to shear. As a result, water is retained within the capillary network during rest, while still being available for flow under applied stress. Such controlled immobilization is particularly relevant in high-fines systems and self-consolidating concretes, where excessive free water rapidly manifests as surface bleeding or laitance.

From an industrial formulation perspective, the efficiency of capillary water immobilization depends strongly on polymer molecular weight, substitution pattern, and solubility kinetics. Cellulose ethers and tailored synthetic copolymers differ in their ability to span pore throats and bridge capillary channels, leading to distinct water retention profiles. Cement chemistry and temperature further influence these interactions by altering pore solution ionic strength, necessitating formulation strategies that are robust across production sites and seasonal conditions.

Industry 4.0 approaches increasingly quantify capillary water immobilization using inline moisture profiling, acoustic emission analysis, and rheological digital twins that correlate water mobility with bleeding risk. Machine learning models trained on these datasets allow predictive adjustment of VMA dosage in real time, reducing reliance on empirical bleed tests and supporting rapid, reproducible technology transfer from laboratory design to full-scale concrete production.

Table 4.2.1: Comparative Influence of VMA Chemistry on Capillary Water Immobilization

VMA Chemistry Type	Dominant Interaction with Water	Immobilization Efficiency	Sensitivity to Cement Chemistry	Industrial Use Case
Cellulose ethers	Hydrogen bonding and chain entanglement	High	Moderate to high	SCC, repair mortars

Natural polysaccharides	Physical gel formation	Moderate	High	Low-shear applications
Synthetic water-soluble polymers	Tunable hydrogen bonding	High and controllable	Low to moderate	Ready-mix, pumping
Hybrid VMA systems	Synergistic water structuring	Very high	Low	High-performance concretes

4.2.2 Aggregate Suspension and Paste Cohesion

Aggregate suspension in fresh concrete is governed by the ability of the cement paste to transmit shear stresses uniformly and counteract gravitational settling of coarse particles. In inadequately stabilized systems, density differences between aggregates and the surrounding paste lead to rapid particle migration, resulting in segregation and non-uniform structural performance. Viscosity-modifying admixtures enhance aggregate suspension by increasing the cohesive strength of the paste phase, allowing it to behave as a continuous load-bearing medium rather than a low-viscosity carrier fluid.

At the microstructural level, VMAs reinforce paste cohesion through polymer-induced thickening of the pore solution and the formation of transient physical networks spanning between cement particles and fines. These networks increase the yield resistance against particle settling without inducing rigid flocculation. Unlike mechanical stabilization achieved solely by high fine content, VMA-driven cohesion remains effective under dynamic shear conditions encountered during pumping, casting, and vibration, which is critical for industrial-scale concrete placement.

From a formulation and manufacturing perspective, aggregate suspension efficiency is highly sensitive to particle size distribution, aggregate shape, and surface roughness. Angular or recycled aggregates with higher specific surface area impose greater demands on paste cohesion, often necessitating tailored VMA chemistries or synergistic use with fillers. Advanced production facilities increasingly deploy inline density gradient monitoring and mixer power analytics to quantify suspension quality in real time, enabling adaptive control of paste rheology across varying aggregate sources.

Industry 4.0 integration further enhances suspension control by linking rheological parameters to digital twins of concrete flow and placement behavior. Machine learning models correlate paste cohesion metrics with segregation indices measured on site, allowing predictive optimization of VMA dosage before defects occur. Such data-driven methodologies significantly improve robustness during technology transfer, ensuring consistent aggregate suspension performance across plants, mix designs, and application environments.

Table 4.2.2: Comparative Effects of Paste Cohesion on Aggregate Suspension Performance

Paste Cohesion Level	Aggregate Suspension Behavior	Segregation Risk	Typical Industrial Scenario
Low	Rapid aggregate settling	High	Lean mixes, high water content
Moderate	Partial suspension under static conditions	Medium	Conventional vibrated concrete
High (VMA-controlled)	Stable suspension under static and dynamic shear	Low	SCC, pumped concrete
Excessive	Restricted flow and poor finishability	Low but impractical	Overdosed VMA systems

4.3 Interaction with Other Admixtures

Viscosity-modifying admixtures function within complex multicomponent chemical environments where their performance is inherently linked to the presence of other admixtures. In industrial concrete formulations, VMAs must be engineered to complement dispersing, air-entraining, and set-controlling agents without compromising robustness or processing efficiency. The combined rheological response is therefore the result of simultaneous polymer adsorption, pore solution modification, and dynamic microstructural evolution under shear.

From a manufacturing and technology transfer perspective, uncontrolled admixture interactions represent a major source of variability during scale-up and cross-plant deployment. Differences in cement chemistry, sulfate balance, and temperature amplify these interactions, often causing deviations between laboratory-optimized and plant-scale behavior. Industry 4.0 tools, including inline rheology monitoring and AI-assisted formulation engines, are increasingly applied to detect interaction-induced drift and automatically correct admixture ratios during production.

4.3.1 Synergy and Antagonism with PCE Superplasticizers

Polycarboxylate ether superplasticizers and viscosity-modifying admixtures are jointly responsible for defining the flow–stability balance in modern high-performance concrete systems. PCEs primarily reduce yield stress by dispersing cement particles through electrostatic repulsion and steric hindrance, while VMAs selectively increase plastic viscosity by structuring the pore solution and forming reversible polymer networks. When correctly formulated, this complementary action produces concretes with high deformability under shear and strong resistance to bleeding and segregation at rest.

Synergistic behavior is achieved when the adsorption domains and functional roles of the two polymer families are effectively decoupled. Contemporary VMA chemistries are increasingly designed to remain predominantly in the aqueous phase, minimizing competition with PCE backbones for cement surface sites. This allows PCE molecules to fully express their dispersing efficiency, while VMAs govern water retention and viscosity independently. Such synergy is essential in self-consolidating concrete, pumped concrete, and mixtures containing high fines or recycled constituents, where both dispersion and stability are simultaneously required.

Antagonism arises when competitive adsorption or unfavorable polymer–polymer interactions occur in the pore solution. High-charge-density PCEs can displace weakly adsorbed VMA chains from cement surfaces, leading to a sudden loss of viscosity control and increased segregation risk. Conversely, certain high-molecular-weight VMAs may partially shield PCE side chains or increase solution viscosity to a level that limits particle mobility, reducing superplasticizer efficiency and increasing pumping pressure. These effects are strongly dependent on calcium ion concentration, sulfate balance, and C_3A content, making cement-specific interaction mapping a critical industrial requirement.

From a technology transfer and manufacturing perspective, Industry 4.0 tools are increasingly employed to manage PCE–VMA compatibility. Inline rheological sensing, mixer power signature analysis, and AI-assisted pattern recognition enable early detection of antagonistic trends during batching and mixing. Digital rheological twins linked to MES platforms allow predictive adjustment of admixture ratios in real time, significantly reducing batch variability and ensuring robust performance across plants, cement sources, and environmental conditions.

Table 4.3.1: Comparative Overview of PCE–VMA Interaction Regimes

Interaction Regime	Dominant Mechanism	Rheological Signature	Practical Implication
Optimal synergy	Dispersed particles with structured water	Low yield stress, elevated plastic viscosity	High flowability with stability
PCE-dominated	Excessive dispersion	Very low viscosity, low thixotropy	Segregation and bleeding risk
VMA-dominated	Overstructured pore solution	High viscosity, limited dispersion	Reduced pumpability
Competitive adsorption	Polymer displacement	Unstable, time-dependent rheology	Batch inconsistency

4.3.2 Sensitivity to Accelerators, Retarders, and Air Entrainers

Viscosity-modifying admixtures exhibit pronounced sensitivity to chemical agents that alter early hydration kinetics and interfacial phenomena. Accelerators, retarders, and air-entraining agents modify ionic strength, calcium availability, and surface activity in the pore solution, thereby influencing VMA solubility, chain conformation, and network formation. In industrial formulations, these interactions directly affect fresh-state robustness, early strength development, and rheological stability during transport and placement.

Interaction with Set Accelerators

Chemical accelerators such as calcium nitrate, calcium formate, alkanolamines, and alkali thiocyanates increase early calcium ion concentration and accelerate C_3S and C_3A hydration. Elevated Ca^{2+} levels can partially collapse VMA polymer coils through charge screening, reducing their thickening efficiency and shortening open time. Polysaccharide-based and anionic synthetic VMAs are particularly sensitive, showing reduced viscosity contribution at equivalent dosages under accelerated systems.

From a manufacturing standpoint, this interaction often manifests as loss of segregation resistance during cold-weather concreting where accelerator dosages are high. Advanced formulation strategies rely on calcium-tolerant VMA architectures or delayed-activation VMAs that develop viscosity only after initial accelerator consumption. Industry 4.0 implementation using inline ion-selective sensors and early-age rheology tracking enables dynamic compensation by adjusting VMA or accelerator dosage during batching, maintaining consistent fresh performance across seasonal conditions.

Interaction with Retarders and Set-Controlling Agents

Retarders including lignosulfonates, hydroxycarboxylic acids, phosphonates, and sugars extend induction periods and reduce early ion release. In the presence of VMAs, this delayed hydration environment can amplify viscosity build-up due to prolonged polymer persistence in the pore solution. While this can enhance static stability and water retention, excessive retardation combined with high VMA levels may lead to overly thixotropic systems with slow structural breakdown under shear.

Industrial experience shows that synthetic VMAs with controlled molecular weight distributions are less prone to over-structuring in retarded systems than natural polymers. Digital rheology models calibrated with calorimetry and torque evolution data are increasingly used to predict such interactions before plant trials. These models support technology transfer by enabling cement-specific tuning of retarder–VMA combinations, reducing the risk of delayed setting or placement difficulties during scale-up.

Interaction with Air-Entraining Agents

Air-entraining agents introduce stabilized microbubbles through surfactant adsorption at the air–water interface. VMAs influence both bubble size distribution and stability by increasing pore solution viscosity and modifying drainage kinetics. Moderate VMA dosages typically enhance air void stability, improving freeze–thaw durability. However, excessive viscosity can inhibit bubble formation during mixing, leading to lower entrained air content and poor reproducibility.

Polymer–surfactant interactions are highly chemistry-dependent. Some VMAs adsorb at the air–water interface, competing with air-entraining agents and altering foam stability. These effects are magnified in low-water or high-powder systems such as self-consolidating or 3D-printed concretes. Inline air content sensors combined with mixer energy signatures allow real-time correction of air-entrainer dosage, aligning with Industry 4.0 objectives for closed-loop quality control.

Industrial Implications and Control Strategies

The combined use of VMAs with accelerators, retarders, and air-entraining agents requires formulation strategies that prioritize compatibility windows rather than single-admixture optimization. Cross-plant deployment benefits from digital fingerprints linking cement chemistry, admixture package, and rheological response. Machine-learning-assisted formulation platforms are increasingly adopted to predict interaction sensitivity and recommend dosage envelopes that remain robust under raw material variability.

Table 4.3.2: Interaction of VMAs with Set-Controlling and Air-Entraining Admixtures

Admixture Type	Dominant Interaction Mechanism	Rheological Effect	Industrial Consequence
Accelerators	Calcium-induced polymer charge screening	Reduced viscosity efficiency	Higher segregation risk in cold weather
Retarders	Prolonged polymer persistence	Increased thixotropy	Slower placement, delayed setting
Air-entrainers	Polymer–surfactant interaction	Modified bubble stability	Variability in air content

Combined systems	Multicomponent ionic effects	Time-dependent rheology	Need for digital dosage control

5. Formulation Strategies for VM Admixture Producers

Formulation strategies for viscosity-modifying admixtures represent the critical bridge between polymer chemistry fundamentals and reliable industrial production. Successful commercialization depends on informed formulation choices and disciplined quality control frameworks that deliver consistent performance under variable raw material, processing, and application conditions. Emphasis is placed on rigorous selection and qualification of raw materials, including polymer purity, impurity control, and supply chain robustness, alongside differentiated design and stability requirements for liquid and powder VMA formats. Equal attention is given to scalable manufacturing routes supported by modern quality systems and Industry 4.0 concepts such as digital monitoring, traceability, and data-driven process optimization. The integrated perspective of material science and digital process management provides a practical pathway for converting laboratory-developed polymer concepts into robust, commercially viable admixtures capable of meeting the demands of diverse concrete production environments.

5.1 Raw Material Selection and Qualification

Raw material selection defines the intrinsic performance envelope of viscosity-modifying admixtures and governs their robustness during manufacturing, storage, and field application. For industrial VMA producers, qualification protocols must extend beyond nominal supplier specifications to include performance-relevant chemical attributes, interaction sensitivity with cement systems, and long-term supply reliability. Minor variations in polymer structure or impurity profile often translate into disproportionate rheological effects in low-water or high-performance concrete formulations.

Modern qualification frameworks increasingly integrate laboratory analytics with application-driven testing and digital data management. Molecular characterization, electrolyte tolerance screening, and rheological fingerprinting are combined to establish acceptance windows rather than fixed targets. Industry 4.0 platforms enable centralized tracking of raw material performance across plants, linking incoming material data to finished product viscosity, customer feedback, and field performance trends.

Polymer Architecture and Molecular Consistency

The molecular architecture of VMA polymers, including backbone chemistry, functional group density, and molecular weight distribution, directly controls water structuring and shear-dependent viscosity development. Narrow distribution synthetic polymers provide superior batch-to-batch consistency, while broader distributions may offer improved tolerance to cement variability at the expense of precision. Qualification protocols therefore balance molecular uniformity with functional robustness.

Advanced producers routinely employ size-exclusion chromatography coupled with multi-angle light scattering and charge titration to establish polymer fingerprints. These data are correlated with rheological response in standardized cement pastes under controlled ionic environments. Digital libraries of polymer fingerprints support rapid identification of off-spec materials and facilitate safe substitution during supply disruptions.

Impurity Profile and Electrolyte Sensitivity

Impurities such as residual salts, unreacted monomers, and low-molecular-weight fractions significantly influence VMA performance. Elevated sodium, potassium, or sulfate levels increase pore solution ionic strength and may suppress polymer hydration or promote competitive adsorption with dispersing admixtures. Even trace levels of multivalent cations can alter polymer conformation and reduce viscosity efficiency.

Industrial qualification therefore includes ash content limits, ion chromatography screening, and controlled reconstitution tests in synthetic pore solutions. Results are processed into electrolyte sensitivity indices that define compatibility boundaries with different cement chemistries. These indices are increasingly embedded into formulation software to guide admixture dosing strategies during plant production.

Water Quality and Auxiliary Raw Materials

Water used in liquid VMA formulations is an active component rather than an inert carrier. Dissolved hardness ions, organic load, and microbial content affect polymer solubility and long-term stability. Similarly, auxiliary raw materials such as neutralizing agents, preservatives, and defoamers must be selected for chemical compatibility and regulatory compliance without compromising rheological function.

Producers apply standardized water qualification protocols combined with real-time conductivity and total organic carbon monitoring. Auxiliary materials are screened not only for immediate effect but also for long-term interaction with polymer chains under storage conditions. Digital batch records capture these variables, enabling traceability and root-cause analysis in the event of performance drift.

Supplier Qualification and Change Management

Supplier qualification is a continuous process that extends through the lifecycle of the raw material. Initial approval includes technical audits, variability assessment, and application testing across representative cement systems. Ongoing monitoring focuses on consistency, change notification discipline, and responsiveness to corrective actions.

Change management is increasingly supported by digital workflows linking supplier notifications to internal risk assessment models. These models simulate the impact of raw material changes on viscosity development and interaction behavior, allowing preemptive mitigation. Such structured approaches significantly reduce the risk of unexpected field issues during technology transfer and cross-plant deployment.

5.1.1 Polymer Purity, Salt Content, and Bioburden Control

Polymer purity is a primary determinant of viscosity efficiency and reproducibility in viscosity-modifying admixtures. In industrial formulations, impurities originating from synthesis, neutralization, or downstream processing alter polymer–water interactions and disrupt predictable rheological behavior. Even low concentrations of residual monomers, oligomers, or process aids may influence hydration kinetics and shear response, particularly in low water-to-cement ratio systems where the pore solution is highly concentrated.

High polymer purity ensures consistent chain conformation and minimizes uncontrolled side interactions with cement phases and co-admixtures. Producers increasingly specify purity targets in terms of molecular weight distribution integrity and extractable fraction rather than simple solids content. Advanced analytical workflows link these metrics to concrete rheology outcomes, allowing acceptance criteria to be based on performance sensitivity rather than arbitrary chemical limits.

Salt Content and Ionic Load Management

Residual salt content represents one of the most critical hidden variables in VMA performance. Sodium, potassium, calcium, and sulfate ions increase pore solution ionic strength, compress polymer electrical double layers, and reduce effective hydrodynamic volume. The consequence is a measurable loss of viscosity contribution at constant dosage, often accompanied by increased sensitivity to cement chemistry variability.

Industrial qualification protocols therefore impose strict ash content and ion-specific limits, supported by ion chromatography and conductivity-based screening. Data are processed into ionic load coefficients that quantify the expected viscosity loss per unit salt content. These coefficients are increasingly integrated into digital formulation platforms, enabling real-time dosage correction when raw material salt levels fluctuate within allowable ranges.

Bioburden Control and Microbial Stability

Bioburden is a critical risk factor for VMAs based on polysaccharides or bio-derived polymers, as well as for aqueous synthetic polymer solutions. Microbial growth leads to enzymatic chain scission, gradual molecular weight reduction, and irreversible viscosity drift during storage. Such degradation often remains undetected until field performance deviates, making prevention more effective than remediation.

Industrial best practice combines controlled water quality, validated preservative systems, and formulation pH management to suppress microbial activity. Routine microbiological screening is complemented by indirect indicators such as viscosity trend analysis and molecular weight monitoring. Industry 4.0 tools enable continuous data capture from storage tanks, allowing early detection of degradation patterns through machine-learning-based anomaly recognition.

Integrated Control Strategies for Manufacturing Robustness

Effective control of purity, salt content, and bioburden requires an integrated approach spanning raw material sourcing, formulation design, and manufacturing operations. Single-point testing is insufficient given the dynamic nature of these variables. Instead, producers establish multidimensional acceptance windows supported by historical data and predictive analytics.

Digital batch records link incoming polymer lots to downstream viscosity performance and shelf-life behavior. When deviations occur, root-cause analysis is accelerated through automated correlation of impurity profiles, storage conditions, and microbial indicators. Such systems support reliable scale-up, cross-plant reproducibility, and confident technology transfer to diverse markets.

5.1.2 Supply Chain Risk Management

Supply chain risk management has become an intrinsic element of viscosity-modifying admixture formulation, directly influencing product continuity, performance consistency, and customer confidence. VMAs frequently rely on specialty polymers, fermentation-derived intermediates, and regionally concentrated production assets, making them vulnerable to logistical disruption, regulatory change, and geopolitical instability. For admixture producers, unmanaged supply risk rapidly translates into forced reformulation, increased variability, and delayed technology transfer.

Effective risk management begins with a granular understanding of raw material criticality. Materials are classified not only by spend or volume, but by formulation sensitivity, substitution tolerance, and interaction criticality with cement systems. High-impact polymers with narrow performance windows require fundamentally different mitigation strategies than auxiliary components with broad functional redundancy. Digital material criticality matrices are increasingly applied to quantify these relationships and guide resource allocation.

Dual Sourcing and Functional Equivalence Qualification

Dual sourcing remains a cornerstone of risk mitigation, yet functional equivalence cannot be assumed for polymers with similar datasheet properties. Minor differences in molecular weight distribution, charge density, or residual salt content often produce non-linear effects on rheology and compatibility. Industrial best practice therefore requires parallel qualification of alternate sources using identical rheological and interaction protocols rather than relying on supplier declarations.

Advanced producers maintain pre-qualified alternates with defined substitution rules embedded in formulation documentation. These rules are supported by digital fingerprints linking supplier-specific polymer characteristics to expected viscosity response. Such preparedness allows rapid switching without triggering customer re-approval cycles, preserving supply continuity during disruptions.

Geographic and Regulatory Exposure Management

Geographic concentration of raw material production introduces correlated risks related to energy availability, environmental regulation, and transportation infrastructure. Regulatory divergence across regions further complicates sourcing, particularly for bio-based polymers, preservatives, and residual monomer limits. These factors directly affect formulation feasibility and market access.

Risk-aware producers integrate regulatory intelligence into their sourcing strategies, tracking emerging restrictions and aligning raw material choices with long-term compliance trajectories. Digital dashboards combining geographic, regulatory, and logistics data enable scenario modeling, allowing proactive reformulation before constraints become critical.

Inventory Strategy and Demand Volatility

Inventory buffering for VMAs must balance supply security against degradation risk, particularly for aqueous formulations sensitive to time and temperature. Excess inventory increases the likelihood of viscosity drift and microbial growth, while insufficient stock exposes customers to supply interruptions.

Data-driven inventory models leveraging demand forecasting, shelf-life analytics, and storage condition monitoring are increasingly deployed. These models optimize safety stock levels dynamically, aligning with Industry 4.0 principles by linking production planning, quality data, and logistics in a closed-loop system.

Digitalization and Predictive Risk Analytics

Digitalization transforms supply chain risk management from reactive to predictive. Integration of supplier performance data, raw material variability trends, and formulation sensitivity models enables early warning of emerging risks.

Machine-learning algorithms detect weak signals such as increasing variability in impurity profiles or delivery lead times, prompting preventive action.

Digital twins of the supply chain, linked to formulation and production data, allow simulation of disruption scenarios and evaluation of mitigation strategies. This capability significantly enhances resilience during scale-up and global deployment, supporting consistent product performance across regions and cement markets.

5.2 Liquid and Powder VMA Formulations

Liquid and powder viscosity-modifying admixtures are engineered to address distinct logistical, operational, and performance requirements across global concrete markets. The formulation approach must therefore consider not only rheological efficiency in cementitious systems, but also stability during storage, transport, and dosing under variable climatic and production conditions. Differences in physical form impose fundamentally different constraints on polymer hydration behavior, degradation pathways, and quality control strategies.

From a technology transfer perspective, the same polymer chemistry often behaves differently when formulated as a liquid or as a powder, requiring format-specific optimization rather than direct translation. Industry 4.0 tools increasingly support this differentiation by linking formulation parameters to real-time stability and performance data across manufacturing sites.

5.2.1 Solubilization, hydration kinetics, and shelf stability

Solubilization behavior governs the initial effectiveness and reproducibility of viscosity-modifying admixtures under industrial mixing conditions. For liquid formulations, polymer dissolution must proceed without incomplete hydration, localized gel formation, or phase separation, as these defects translate directly into inconsistent rheology during concrete batching. Polymer backbone polarity, degree of neutralization, and counterion selection determine water affinity and dissolution rate, making them primary formulation levers during product development.

Hydration kinetics must be matched to realistic mixing and transport timelines. VMAs that hydrate too rapidly may generate transient viscosity spikes, increasing mixer torque and impairing dispersion of other admixtures. Conversely, slow hydration delays viscosity development, leaving mixtures vulnerable to segregation during early handling. Industrial optimization relies on time-resolved rheological profiling and conductivity monitoring, converting hydration behavior into quantifiable rate parameters that guide formulation adjustments. These data-driven approaches are increasingly embedded into digital formulation platforms to support rapid scale-up and cross-plant replication.

Powder VMAs introduce distinct solubilization challenges linked to wetting efficiency and particle morphology. Spray-dried or granulated polymers are engineered to control water ingress and swelling, avoiding agglomeration while ensuring rapid dispersion in concrete mixers. Particle size distribution, surface treatment, and residual moisture content are tightly controlled, as they directly influence dissolution kinetics and early-age rheology. Validation protocols include high-shear dispersion tests and simulated mixer residence time studies to ensure reliable performance under field conditions.

Shelf stability is governed by both chemical and physical factors that evolve during storage and transport. In liquid VMAs, slow hydrolysis, oxidative processes, or associative polymer rearrangement can alter viscosity over time, even in the absence of microbial activity. Temperature cycling accelerates these mechanisms, making thermal stability testing a critical qualification step. For powder formulations, moisture uptake and thermal exposure affect flowability and dissolution behavior, necessitating robust packaging and environmental control.

Industry 4.0 concepts increasingly support shelf-life management through continuous data acquisition and predictive analytics. Inline viscosity sensors, temperature logging, and digital batch records generate datasets used to model long-term stability and forecast drift before it becomes critical. These tools enable proactive formulation refinement and reduce the risk of performance loss during extended storage or global distribution.

5.2.2 Anti-microbial strategies and viscosity drift control

Anti-microbial protection is a critical design element for viscosity-modifying admixtures, particularly for aqueous formulations and bio-derived polymer systems. Uncontrolled microbial growth leads to enzymatic chain scission, progressive reduction in molecular weight, and irreversible loss of viscosity efficiency. In industrial practice, such degradation often manifests as gradual rheology drift that escapes routine quality checks until field performance is compromised.

Effective anti-microbial strategies are based on prevention rather than remediation. Water quality control represents the first barrier, as dissolved organic carbon and microbial load strongly influence contamination risk. Deionized or treated process water, combined with closed storage and hygienic handling, significantly reduces initial bioburden. Formulation pH is adjusted to levels unfavorable for microbial proliferation while maintaining polymer stability and regulatory compliance.

Preservative systems are selected to provide broad-spectrum activity without interfering with polymer hydration or cement compatibility. Single-component biocides are increasingly replaced by synergistic preservative blends that lower active concentration while extending protection duration. Compatibility testing includes long-term viscosity monitoring and molecular weight analysis to confirm that antimicrobial agents do not induce secondary degradation or associative effects. Regulatory trends toward reduced biocide usage further reinforce the need for optimized, low-dosage systems.

Viscosity drift control extends beyond microbial stability and encompasses physicochemical mechanisms such as slow hydrolysis, oxidative chain modification, and time-dependent polymer association. These processes are strongly influenced by temperature, shear history, and formulation composition. Industrial monitoring therefore relies on viscosity trend analysis rather than single-point specification checks, capturing gradual changes that affect performance robustness.

Industry 4.0 tools play a central role in drift management. Inline and at-line viscometers, combined with temperature and residence time logging, generate continuous datasets analyzed using statistical process control and machine-learning algorithms. Early-stage deviations from expected viscosity trajectories trigger corrective actions such as controlled recirculation, temperature adjustment, or targeted formulation correction. For powder VMAs, humidity and thermal exposure are tracked digitally to prevent dissolution performance drift linked to storage conditions.

An integrated approach combining hygienic design, validated preservative systems, and digital monitoring enables long-term stability without excessive chemical intervention. Such strategies support consistent product quality, simplify technology transfer, and reduce field variability across diverse climatic and logistical environments.

5.3 Manufacturing Scale-Up and Quality Control

Manufacturing scale-up of viscosity-modifying admixtures represents a critical transition where laboratory-optimized formulations are exposed to industrial realities such as variable shear fields, extended residence times, and thermal gradients. These factors directly influence polymer conformation, hydration state, and long-term stability. Quality control systems must therefore evolve from static end-point testing toward process-aware, data-driven frameworks that capture formulation behavior throughout production.

Successful scale-up depends on preserving functional polymer attributes while accommodating increased batch size, equipment geometry, and energy input. Small deviations in mixing sequence or shear history can alter apparent viscosity and hydration kinetics, creating discrepancies between pilot and commercial material. Industry 4.0 principles increasingly underpin scale-up strategies by integrating process analytics, digital twins, and real-time quality feedback into manufacturing operations.

Process Sensitivity During Scale-Up

Polymer-based VMAs exhibit strong sensitivity to shear rate, temperature, and dilution pathway. High local shear may induce partial chain degradation or excessive disentanglement, while insufficient energy leads to incomplete hydration and heterogeneity. Thermal exposure during production or storage accelerates hydrolysis and associative rearrangement, influencing both immediate viscosity and shelf stability.

Industrial scale-up protocols therefore include controlled energy mapping across mixers and reactors, combined with temperature profiling and residence time distribution analysis. Data are used to define acceptable operating envelopes rather than fixed setpoints, allowing robust transfer across plants with different equipment configurations.

Quality Control Beyond Conventional Specifications

Traditional quality control relying solely on solids content and single-point viscosity measurements is insufficient for VMAs with complex time-dependent behavior. Modern QC frameworks incorporate dynamic rheology, hydration rate assessment, and stability indicators that better reflect field performance. These metrics are correlated with concrete application data to ensure relevance.

Inline sensing technologies increasingly supplement laboratory testing. Viscosity, conductivity, and density measurements captured during production provide continuous assurance of formulation consistency. When integrated into manufacturing execution systems, these data streams enable immediate corrective action and prevent off-spec material from progressing downstream.

Digitalization and Traceability

Digital batch records form the backbone of advanced quality control. Each production lot is linked to raw material identity, process conditions, and real-time sensor data, creating full traceability from supplier to customer. This digital continuity supports rapid root-cause analysis when deviations occur and accelerates corrective learning across sites.

Machine-learning algorithms analyze historical production data to identify patterns associated with subtle quality drift. Predictive alerts allow intervention before specifications are breached, reducing waste and improving overall equipment effectiveness. Such digitalization supports consistent global product quality and simplifies technology transfer during capacity expansion.

5.3.1 Batch Versus Continuous Processing

The choice between batch and continuous processing fundamentally influences the consistency, scalability, and cost structure of viscosity-modifying admixture production. Polymer-based VMAs are particularly sensitive to processing history, as shear intensity, residence time, and thermal exposure directly affect polymer hydration, molecular integrity, and final rheological performance. Selecting the appropriate processing mode therefore requires alignment between formulation sensitivity, production volume, and quality assurance capability.

Batch Processing Characteristics and Limitations

Batch processing remains prevalent for specialty VMAs and multi-product facilities due to its flexibility and low changeover barrier. It allows precise control over addition sequence, hydration time, and post-adjustment, which is advantageous during product development and small-volume production. However, batch-to-batch variability arises from operator-dependent execution, inconsistent shear distribution, and sensitivity to raw material addition rate.

Scale-up in batch systems often amplifies these limitations. Larger vessels introduce non-uniform energy input and temperature gradients that alter polymer dissolution and associative behavior. Industrial mitigation strategies include standardized addition protocols, automated dosing, and defined mixing energy targets rather than fixed mixing times. Digital operator guidance and recipe enforcement systems reduce human variability and improve reproducibility.

Continuous Processing Advantages and Design Considerations

Continuous processing offers superior consistency by stabilizing shear history and residence time, making it increasingly attractive for high-volume VMA grades. Steady-state operation minimizes transient effects associated with batch start-up and shutdown, resulting in narrower viscosity distributions and improved shelf-life predictability.

Designing continuous processes for VMAs requires careful management of polymer hydration kinetics. Inline dilution, staged polymer addition, and controlled energy input prevent localized over-shearing or incomplete dissolution. Residence time distribution analysis and modular reactor design are applied to ensure sufficient hydration without excessive thermal exposure. These parameters are validated using pilot-scale continuous units before full industrial deployment.

Quality Control and Industry 4.0 Integration

Quality control strategies differ markedly between batch and continuous systems. Batch processes rely on lot-based release testing supplemented by in-process checks, while continuous systems shift emphasis toward real-time monitoring and statistical control. Inline viscosity measurement, pressure drop analysis, and power consumption monitoring provide continuous insight into formulation consistency.

Industry 4.0 frameworks integrate these sensors with manufacturing execution systems and digital twins. Predictive models correlate process conditions with viscosity outcomes, enabling automatic adjustment of flow rates, dilution ratios, or energy input. Such closed-loop control significantly reduces off-spec production and supports reliable cross-plant technology transfer.

Comparative Processing Assessment

The choice between batch and continuous processing constitutes a core manufacturing strategy in viscosity-modifying admixture production, directly influencing rheological reproducibility, production economics, and readiness for digital manufacturing. Batch operations enable formulation agility and rapid grade switching, which is advantageous for diversified VMA portfolios, yet they remain inherently sensitive to raw material variability, operator intervention, and scale-dependent mixing history. Continuous processing, once formulation and residence time distributions are stabilized, delivers markedly improved product uniformity, tighter viscosity control, and seamless integration of inline sensors and closed-loop control architectures aligned with Industry 4.0 principles. Evaluating processing routes through the lenses of scale-up risk, quality assurance philosophy, and digital maturity allows producers to rationally match manufacturing configuration with market volume, performance criticality, and long-term automation strategy.

Table 5.3.1: Comparison of Batch and Continuous Processing for VMA Production

Parameter	Batch Processing	Continuous Processing
Flexibility	High, suitable for multiple grades	Lower, optimized for dedicated products
Batch-to-batch variability	Higher without automation	Intrinsically low
Scale-up complexity	Moderate to high	High initial, stable once optimized
Quality control	Lot-based testing	Real-time process control
Industry 4.0 compatibility	Operator-guided digitalization	Full closed-loop automation

5.3.2 Inline Viscosity Monitoring and Digital Batch Records

Role of Inline Viscosity Measurement in Industrial VMA Manufacturing

Inline viscosity monitoring has become a core element in the industrial production of polymer-based viscosity-modifying admixtures, particularly where hydration kinetics and shear sensitivity govern performance. Unlike laboratory rheometry, inline systems capture viscosity evolution under true processing conditions, including actual shear fields, temperature profiles, and dilution histories. This real-time visibility allows manufacturers to detect deviations linked to incomplete polymer hydration, local over-shearing, or raw material variability before these effects are locked into the finished product.

From an industrial formulation perspective, inline viscosity data provide a functional fingerprint rather than a single release value. Continuous tracking of apparent viscosity, yield stress proxies, or pressure drop across defined flow elements enables correlation with polymer molecular integrity and associative network development. These measurements are especially valuable during scale-up, where geometric differences between pilot and production equipment alter hydrodynamic conditions in ways that cannot be fully replicated offline.

Sensor Technologies and Deployment Strategies

Several sensor technologies are applied for inline viscosity monitoring in VMA production, each with specific advantages and constraints. Vibrational and torsional resonator sensors offer robustness and low maintenance, making them suitable for high-solids aqueous polymer systems. Capillary-based pressure drop sensors provide indirect viscosity measurement under defined shear rates, aligning well with process-relevant flow conditions. Rotational inline rheometers, while more complex, deliver richer rheological data and are increasingly used in high-value formulations.

Sensor placement is critical. Locating measurement points after key hydration or dilution stages ensures that viscosity data reflect meaningful polymer conformation states rather than transient mixing artifacts. Redundant sensing at different process locations supports fault detection and improves confidence in data integrity.

Calibration protocols are increasingly automated and linked to reference materials to maintain long-term accuracy under industrial operating conditions.

Digital Batch Records as a Quality Backbone

Digital batch records extend inline monitoring from data collection to structured knowledge capture. Each production lot is digitally linked to time-stamped process variables, including viscosity profiles, temperature history, energy input, and raw material identifiers. This comprehensive dataset replaces traditional paper-based records and transforms quality control into a traceable, auditable, and analyzable system.

In polymer admixture manufacturing, digital batch records enable rapid root-cause analysis when viscosity drift or field performance issues arise. Instead of relying on post hoc laboratory tests, engineers can reconstruct the full process trajectory and identify subtle deviations in hydration sequence or shear exposure. This capability is particularly valuable for multi-site operations, where harmonizing product quality across different plants is a persistent challenge.

Industry 4.0 Integration and Predictive Control

When inline viscosity monitoring and digital batch records are integrated into Industry 4.0 architectures, quality control evolves from reactive to predictive. Manufacturing execution systems aggregate sensor data and feed them into machine-learning models trained on historical production outcomes. These models identify patterns that precede off-spec viscosity or stability loss, enabling early intervention.

Closed-loop control strategies are increasingly implemented, where viscosity signals directly adjust dilution rates, polymer feed, or mixing energy in real time. Digital twins of the hydration and mixing process simulate the impact of corrective actions before execution, reducing the risk of overcorrection. Such systems support consistent product quality, minimize waste, and accelerate technology transfer during capacity expansion or localization of production.

Data Governance and Regulatory Considerations

Robust data governance underpins the industrial use of digital batch records. Secure data storage, version control, and audit trails are essential to meet customer and regulatory expectations, particularly in construction chemicals supplied to regulated infrastructure projects. Standardized data formats facilitate integration with enterprise quality systems and external audits.

From a technology transfer standpoint, digital batch records simplify qualification of new plants or contract manufacturers. Process windows, rather than fixed recipes, are embedded in the digital framework, allowing controlled flexibility while preserving functional performance. This approach aligns with modern quality-by-design principles increasingly adopted across the polymer additives industry.

Table 5.3.2: Representative Inline Monitoring Parameters for VMAs

Parameter Monitored	Industrial Relevance	Typical Use in Control Strategy
Apparent viscosity (inline)	Polymer hydration and integrity	Release prediction, closed-loop dilution
Pressure drop	Flow consistency, shear sensitivity	Early fouling or degradation detection
Temperature	Hydrolysis and stability risk	Thermal envelope enforcement
Power consumption	Mixing efficiency	Energy-based scale-up validation
Conductivity	Dilution accuracy	Raw material and water balance control

6. Compatibility with Cementitious Systems

Compatibility of VMAs with cementitious systems is no longer a static material property but a dynamic interaction between polymer chemistry, binder design, and process variability. Robust performance increasingly depends on adaptive formulation strategies supported by real-time data and digital quality frameworks. This systems-level view is essential for reliable application across modern, low-carbon, and resource-variable concrete technologies.

6.1 Fundamental Chemical Interactions with Cement Phases

Alkaline Pore Solution Effects on Polymer Stability

Viscosity-modifying admixtures operate immediately within cement pore solutions characterized by pH values typically exceeding 12.5 and high ionic strength. Such conditions impose severe chemical stress on polymer backbones, particularly for hydrolytically sensitive linkages such as esters, amides, and glycosidic bonds. Alkaline hydrolysis alters molecular weight distribution and reduces effective chain length, leading to diminished thickening efficiency and accelerated viscosity decay. Industrially relevant VMAs therefore rely on alkali-stable backbones or protective substitution patterns that suppress nucleophilic attack under sustained high-pH exposure.

From a formulation standpoint, polymer stability must be evaluated over realistic residence times that reflect transport, pumping, and placement delays. Short laboratory exposure tests often underestimate degradation kinetics that become relevant at jobsite scale.

Calcium Ion Complexation and Polymer Conformation

Calcium ions released during early hydration exert a dominant influence on VMA behavior. Functional groups such as carboxylates, hydroxyls, and amides interact with Ca^{2+} through coordination or electrostatic association. Moderate interaction can enhance network formation and viscosity build-up, while excessive complexation induces coil contraction, interchain association, or even localized precipitation.

Calcium tolerance is therefore a critical design parameter. Polymers engineered with controlled charge density and steric shielding maintain extended conformations in calcium-rich environments, preserving effective hydrodynamic volume. This balance directly controls plastic viscosity development and segregation resistance in fresh concrete.

Competitive Adsorption on Clinker and Hydrate Surfaces

Cement phases present a wide range of adsorption sites that compete for polymer attachment. C3A and early aluminate hydrates exhibit strong affinity for anionic and polar polymer segments, often outcompeting silicate phases. Adsorption reduces the free polymer fraction available for bulk solution thickening, creating apparent under-dosage effects even when total VMA content is unchanged.

Industrial compatibility screening increasingly includes adsorption profiling against individual clinker phases or synthetic hydrates. Such data guide polymer architecture selection toward reduced surface affinity when bulk rheology control is the primary objective.

Sulfate Balance and Early Hydration Dynamics

Dissolved sulfates regulate C3A reactivity and indirectly influence VMA availability. Proper sulfate balance maintains aluminate phases in a passivated state, limiting premature polymer fixation. Sulfate deficiency accelerates aluminate hydration and increases polymer adsorption, while sulfate excess modifies ionic strength and can delay viscosity development.

For manufacturing and field application, VMA performance must be validated across the expected sulfate variability window of supplied cements. Static compatibility claims based on a single cement source are insufficient for robust commercial deployment.

Interactions with Superplasticizers at the Molecular Level

Polycarboxylate ether superplasticizers interact with cement surfaces through similar functional motifs as many VMAs, leading to competitive adsorption and dynamic displacement. Depending on molecular architecture, this interaction may suppress or amplify VMA effectiveness. Systems designed without considering co-admixture chemistry frequently exhibit delayed viscosity loss or unpredictable thixotropic behavior.

Advanced formulation practice treats VMA and superplasticizer as a coupled system. Co-optimization of charge density, side-chain length, and addition sequence reduces antagonistic interactions and stabilizes fresh-state rheology.

Industry 4.0 Perspective on Chemical Compatibility

Digitalization enables chemical compatibility to be managed as a measurable process variable rather than an empirical risk. Inline conductivity, temperature, and viscosity data are correlated with cement chemistry inputs to predict polymer–ion interaction intensity. Digital twins incorporating pore solution chemistry simulate polymer conformation changes under varying cement compositions, supporting adaptive dosing strategies in real time.

Such data-driven frameworks are increasingly applied during technology transfer to new regions, where cement chemistry variability is unavoidable and rapid formulation adjustment is commercially critical.

Table 6.1: Key Interaction Mechanisms and Industrial Implications

Interaction Mechanism	**Cement Phase Driver**	**Industrial Consequence**
Alkaline hydrolysis	High OH^- concentration	Viscosity decay, shelf-life loss
Calcium complexation	C3S, C2S hydration	Coil contraction or network formation
Polymer adsorption	C3A, AFt, AFm phases	Reduced free polymer fraction
Sulfate modulation	Gypsum, anhydrite	Early-age viscosity instability
Competitive adsorption	PCE superplasticizers	Non-linear rheological response

6.2 Portland Cement Variability and Its Impact on VMA Performance

Portland cement is not a chemically uniform raw material but a statistically controlled product whose phase composition and minor constituents vary with raw meal chemistry, kiln operation, and sulfate adjustment practice. For viscosity-modifying admixtures, such variability directly influences polymer solubility, adsorption behavior, and effective thickening efficiency. Industrial robustness of VMAs therefore depends less on optimal performance with a reference cement and more on tolerance to realistic compositional fluctuation across supply regions and production campaigns.

From a manufacturing and application standpoint, cement variability manifests as shifts in pore solution chemistry during the first minutes of hydration. These early-age conditions govern polymer conformation and availability, establishing rheological behavior that persists through transport, placement, and finishing. Understanding and anticipating these effects is essential for stable field performance.

6.2.1 Alkali Content, C3A Reactivity, and Sulfate Balance

Alkali Content and Polymer Stability

Alkali oxides present in cement, primarily Na_2O and K_2O, dissolve rapidly and elevate pore solution alkalinity beyond pH 13 in some systems. Elevated alkali levels increase ionic strength and accelerate hydrolytic stress on alkali-sensitive polymer linkages. Even alkali-stable VMAs experience altered hydration kinetics under such conditions, as charge screening reduces polymer chain expansion and lowers effective hydrodynamic volume.

Industrial evaluation of alkali sensitivity requires testing across a representative alkali window rather than a single cement grade. VMAs formulated with controlled charge density and sterically protected functional groups demonstrate reduced sensitivity to alkali-driven conformation changes, maintaining viscosity contribution despite ionic strength fluctuations.

C3A Content and Adsorption Dynamics

Tricalcium aluminate content varies significantly between Portland cements and exerts a disproportionate effect on admixture compatibility. High C3A levels generate abundant positively charged sites during early hydration, promoting rapid adsorption of anionic or polar polymer segments. For VMAs, this adsorption depletes the free polymer fraction in the pore solution, reducing bulk thickening efficiency and increasing apparent dosage demand.

The rate of C3A reaction is equally important. Highly reactive aluminates can immobilize polymers within minutes, creating time-dependent viscosity loss that may only become evident during transport. Industrial mitigation strategies include polymer architectures with reduced aluminate affinity and optimized addition sequences that delay exposure during peak aluminate activity.

Sulfate Balance as a Modulating Factor

Sulfate addition through gypsum or anhydrite regulates C3A hydration and indirectly controls VMA availability. Adequate sulfate levels maintain aluminate phases in a passivated state, limiting premature polymer adsorption and supporting stable viscosity development. Sulfate deficiency accelerates aluminate dissolution and intensifies polymer fixation, while sulfate excess modifies pore solution chemistry and may delay polymer hydration.

For VMA qualification, sulfate balance sensitivity must be evaluated using cements with controlled sulfate variation rather than nominally identical compositions. Industrial experience shows that VMAs tolerant to moderate sulfate imbalance exhibit significantly improved robustness across multiple cement sources.

Combined Effects and Practical Implications

In practice, alkali content, C3A reactivity, and sulfate balance act simultaneously rather than independently. High-alkali, high-C3A cements with marginal sulfate control represent the most challenging scenario for VMA performance. Under such conditions, polymers experience strong adsorption forces, rapid conformation changes, and elevated degradation risk.

Industry 4.0-enabled batching systems increasingly incorporate cement chemistry metadata, allowing adaptive VMA dosing based on alkali and aluminate indicators. Such approaches shift compatibility management from empirical safety margins toward data-driven control.

Table 6.2.1: Key Cement Variability Parameters Affecting VMA Performance

Cement Parameter	Primary Mechanism	Impact on VMA Function
Alkali content	Ionic strength, hydrolysis	Reduced chain expansion, viscosity loss
C3A content	Polymer adsorption	Depletion of free polymer
C3A reactivity	Early fixation kinetics	Time-dependent viscosity decay
Sulfate balance	Aluminate passivation	Stability or instability of viscosity build-up

6.3 Interaction with Supplementary Cementitious Materials

Supplementary cementitious materials significantly modify the chemical and physical environment in which viscosity-modifying admixtures operate. Beyond clinker dilution, SCMs alter pore solution composition, surface chemistry, and fine particle population, all of which influence polymer availability and rheological efficiency. As SCM usage increases to meet decarbonization targets, VMA compatibility with these materials has become a primary determinant of formulation robustness rather than a secondary optimization parameter.

From an industrial perspective, SCM effects on VMAs are governed by three coupled mechanisms: surface adsorption, ion release, and fines-driven water demand. The relative importance of each mechanism depends on SCM type, fineness, and replacement level.

6.3.1 Fly Ash, Slag, Limestone, and Calcined Clay Systems

Fly Ash Systems

Fly ash introduces spherical particles with variable glass chemistry and residual carbon content. Low-calcium fly ash primarily affects VMAs through surface adsorption and dilution of clinker-derived calcium, often leading to delayed viscosity development. High-calcium fly ash releases additional Ca^{2+} and alkalis, intensifying polymer complexation and adsorption effects similar to reactive Portland cements.

Unburned carbon presents a secondary compatibility risk by adsorbing organic polymers non-selectively, reducing effective VMA concentration. Industrial qualification therefore includes loss-on-ignition sensitivity testing to establish safe operating windows across ash sources.

Ground Granulated Blast Furnace Slag

Slag-rich systems exhibit lower early calcium availability and modified pore solution chemistry, which can delay polymer hydration and network formation. VMAs formulated for pure Portland cement may show reduced initial viscosity in slag blends, followed by rapid viscosity rise as slag activation progresses.

Industrial mitigation strategies include polymers with faster intrinsic hydration kinetics or staged addition concepts that align polymer activation with slag reactivity. Compatibility screening must extend beyond early-age rheology to capture delayed structuration effects relevant to pumping and finishing.

Limestone Fillers

Fine limestone additions increase total surface area and introduce positively charged calcium carbonate surfaces that readily adsorb polar polymer segments. While limestone generally improves rheological stability through particle packing effects, excessive adsorption can increase apparent VMA demand and suppress thickening efficiency.

Optimized VMAs for limestone-rich systems employ reduced surface affinity and broader molecular weight distributions to maintain free polymer availability despite elevated filler content. These design choices are increasingly important in cements with high limestone substitution levels.

Calcined Clay and LC^3 Systems

Calcined clays present the most aggressive environment for VMAs due to high specific surface area, platelet morphology, and reactive alumina content. Strong polymer adsorption onto metakaolin surfaces rapidly depletes free polymer and destabilizes viscosity control if formulations are not specifically adapted.

In LC^3 systems, synergistic interactions between calcined clay and limestone further complicate polymer behavior by modifying aluminate availability and pore solution chemistry. VMAs intended for such systems require steric shielding, controlled charge density, and enhanced calcium tolerance to maintain functional performance.

Table 6.3.1: SCM-Specific Interaction Mechanisms and VMA Implications

SCM Type	Dominant Interaction Mechanism	VMA Performance Risk
Low-Ca fly ash	Surface adsorption, dilution	Delayed viscosity build-up
High-Ca fly ash	Calcium release, adsorption	Over-structuration or loss of flow
Slag	Delayed ion availability	Time-dependent viscosity rise
Limestone	Filler adsorption	Increased dosage demand
Calcined clay	Strong adsorption, alumina reactivity	Severe polymer depletion

Industry 4.0 Approach to SCM Compatibility

Digital formulation platforms increasingly integrate SCM characterization data, including fineness, surface area, and chemical reactivity, into predictive rheology models. Inline viscosity and torque measurements during batching provide real-time feedback on SCM-induced variability. Adaptive dosing algorithms adjust VMA input dynamically to compensate for SCM source fluctuations, reducing reliance on conservative over-dosage.

Such data-driven compatibility management is essential for consistent performance in modern, SCM-intensive concrete technologies and supports rapid technology transfer across regions with diverse raw material availability.

6.4 SCM-Rich and Low-Clinker Cement Systems

SCM-rich and low-clinker cement systems represent a structural shift in binder design driven by decarbonization, resource efficiency, and regulatory pressure. These systems fundamentally alter early-age hydration pathways, pore solution chemistry, and fine particle population, creating operating conditions for viscosity-modifying admixtures that differ markedly from conventional Portland cement. VMA performance in such binders is therefore governed by robustness to delayed calcium availability, elevated surface adsorption demand, and extended dormant periods.

From an industrial formulation perspective, the primary challenge is no longer maximizing viscosity at a fixed dosage, but maintaining predictable rheological evolution across extended time windows and variable activation mechanisms.

6.4.1 Altered Pore Solution Chemistry and Polymer Availability

SCM-rich and low-clinker cement systems generate pore solutions that differ fundamentally from those of conventional Portland cement during the critical early hydration window. Reduced clinker content lowers the immediate release of Ca^{2+} and OH^{-} ions, while the relative contribution of alkalis, aluminates, and SCM-derived species increases. For viscosity-modifying admixtures, this shift directly controls polymer solubility, conformation, and effective availability in the aqueous phase.

In practical terms, the initial chemical environment often lacks sufficient calcium to promote polymer association mechanisms that underpin viscosity development in traditional systems. VMAs formulated with strong calcium dependence therefore exhibit delayed or insufficient thickening during the first mixing and transport stages.

Calcium Availability and Polymer Structuring

Calcium ions play a dual role in VMA performance by screening electrostatic repulsion and enabling associative network formation. In low-clinker binders, delayed calcium release suppresses these mechanisms, leading to reduced hydrodynamic volume and lower apparent viscosity. As hydration or SCM activation progresses, a sudden increase in calcium concentration can trigger rapid polymer association, resulting in abrupt viscosity rise.

Industrial formulation must therefore balance early-stage functionality with controlled later structuration. Polymers with partial calcium independence or graded interaction strength provide smoother rheological evolution across time.

Ionic Strength and Charge Screening Effects

Despite reduced calcium levels, SCM-rich systems often exhibit elevated ionic strength due to soluble alkalis and alumina species. High ionic strength compresses the electrical double layer surrounding polymer chains, promoting coil contraction even in the absence of strong calcium complexation. This effect further reduces effective polymer volume and weakens viscosity contribution.

Charge density optimization and steric shielding are key design tools to mitigate ionic strength sensitivity. Such molecular strategies maintain polymer extension and hydration under variable pore solution compositions.

Aluminate and Silicate Species Interactions

Reactive alumina released from calcined clays and certain slags introduces additional adsorption and complexation pathways. Aluminate species interact strongly with polar functional groups, increasing the risk of polymer fixation onto solid surfaces or early hydrates. Silicate species, while less aggressive, contribute to evolving pore solution composition that affects polymer hydration kinetics.

Compatibility assessment must therefore include SCM-specific pore solution simulants rather than relying solely on Portland cement filtrates. This approach improves predictive accuracy for polymer availability under real binder conditions.

Temporal Evolution of Polymer Availability

A defining feature of low-clinker systems is the pronounced time dependence of pore solution chemistry. Polymer availability at mixing, during transport, and at placement can differ substantially. VMAs that appear compatible in short-duration laboratory tests may fail under extended holding conditions typical of ready-mix operations.

Time-resolved rheology and staged ion addition experiments are increasingly used to map polymer availability against evolving chemical conditions. These data inform both polymer design and adaptive dosing strategies.

Industry 4.0-Enabled Control of Polymer Availability

Digitalization enables continuous estimation of pore solution evolution using indirect process indicators such as temperature rise, conductivity, and early calorimetric response. These signals are correlated with polymer availability models to predict viscosity development in real time. Manufacturing execution systems can adjust VMA dosage or dilution dynamically to compensate for delayed calcium release or unexpected ionic strength shifts.

Such data-driven control supports consistent rheological performance across SCM-rich cement systems and accelerates technology transfer in low-carbon concrete production.

Table 6.4.1: Key Chemical Drivers of Polymer Availability

Chemical Driver	Effect on Polymer	Rheological Consequence
Reduced Ca^{2+}	Limited association	Low early viscosity
High ionic strength	Coil contraction	Reduced thickening efficiency
Reactive aluminates	Adsorption, fixation	Polymer depletion
Time-dependent chemistry	Delayed activation	Abrupt viscosity rise

6.4.2 Fine Content Amplification and Adsorption Effects

SCM-rich and low-clinker cement systems are intrinsically characterized by elevated fine content arising from supplementary materials, interground limestone, and increased filler usage. This amplification of fines fundamentally alters the interaction landscape for viscosity-modifying admixtures by expanding the total available surface area for non-productive polymer adsorption. As a result, the effective concentration of free polymer in the pore solution is often significantly lower than the nominal dosage, particularly during the early stages of mixing.

From an industrial formulation perspective, fine content effects frequently dominate over intrinsic polymer chemistry when SCM replacement levels exceed moderate thresholds. Failure to account for this mechanism is a primary cause of apparent under-dosage and unstable rheological behavior in low-clinker concretes.

Surface Area, Mineralogy, and Polymer Depletion

The adsorption propensity of VMAs is governed not only by total surface area but also by mineralogical identity and surface charge. Calcined clays, metakaolin, and ultra-fine limestone present high-energy surfaces with strong affinity for polar and anionic polymer segments. Platelet-shaped particles further increase effective contact probability, accelerating polymer depletion from the aqueous phase.

In contrast, coarser or glassy SCMs exhibit lower adsorption intensity but still contribute cumulatively to polymer loss at high replacement levels. Industrial evaluation must therefore include both BET surface area and mineralogical descriptors rather than relying on fineness alone.

Impact on Effective Dosage and Rheological Response

As adsorption increases, the discrepancy between added and available polymer widens. This manifests macroscopically as reduced plastic viscosity, delayed thickening, or erratic thixotropic behavior. Increasing VMA dosage without addressing adsorption often leads to diminishing returns, as additional polymer is preferentially immobilized on solid surfaces.

Robust formulation strategies focus on preserving free polymer fraction through molecular design rather than compensatory over-dosage. Such approaches deliver more stable rheological profiles and improved cost efficiency.

Competitive Adsorption with Other Admixtures

Fine-rich systems intensify competition between VMAs and other chemical admixtures for surface sites. Polycarboxylate ether superplasticizers are particularly affected, as both admixture classes target similar adsorption domains. Changes in addition sequence or mixing energy can shift the balance between dispersion and viscosity control, producing non-linear responses.

Industrial compatibility testing therefore prioritizes multi-admixture systems evaluated under realistic mixing protocols. Static paste tests are insufficient to capture these competitive dynamics.

Strategies for Mitigating Adsorption Losses

Effective mitigation combines polymer architecture optimization and process control. Steric shielding through side-chain design reduces surface affinity, while broader molecular weight distributions maintain hydrodynamic effectiveness despite partial adsorption. In some cases, delayed-release or staged addition concepts preserve polymer availability during critical early mixing stages.

At the process level, adaptive dosing informed by fines content estimation provides a pragmatic control lever. Integration of fines characterization into batching systems supports real-time compensation for raw material variability.

Industry 4.0 Integration for Fine Content Management

Digital batching platforms increasingly incorporate aggregate grading data, SCM fineness, and historical adsorption behavior into predictive rheology models. Inline torque, power consumption, and viscosity signals serve as proxies for effective polymer availability. Machine-learning algorithms correlate these signals with fines content and recommend dosage adjustments dynamically.

Such Industry 4.0-enabled control frameworks are essential for maintaining consistency in SCM-intensive concretes where fine content variability is unavoidable.

Table 6.4.2: Fine Content–Driven Adsorption Effects and Industrial Consequences

Fine Content Driver	Adsorption Mechanism	Impact on VMA Performance
High BET surface area	Increased contact sites	Polymer depletion
Reactive alumina	Strong polar interactions	Rapid adsorption
Platelet morphology	Enhanced collision frequency	Accelerated loss of free polymer
Multi-admixture competition	Shared surface sites	Non-linear rheology

6.4.3 Extended Dormant Periods and Time-Dependent Rheology

SCM-rich and low-clinker cement systems frequently exhibit prolonged dormant periods prior to the onset of accelerated hydration. This altered kinetic profile has direct implications for viscosity-modifying admixtures, as VMAs must sustain functional rheological control over extended time windows without relying on rapid clinker-driven structuration. Time-dependent rheology therefore becomes a primary compatibility criterion rather than a secondary performance attribute.

From an industrial operations perspective, extended dormancy aligns poorly with traditional short-duration laboratory tests and demands evaluation protocols that reflect realistic transport, pumping, and placement timelines.

Dormant Period Length and Polymer Stability

During the dormant phase, limited hydration reduces ion release while mechanical energy input continues through mixing and agitation. VMAs are exposed to sustained shear under chemically evolving conditions, increasing sensitivity to chain disentanglement, gradual hydrolysis, or associative rearrangement. Polymers that perform adequately in rapid-setting systems may exhibit slow viscosity decay or unpredictable recovery when dormancy extends beyond typical laboratory observation periods.

Industrial screening must therefore assess viscosity retention and recovery under continuous low-shear conditions representative of ready-mix handling.

Time-Dependent Structuration and Thixotropy

As hydration eventually accelerates, SCM activation or delayed calcium release can trigger rapid polymer association. This transition often manifests as a sharp increase in plastic viscosity or static yield stress, compromising pumpability and finishability if not properly controlled. The magnitude and rate of this structuration depend on polymer architecture, calcium sensitivity, and prior shear history.

VMAs designed for time-stable rheology exhibit gradual, predictable structuration rather than abrupt transitions. Such behavior is essential for applications involving long haul distances or extended placement schedules.

Interaction with Temperature and Environmental Conditions

Extended dormant periods amplify the influence of temperature on rheological evolution. Modest temperature increases can significantly accelerate SCM activation, shifting the timing of viscosity build-up into critical operational windows. Conversely, low temperatures prolong dormancy and increase the risk of viscosity loss due to sustained shear exposure.

Industrial compatibility evaluation must therefore incorporate temperature-controlled rheology profiles to capture seasonal and regional variability.

Rheological Metrics for Time-Dependent Performance

Single-point measurements such as initial slump or flow spread are insufficient to characterize VMA performance under extended dormancy. Time-resolved rheological parameters, including viscosity decay rate, rebuild kinetics, and static yield stress development, provide a more accurate representation of field behavior.

Advanced rheometers and inline sensing technologies enable continuous monitoring of these parameters, supporting data-driven formulation refinement.

Industry 4.0 Control of Time-Dependent Rheology

Digital platforms integrate rheological time series with process data such as mixing energy, temperature, and holding duration. Predictive models estimate remaining workable time and trigger adaptive VMA dosing or dilution adjustments before critical thresholds are reached. Digital twins simulate rheological evolution under different dormancy scenarios, guiding operational decisions in real time.

Such Industry 4.0-enabled control transforms extended dormancy from a source of variability into a manageable process parameter.

Table 6.4.3: Time-Dependent Rheology Drivers and Operational Implications

Time-Dependent Driver	Polymer Response	Operational Risk
Prolonged dormancy	Viscosity decay	Segregation during transport
Delayed Ca^{2+} release	Abrupt structuration	Pump blockage
Sustained shear	Chain disentanglement	Loss of viscosity
Temperature fluctuation	Shifted activation timing	Unpredictable workability

6.4.4 Interaction with Chemical Activators and Admixture Packages

SCM-rich and low-clinker cement systems frequently depend on chemical activators and complex admixture packages to compensate for reduced clinker reactivity and delayed strength development. These additives substantially modify pore solution chemistry, hydration kinetics, and surface charge distribution, thereby introducing additional interaction pathways for viscosity-modifying admixtures. VMA performance in such systems is therefore governed by the combined chemical environment rather than by binder composition alone.

From an industrial formulation standpoint, failure to consider activator–VMA coupling is a common root cause of unstable rheology and non-reproducible field performance.

Alkali and Sulfate Activators

Alkali-based activators increase pore solution ionic strength and pH, intensifying charge screening and promoting polymer coil contraction. While moderate alkalinity can accelerate polymer hydration, excessive alkali levels often suppress effective thickening by reducing hydrodynamic volume. Sulfate activators, introduced as gypsum, anhydrite, or soluble sulfates, regulate aluminate reactivity and indirectly control polymer adsorption kinetics.

Incompatible sulfate balance can either immobilize VMAs prematurely through accelerated aluminate hydration or delay viscosity development by modifying ion availability. Industrial qualification must therefore include sensitivity analysis across realistic activator dosage ranges.

Calcium-Based Accelerators and Polymer Association

Calcium nitrate, calcium formate, and related accelerators directly increase Ca^{2+} concentration in the pore solution. For VMAs relying on calcium-mediated association, such additives can induce rapid network formation and abrupt viscosity increase. While beneficial for segregation resistance, uncontrolled association compromises pumpability and finishing operations.

Effective formulations employ polymers with graded calcium affinity or incorporate staged addition strategies to decouple accelerator action from VMA structuration during early processing.

Interaction with Superplasticizers

Polycarboxylate ether superplasticizers share adsorption domains with many VMAs, particularly on aluminate-rich surfaces. Competitive adsorption can displace VMAs from the pore solution or alter their spatial distribution, leading to delayed viscosity loss or unexpected thixotropic behavior. The interaction strength depends on side-chain length, charge density, and addition sequence.

Industrial practice increasingly treats VMAs and superplasticizers as a co-designed system. Compatibility mapping across realistic dosage ratios and mixing protocols is essential for stable multi-admixture performance.

Retarders, Accelerators, and Set Control Agents

Organic retarders and set modifiers introduce additional solvation and adsorption effects that can interfere with polymer hydration. In SCM-rich systems, where hydration kinetics are already altered, these effects are amplified. VMAs with slow intrinsic hydration may exhibit prolonged low-viscosity phases followed by sudden structuration once retardation effects dissipate.

Comprehensive compatibility evaluation must therefore encompass the full admixture package rather than isolated components.

Industrial Control and Process Implications

The combined influence of activators and admixture packages creates a narrow operational window for stable VMA performance. Minor deviations in dosage or addition sequence can produce disproportionate rheological responses. Industry 4.0-enabled batching systems mitigate this risk by linking admixture dosing to real-time indicators such as conductivity, temperature, and inline viscosity.

Predictive models correlate activator chemistry and admixture ratios with rheological outcomes, enabling adaptive control and reducing reliance on conservative over-dosage strategies.

Table 6.4.4: Key Activator and Admixture Interactions Affecting VMA Performance

Additive Type	Primary Interaction Mechanism	Rheological Risk
Alkali activators	Ionic strength increase	Coil contraction, viscosity loss
Sulfate activators	Aluminate regulation	Premature adsorption or delay
Calcium accelerators	Polymer association	Abrupt viscosity rise
Superplasticizers	Competitive adsorption	Non-linear rheology
Retarders	Solvation effects	Delayed structuration

6.4.5 Industry 4.0 Strategies for Robust Performance

The complexity of SCM-rich and low-clinker cement systems has accelerated the adoption of Industry 4.0 methodologies to stabilize viscosity-modifying admixture performance under variable chemical and physical conditions. Conventional fixed-dosage approaches are increasingly inadequate, as small fluctuations in binder chemistry, fine content, or activator usage can trigger disproportionate rheological responses. Digitalization enables VMA performance to be managed as a dynamic, data-driven process variable rather than a static formulation attribute.

From an industrial manufacturing and ready-mix perspective, robustness is achieved by coupling real-time sensing with predictive control models that anticipate rather than react to incompatibility.

Real-Time Sensing and Process Analytics

Inline measurement of viscosity, torque, power consumption, temperature, and conductivity provides continuous proxies for polymer availability and hydration state. In SCM-intensive systems, these signals are more informative than conventional batch-wise slump or flow tests, as they capture time-dependent rheological evolution during mixing and transport.

Advanced analytics correlate sensor outputs with historical performance data to distinguish between normal variability and emerging instability. Such correlation is particularly valuable where raw material variability cannot be fully controlled.

Predictive Modeling and Digital Twins

Digital twins of cement hydration and polymer interaction simulate the evolution of pore solution chemistry and rheology under varying SCM compositions and activator regimes. These models incorporate cement chemistry metadata, SCM fineness, and admixture architecture to predict viscosity development over time.

In industrial deployment, digital twins are used to evaluate corrective actions virtually before implementation, reducing the risk of overcompensation and minimizing off-spec production. This approach is especially effective during technology transfer across plants with different raw material supply profiles.

Adaptive Dosing and Closed-Loop Control

Adaptive dosing systems adjust VMA input dynamically based on real-time process feedback rather than predefined recipes. When indicators suggest delayed polymer activation or excessive adsorption, dosage, dilution rate, or addition timing is modified automatically to restore target rheology.

Closed-loop control significantly reduces batch-to-batch variability and supports consistent performance across extended production runs. Such systems are increasingly integrated into manufacturing execution systems for centralized oversight.

Machine Learning for Variability Management

Machine-learning algorithms analyze multivariate datasets encompassing cement chemistry, SCM characteristics, admixture usage, and rheological outcomes. These models identify subtle patterns that precede performance drift, enabling early intervention.

In practice, machine learning supports proactive maintenance of rheological stability, reduces reliance on empirical safety margins, and improves overall material efficiency.

Digital Traceability and Knowledge Transfer

Digital batch records link each production lot to its full process and material history, creating a transparent quality backbone. When performance deviations occur, root-cause analysis is accelerated, and corrective knowledge is rapidly disseminated across sites.

Such digital continuity simplifies qualification of new SCM sources and supports scalable deployment of low-clinker technologies.

Table 6.4.5: Industry 4.0 Tools and Their Functional Roles

Digital Tool	Primary Function	Impact on VMA Robustness
Inline rheology sensors	Real-time viscosity tracking	Early instability detection
Digital twins	Process simulation	Risk-free optimization
Adaptive dosing systems	Dynamic correction	Reduced off-spec output
Machine learning models	Pattern recognition	Predictive control
Digital batch records	Traceability	Faster root-cause analysis

6.4.6 Key Challenges and Design Responses in Low-Clinker Systems

Low-clinker cement systems introduce a fundamentally different chemical and physical environment for viscosity-modifying admixtures, where clinker dilution, elevated SCM content, and alternative activation pathways reshape polymer effectiveness. VMAs developed for conventional Portland cement often exhibit non-linear or unstable behavior when transferred directly into these systems. Robust performance therefore depends on recognizing the dominant challenges and embedding targeted design responses at both formulation and process levels.

Reduced Calcium Availability and Delayed Polymer Anchoring

Clinker reduction lowers the early availability of Ca^{2+} ions that facilitate polymer adsorption and network formation. In slag-rich, fly ash-dominant, or LC^3 binders, delayed calcium release shifts VMA action toward later hydration stages, weakening early-age viscosity build-up and segregation resistance.

Industrial design responses focus on polymer architectures with reduced dependence on divalent cation bridging. Modified polysaccharides and synthetic associative polymers with hydrogen bonding or hydrophobic interaction dominance maintain functionality under low-calcium conditions. Controlled molecular weight distribution further minimizes sensitivity to delayed ion release.

Elevated Fine Surface Area and Competitive Adsorption

Low-clinker systems typically contain higher total fines from SCMs and fillers, dramatically increasing available adsorption sites. VMAs compete with superplasticizers and activators for surface interaction, often leading to reduced free polymer concentration in the pore solution.

Mitigation strategies include surface-selective polymers with steric shielding, as well as optimized addition sequences that temporally separate dispersant adsorption from viscosity build-up. From a manufacturing standpoint, pre-dilution and staged dosing reduce instantaneous adsorption losses.

Altered Alkali and Sulfate Regimes

Alkali activation and sulfate adjustment are common in low-clinker formulations, particularly for slag and calcined clay systems. Elevated pH and non-traditional sulfate sources modify polymer ionization, conformation, and hydration kinetics, occasionally triggering excessive thickening or delayed response.

Design responses emphasize pH-tolerant backbones and side-chain chemistries that maintain controlled expansion across a wide ionic strength range. Buffer-compatible formulations reduce sensitivity to sulfate balance drift during production.

Extended Dormant Periods and Time-Dependent Instability

Many low-clinker binders exhibit prolonged dormant periods followed by rapid acceleration. VMAs must remain stable and effective over extended induction times without premature structuring or viscosity decay.

Time-resilient polymer designs combine slow hydration kinetics with reversible associative interactions. These systems preserve pumpability while ensuring viscosity recovery during placement. Shelf-life stability is reinforced through antioxidant protection and controlled residual moisture levels.

Interaction with Complex Admixture Packages

Low-clinker concretes frequently rely on combined admixture systems, including activators, retarders, accelerators, and high-performance dispersants. Unmanaged interactions can suppress VMA efficiency or induce unpredictable rheological transitions.

Formulation-level compatibility screening under realistic binder chemistries is essential. Industrial practice increasingly adopts matrix-based testing rather than binary compatibility checks, supported by rheological fingerprinting across time and temperature.

Table 6.4.6: Integrated Design and Control Responses

Challenge	Primary Risk	Design Response
Low Ca^{2+} availability	Weak early viscosity	Calcium-independent polymer architectures
High fine content	Polymer depletion	Selective adsorption and staged dosing
High alkalinity	Over-thickening or delay	pH-tolerant backbone chemistry
Long dormant periods	Viscosity drift	Reversible associative networks
Complex admixtures	Incompatibility	Matrix-based compatibility design

Industry 4.0-Enabled Reinforcement

Digital control strategies complement chemical design by stabilizing performance under unavoidable variability. Real-time rheological monitoring, adaptive dosing, and predictive modeling reduce reliance on conservative over-dosing. Digital twins simulate SCM source changes and admixture interactions before plant-scale implementation, accelerating qualification cycles.

Such integration transforms low-clinker VMA deployment from empirical adjustment to controlled technology transfer, supporting scalable decarbonization without compromising fresh concrete stability.

6.5 Rheological Manifestations in Fresh Concrete

Rheological behavior in fresh concrete represents the macroscopic expression of molecular-scale interactions between viscosity-modifying admixtures, cement phases, supplementary cementitious materials, and the evolving pore solution. In low-clinker and SCM-rich systems, these manifestations are amplified by delayed hydration, elevated fine content, and complex ionic environments. Understanding how VMA chemistry translates into measurable flow behavior is essential for formulation design, process control, and reliable field performance.

6.5.1 Yield Stress Development and Structural Build-Up

Yield stress in fresh concrete represents the minimum stress required to initiate flow and is a critical parameter governing segregation resistance, shape stability, and early-age robustness. In systems containing viscosity-modifying admixtures, yield stress emerges from the combined effects of polymer-mediated network formation, particle flocculation dynamics, and the evolving chemistry of the pore solution. The structural build-up process is inherently time-dependent and highly sensitive to binder composition and processing history.

Molecular Origins of Yield Stress in VMA-Modified Systems

VMAs contribute to yield stress by increasing the effective volume fraction of the aqueous phase and by creating transient polymer networks that bridge or confine solid particles. Associative interactions, hydrogen bonding, and entanglement dominate these mechanisms rather than permanent adsorption. The resulting structure resists gravitational separation while remaining reversible under applied shear.

In cementitious systems with adequate early calcium availability, polymer association is reinforced by cation-mediated screening, accelerating network formation. In low-clinker and SCM-rich binders, reduced Ca^{2+} concentration delays this process, shifting yield stress development to later hydration stages unless polymer architecture is specifically adapted.

Kinetics of Structural Build-Up at Rest

Structural build-up at rest reflects the progressive recovery of yield stress following mixing or shear. VMAs modulate this process by controlling the rate of particle re-flocculation and polymer re-association. Rapid build-up enhances shape retention and formwork pressure control, while overly aggressive structuring compromises workability.

Industrial formulations target a balanced build-up profile, characterized by fast initial recovery followed by a plateau that maintains stability during transport and placement. Temperature and water chemistry exert strong influence, particularly in SCM-rich systems with prolonged dormant periods.

Influence of Fine Content and Particle Packing

Fine particles from SCMs and fillers increase contact points and confinement effects, amplifying yield stress at constant VMA dosage. While this enhances segregation resistance, it also raises sensitivity to small dosage variations.

Design strategies include tailoring polymer molecular weight and charge density to decouple yield stress from fine surface area. Staged addition and dilution of VMAs further mitigate abrupt structural build-up in fine-rich formulations.

Shear Sensitivity and Reversibility

A defining requirement for VMAs is the reversibility of yield stress under shear. Structural build-up must collapse under mixing and pumping, then recover predictably once shear is reduced. Excessive chain entanglement or irreversible adsorption leads to hysteresis and poor placement behavior.

Shear cycling tests and time-resolved rheometry are therefore essential for qualification. These methods reveal whether yield stress recovery remains controlled over repeated mixing and transport events.

Table 6.5.1: Industrial Performance Indicators

Indicator	Observation	Practical Implication
Initial yield stress	Early segregation resistance	Casting stability
Build-up rate	Speed of structure recovery	Shape retention
Yield stress plateau	Time-stable structure	Transport robustness
Reversibility	Collapse under shear	Pumpability

Digital Monitoring and Control

Inline torque and power draw measurements provide indirect but continuous indicators of yield stress evolution during mixing. When integrated into digital control systems, these signals enable adaptive VMA dosing and energy adjustment to maintain target structural build-up profiles across variable raw materials.

Such data-driven control is increasingly used to stabilize yield stress in low-clinker concretes, where intrinsic variability would otherwise require conservative over-design.

6.5.2 Plastic Viscosity and Flow Stability

Plastic viscosity governs the resistance of fresh concrete to sustained flow once yielding has occurred and directly determines pumping energy, form filling efficiency, and surface finish quality. In systems incorporating viscosity-modifying admixtures, plastic viscosity reflects the balance between polymer-induced hydrodynamic effects, particle interactions, and shear-dependent structural rearrangements. Unlike yield stress, which controls flow initiation, plastic viscosity dictates flow stability under continuous deformation.

Polymer Contributions to Plastic Viscosity

VMAs increase plastic viscosity primarily by enlarging the effective hydrodynamic volume of the aqueous phase through polymer hydration and chain expansion. Associative interactions between polymer segments further dissipate energy during shear, stabilizing flow against segregation and washout.

Excessive chain length or high association density can, however, elevate plastic viscosity beyond practical limits, increasing pumping pressure and risking incomplete mold filling. Industrial formulation therefore emphasizes controlled molecular weight distributions and tunable associative strength.

Shear-Thinning Behavior and Energy Efficiency

A defining performance target for VMAs is shear-thinning rheology, where apparent viscosity decreases under high shear and recovers rapidly at rest. Shear-thinning reduces energy demand during pumping and mixing while preserving stability during placement.

This behavior arises from reversible disentanglement and alignment of polymer chains under shear. Upon shear cessation, rapid re-association restores viscosity. Polymer architectures with flexible backbones and reversible hydrophobic interactions are particularly effective in achieving this response.

Influence of Binder Composition and Fine Content

SCM-rich and low-clinker systems often exhibit elevated baseline plastic viscosity due to higher fine content and altered particle packing. VMAs interact with these fines to further increase viscosity, making dosage control more critical.

Formulation strategies include lowering polymer association density or introducing steric side chains to reduce sensitivity to fine surface area. In practice, plastic viscosity is optimized jointly with superplasticizer selection to avoid antagonistic effects.

Flow Stability under Dynamic Conditions

Flow stability describes the ability of concrete to maintain uniform rheology during prolonged shear, transport, and placement. VMAs contribute to stability by preventing particle migration and phase separation under flow.

Inconsistent flow stability often indicates partial polymer degradation, excessive adsorption, or incompatibility with dispersants. Industrial diagnostics rely on pressure loss monitoring during pumping and time-resolved rheological measurements.

Table 6.5.2: Industrial Metrics and Correlation

Parameter	Measurement Proxy	Operational Impact
Plastic viscosity	Rheometer slope	Pumping pressure
Shear-thinning index	Multi-shear tests	Energy efficiency
Viscosity recovery	Step-shear tests	Placement stability
Flow uniformity	Pressure consistency	Defect reduction

Industry 4.0-Enabled Optimization

Inline monitoring of power draw, pressure drop, and flow rate provides continuous insight into plastic viscosity during production and placement. These signals feed predictive models that adjust VMA dosage or mixing energy in real time.

Digital twins simulate the impact of formulation changes on pumping and placement behavior, reducing trial-and-error during scale-up. Such tools are increasingly essential for maintaining flow stability in variable SCM-rich systems.

6.5.3 Time-Dependent Rheology and Thixotropy

Time-dependent rheology in fresh concrete reflects the continuous evolution of microstructure under the combined influence of hydration reactions, particle interactions, and polymer reorganization. Viscosity-modifying admixtures play a central role in governing thixotropic behavior by introducing reversible structural elements that respond dynamically to shear history and rest periods. In modern SCM-rich and low-clinker systems, these effects become more pronounced due to extended dormant periods and delayed strength development.

Mechanistic Basis of Thixotropy in VMA-Modified Concrete

Thixotropy arises from the breakdown and reformation of weak physical networks during alternating shear and rest. VMAs contribute by forming transient polymer networks through entanglement, hydrogen bonding, and associative interactions. Under shear, these networks partially collapse, reducing viscosity. Upon rest, rapid re-association restores structure.

The reversibility of these interactions is essential. Excessive adsorption or irreversible flocculation leads to progressive stiffening rather than controlled thixotropic recovery.

Structural Build-Up and Breakdown Cycles

Repeated shear cycles during mixing, transport, and placement impose alternating breakdown and recovery phases. VMAs must maintain consistent recovery kinetics across these cycles without cumulative viscosity loss or gain.

In SCM-rich systems, slower hydration and altered ionic strength extend recovery times. Polymer design therefore targets stable recovery rates that are insensitive to modest temperature and chemistry fluctuations.

Impact on Buildability and Placement Technologies

Time-dependent rheology directly affects buildability in applications such as slip-forming, layered casting, and additive manufacturing. Controlled thixotropy allows concrete to support subsequent layers while remaining pumpable.

VMAs designed for such applications emphasize rapid yield stress recovery with moderate plastic viscosity increase. Overly aggressive thixotropy increases placement resistance and surface defects.

Temperature and Time Sensitivity

Temperature accelerates both hydration and polymer dynamics, modifying thixotropic response. In hot conditions, rapid recovery can reduce open time, while in cold environments, delayed recovery increases segregation risk.

Industrial formulations incorporate temperature-resilient polymer chemistries and, when necessary, combine VMAs with retarders or accelerators to stabilize time-dependent behavior.

Table 6.5.3: Measurement and Industrial Interpretation

Test Method	Thixotropic Indicator	Practical Insight
Loop rheometry	Hysteresis area	Structural reversibility
Step-shear tests	Recovery rate	Transport stability
Rest-period tests	Yield stress gain	Buildability
Pumping simulations	Pressure recovery	Placement reliability

Digital Control of Time-Dependent Behavior

Inline rheological proxies such as torque recovery and power draw trends enable continuous assessment of thixotropic evolution during production. These data feed adaptive control systems that adjust VMA dosage or mixing energy to maintain target recovery profiles.

Machine-learning models trained on historical shear-rest cycles predict thixotropic drift before field performance is affected, supporting proactive correction.

6.5.4 Interaction with Mixing, Transport, and Placement

The rheological performance imparted by viscosity-modifying admixtures is inseparable from the mechanical history imposed during mixing, transport, and placement. VMAs respond dynamically to shear intensity, shear duration, and rest intervals, translating process conditions into measurable changes in yield stress, plastic viscosity, and thixotropic recovery. Industrial robustness therefore requires that VMA chemistry be aligned with realistic processing pathways rather than idealized laboratory protocols.

Mixing Energy and Shear History

During mixing, VMAs are exposed to high and often non-uniform shear fields that influence polymer hydration, chain conformation, and network formation. Adequate energy is required to fully disperse and hydrate polymer chains, particularly for high-molecular-weight or associative systems. Insufficient mixing leads to incomplete activation and heterogeneity, while excessive shear risks partial chain scission or persistent disentanglement.

Industrial practice increasingly specifies mixing energy input rather than fixed mixing time, using power draw or torque as control parameters. This approach improves reproducibility across mixers of different geometry and scale.

Transport-Induced Structural Evolution

Transport by truck mixers or pumps subjects concrete to prolonged, moderate shear interspersed with rest periods. Under these conditions, VMA-induced structures repeatedly break down and reform. Well-designed systems exhibit stable recovery kinetics, maintaining flowability without progressive thickening or thinning.

Temperature rise during transport further accelerates polymer dynamics and hydration reactions. Formulations must therefore tolerate combined thermal and mechanical aging to preserve rheological stability upon arrival at the placement site.

Pumping and High-Shear Placement

Pumping imposes high shear rates that temporarily suppress viscosity and yield stress. VMAs must collapse predictably under these conditions to limit pressure build-up, then recover rapidly after discharge to prevent segregation.

Associative polymers with reversible interactions provide superior performance by decoupling pumpability from placement stability. Excessive irreversible adsorption or chain degradation manifests as increased pumping pressure or loss of post-pump stability.

Placement, Finishing, and Rest-Phase Recovery

After placement, shear is minimal and structural recovery dominates. VMAs control the rate at which yield stress and viscosity rebuild, influencing surface finish, formwork pressure, and resistance to bleeding.

In applications such as slip-forming or layered casting, recovery must be sufficiently rapid to support subsequent operations without compromising interlayer bonding. Polymer design and dosage are therefore tailored to the specific placement technique.

Table 6.5.4: Process-Specific Performance Mapping

Process Stage	Dominant Shear Regime	VMA Performance Requirement
Mixing	High, non-uniform	Complete hydration without degradation
Transport	Moderate, cyclic	Stable recovery kinetics
Pumping	Very high	Shear collapse with low pressure
Placement	Low to none	Rapid, controlled structural build-up

Industry 4.0 Process Integration

Modern production facilities integrate rheological proxies such as power consumption, mixer torque, and pumping pressure into digital control platforms. These data streams are used to map rheological response to process history and to adjust VMA dosage, dilution, or mixing energy in real time.

Digital twins simulate mixing and transport scenarios to predict rheological evolution before execution. Such tools reduce trial-and-error during formulation transfer and enhance consistency across sites and equipment configurations.

6.5.5 Measurement and Industrial Relevance

Reliable deployment of viscosity-modifying admixtures in modern concrete production depends on the ability to translate laboratory rheological descriptors into operationally meaningful indicators. Measurement strategies must capture the time-dependent, shear-sensitive behavior of fresh concrete while remaining compatible with industrial production constraints. The relevance of any rheological parameter is therefore defined by its predictive value for mixing, transport, placement, and in-situ performance.

Laboratory Rheology and Its Limitations

Rotational rheometers provide fundamental parameters such as yield stress and plastic viscosity under controlled conditions. These measurements are indispensable for formulation screening and comparative evaluation of VMA chemistries. However, laboratory shear protocols often fail to replicate the complex and non-uniform shear histories encountered in industrial mixers, pumps, and formwork.

Consequently, absolute rheological values are less informative than trends and relative changes when transferring results to plant-scale operations. Time-resolved testing and cyclic shear protocols improve relevance by mimicking realistic process conditions.

Process-Oriented Measurement Proxies

Industrial environments favor indirect but continuous indicators of rheological behavior. Mixer torque, power consumption, pressure loss during pumping, and flow rate stability serve as practical proxies for yield stress and plastic viscosity. These signals integrate the effects of polymer behavior, particle interactions, and temperature in real time.

Such proxies are particularly valuable in SCM-rich systems, where rapid chemical evolution complicates interpretation of single-point laboratory tests.

Correlation with Field Performance

Measurement relevance is established by correlating rheological indicators with observable field outcomes such as segregation resistance, pumpability, form filling, and surface finish. VMAs that deliver similar laboratory rheology may differ significantly in field behavior due to differences in shear reversibility or time-dependent recovery.

Industrial qualification therefore emphasizes performance envelopes rather than fixed numerical targets, allowing controlled variability without compromising application requirements.

Standardization and Repeatability Challenges

Concrete rheology lacks universal standardization, and results depend strongly on instrument geometry, testing protocol, and sample preparation. For industrial consistency, internal reference methods and calibration against field benchmarks are often more effective than strict reliance on external standards.

Repeatability improves when rheological testing is embedded within a broader quality framework that includes raw material characterization and process monitoring.

Table 6.5.5: Measurement Methods and Industrial Insight

Measurement Approach	Rheological Insight	Industrial Relevance
Rotational rheometry	Yield stress, viscosity	Formulation screening
Loop and cyclic tests	Thixotropy	Transport stability
Mixer torque	Structural build-up	Process control
Pump pressure loss	Flow resistance	Pumpability assurance
Slump flow trends	Integrated response	Site verification

Industry 4.0-Enabled Measurement Integration

Digital platforms consolidate laboratory data, inline process measurements, and field performance records into unified datasets. Machine-learning models use these data to refine correlations between measured rheology and operational outcomes, progressively improving predictive accuracy.

Such integration transforms rheological measurement from a diagnostic tool into an active control variable, supporting consistent VMA performance across variable materials and production environments.

6.5.6 Industry 4.0-Enabled Rheology Control

The integration of Industry 4.0 concepts into fresh concrete rheology management is redefining how viscosity-modifying admixtures are selected, dosed, and validated in real time. Traditional empirical approaches based on laboratory trial batches are increasingly inadequate for modern concrete plants operating under variable raw material quality, fluctuating ambient conditions, and tight performance tolerances. Digitalization enables continuous feedback between material behavior and process control, allowing VMAs to be used as adaptive rheology regulators rather than static formulation components.

From a formulation perspective, VMAs exhibit non-linear and time-dependent effects on yield stress, plastic viscosity, and thixotropic rebuild. These effects are strongly coupled to cement chemistry, fines content, and dispersant state. Industry 4.0 frameworks translate these complex interactions into actionable control parameters by combining sensor data, material models, and automated dosing logic. As a result, rheological consistency can be maintained despite upstream variability in cement fineness, supplementary cementitious material reactivity, or moisture content of aggregates.

Sensor-Based Rheology Monitoring in Fresh Concrete

Advanced sensing technologies form the backbone of digital rheology control. Inline and at-line sensors increasingly provide indirect but robust indicators of fresh concrete flow behavior. Torque measurements in mixers, power consumption profiles, vibration response during discharge, and pressure differentials in pumping lines are correlated with apparent viscosity and yield stress. When properly calibrated, these signals allow continuous inference of rheological state without interrupting production.

For VMAs, sensor-based monitoring is particularly valuable because their impact is often subtle and masked by superplasticizer action at low shear rates. Digital systems can detect early signs of over-thickening, segregation resistance loss, or excessive thixotropic buildup that would otherwise only become evident during placement. This enables corrective micro-dosing of VMAs or dispersants within the same batch cycle, reducing scrap and rework.

Data-Driven VMA Dosage Optimization

Machine learning and multivariate statistical models are increasingly applied to map the relationship between raw material descriptors, admixture chemistry, and measured rheological responses. Cement phase composition, alkali content, sulfate balance, and SCM mineralogy are incorporated alongside VMA molecular attributes such as charge density, molecular weight distribution, and substitution pattern.

These models allow predictive adjustment of VMA dosage to achieve target rheological windows under changing conditions. For example, higher C3A reactivity or elevated fines content may trigger algorithmic recommendations for reduced VMA concentration or modified addition sequence. Over time, self-learning systems refine these correlations, improving robustness across seasons, suppliers, and production sites. This approach supports technology transfer by embedding formulation know-how into digital workflows rather than relying solely on expert operators.

Closed-Loop Rheology Control Architectures

Closed-loop control represents the most mature application of Industry 4.0 to VMA-enabled rheology. In such architectures, real-time sensor inputs are continuously compared against predefined rheological setpoints linked to placement method, formwork geometry, and reinforcement density. Deviations trigger automated adjustments in admixture dosing, mixing energy, or water correction within permissible limits.

For VMAs, closed-loop control mitigates risks associated with overdosing, such as pumpability loss or delayed surface finishing. It also compensates for batch-to-batch variability in polymer performance due to temperature-dependent hydration kinetics or shear history. Importantly, these systems require well-characterized VMA response curves and robust compatibility data with cementitious systems, reinforcing the need for systematic rheological profiling during product development.

Digital Twins for Fresh Concrete Rheology

Digital twin concepts extend rheology control beyond the plant into simulation environments that mirror real production conditions. A digital twin integrates material models for cement hydration, admixture adsorption, and polymer-induced structuring of the pore solution. VMAs are represented as dynamic rheology modifiers whose effectiveness evolves with time, shear, and ionic environment.

By simulating fresh concrete behavior before production, digital twins support proactive formulation adjustments and scenario testing. This is particularly valuable for complex applications such as self-consolidating concrete or 3D-printed cementitious materials, where rheological margins are narrow. For VMA suppliers, digital twins facilitate technology transfer by providing customers with predictive tools that reduce commissioning time and trial costs.

Practical Implementation Considerations

Despite clear benefits, implementation of Industry 4.0-enabled rheology control requires careful alignment between hardware, software, and admixture chemistry. VMAs must exhibit reproducible performance and low sensitivity to minor dosage fluctuations to be suitable for automated control. Data integrity, sensor calibration, and cybersecurity also become critical factors in industrial environments.

Successful deployment typically follows a phased approach, starting with data acquisition and visualization, progressing to advisory systems, and ultimately achieving autonomous control. Throughout this progression, collaboration between admixture formulators, concrete producers, and automation specialists is essential to ensure that rheological targets are technically meaningful and operationally achievable.

Table 6.5.6: Representative Industry 4.0 Tools for VMA Rheology Control

Digital Element	Function in Rheology Control	Relevance to VMAs
Mixer torque sensors	Indirect viscosity and yield stress tracking	Detect over-thickening and thixotropy
Machine learning models	Predictive dosage optimization	Adapt VMA levels to cement variability

Closed-loop dosing systems	Real-time correction	Prevent segregation or pump blockage
Digital twins	Scenario simulation and training	Accelerate formulation transfer

6.6 Recycled Constituents and Non-Traditional Aggregates

The increasing incorporation of recycled constituents and non-traditional aggregates into concrete formulations introduces a new level of rheological complexity that directly impacts the effectiveness and robustness of viscosity-modifying admixtures. Recycled concrete aggregates, crushed masonry, industrial by-products, and manufactured sands exhibit wider variability in particle morphology, surface chemistry, and water interaction compared to natural aggregates. These characteristics amplify the sensitivity of fresh concrete rheology to fine content and moisture fluctuations, making VMA selection and control a critical formulation variable rather than a secondary stabilizer.

From an industrial formulation standpoint, VMAs must compensate for heterogeneity while maintaining pumpability, segregation resistance, and surface finish. The challenge lies in managing variability without resorting to excessive admixture dosages that compromise robustness or cost efficiency. Digital material characterization and adaptive dosing strategies aligned with Industry 4.0 principles increasingly serve as enabling tools to stabilize rheology when recycled constituents are used at scale.

6.6.1 Fine Content Variability

The incorporation of recycled constituents and non-traditional aggregates introduces fine fractions that are markedly different from conventional quarry-derived fillers in both origin and functionality. These fines are generated through crushing, abrasion, and the liberation of adhered cement paste, resulting in particles with high angularity, elevated specific surface area, and heterogeneous mineralogical composition. Such characteristics intensify water demand and disrupt particle packing, leading to pronounced shifts in yield stress and plastic viscosity even at modest replacement levels.

From a polymer–particle interaction perspective, fine content variability directly governs VMA adsorption efficiency and structuring behavior within the pore solution. Recycled concrete fines often contain partially hydrated clinker phases, calcite, and amorphous calcium silicate hydrates that compete for polymer adsorption sites. This competition reduces the effective concentration of free VMA chains available for solution structuring, creating batch-to-batch inconsistency in segregation resistance and flow retention. The effect is amplified in systems with high sulfate or alkali availability, where electrostatic screening alters polymer conformation.

Industrial formulation strategies increasingly move beyond simple mass-based fines limits toward functional classification based on surface area, zeta potential, and mineral reactivity. VMAs with moderate molecular weight and controlled substitution patterns demonstrate improved tolerance to fines-rich systems by balancing adsorption and solution thickening mechanisms. In practice, co-optimization with superplasticizers is essential, as dispersant-fines interactions strongly influence the apparent VMA demand and the risk of over-structuration at low shear.

Digital characterization tools aligned with Industry 4.0 concepts support stabilization of rheology under fine content variability. Inline monitoring of mixer torque and energy consumption provides indirect yet reliable indicators of fines-induced rheological shifts. When combined with historical raw material datasets, predictive models can recommend incremental VMA dosage adjustments or alternative polymer architectures before visible loss of workability occurs. Such data-driven approaches embed empirical know-how into automated decision frameworks, enhancing robustness across recycled material supply chains.

Table 6.6.1: Practical Implications for VMA Selection

Fine Characteristic	Rheological Effect	VMA Design Response
High specific surface	Increased water and polymer demand	Moderate molecular weight, low charge density
Reactive cementitious fines	Competitive adsorption	Balanced adsorption and solution structuring

High angularity	Elevated yield stress	Enhanced lubrication through controlled thickening
Variable mineralogy	Inconsistent flow retention	Data-driven adaptive dosing

6.6.2 Moisture and Absorption-Driven Instability

Recycled and non-traditional aggregates exhibit moisture behavior that deviates fundamentally from that of dense natural aggregates, creating a primary source of rheological instability in fresh concrete. Elevated porosity, microcracked surfaces, and adhered cement paste lead to high absorption capacity and slow moisture equilibration. As a result, the effective water available to lubricate the cementitious matrix becomes time dependent, changing continuously during mixing, transport, and placement. Even when aggregates are nominally pre-saturated, residual absorption kinetics remain active under shear.

From a rheological standpoint, these dynamic moisture exchanges directly influence the performance envelope of viscosity-modifying admixtures. VMAs rely on a stable free-water phase and predictable ionic environment to develop controlled solution structuring and particle network stabilization. Rapid aggregate water uptake during early mixing increases apparent viscosity and accelerates thixotropic rebuild, while delayed water release during haulage can locally dilute the pore solution, weakening segregation resistance and bleeding control. Such opposing effects complicate dosage selection and reduce robustness when conventional fixed-dosage strategies are applied.

Industrial formulation approaches increasingly integrate absorption kinetics rather than static water absorption values into mix design logic. VMAs with reversible associative mechanisms and limited sensitivity to short-term water fluctuations demonstrate superior performance in moisture-unstable systems. These polymers maintain rheological continuity across shear cycles, reducing the risk of pump pressure spikes and finishing delays. Co-formulation with superplasticizers must be carefully balanced, as dispersant efficiency is also moisture dependent and can exacerbate instability if water availability shifts abruptly.

Industry 4.0-enabled control architectures provide practical mitigation pathways. Real-time aggregate moisture sensors, mixer power signatures, and discharge flow indicators feed predictive models that estimate effective water evolution during processing. These models enable adaptive adjustment of VMA and water-reducing admixture dosages within defined safety margins. By acting on predicted rheological drift rather than observed failure, digital control systems stabilize fresh concrete behavior despite high absorption variability inherent to recycled materials.

Table 6.6.2: Industrial Control Implications

Moisture Phenomenon	Rheological Consequence	VMA-Oriented Mitigation
Rapid early absorption	Sudden viscosity increase	Reversible VMA structuring
Time-dependent release	Loss of cohesion and bleeding	Stable polymer networks
Moisture heterogeneity	Batch inconsistency	Sensor-driven adaptive dosing
Temperature sensitivity	Accelerated water exchange	Integrated thermal and moisture control

6.7 Adaptive Dosing and Variability Mitigation Strategies

Concrete production environments increasingly operate under conditions of high raw material variability, narrow performance tolerances, and accelerated delivery schedules. Under such constraints, fixed admixture dosages are structurally insufficient to ensure stable fresh concrete rheology. Adaptive dosing strategies reposition viscosity-modifying admixtures from passive stabilizers to active control agents that respond dynamically to fluctuations in cement chemistry, aggregate properties, moisture state, and processing history. This approach aligns admixture technology with modern production realities and supports scalable use of recycled and alternative constituents.

Adaptive dosing frameworks combine material science understanding with digital control logic. VMAs are selected not only for peak performance under ideal conditions but also for predictable response gradients under deviation. The objective is to maintain rheological behavior within a defined operational window rather than targeting a single nominal value. Such strategies reduce sensitivity to disturbances and improve consistency across batches, plants, and seasons.

6.7.1 Sources of Variability Requiring Adaptive Control

Fresh concrete rheology is governed by a dense network of interacting material and process variables, many of which fluctuate outside the tolerance range assumed in conventional mix designs. These fluctuations translate directly into unstable yield stress, plastic viscosity, and thixotropic rebuild, making adaptive control a prerequisite rather than an optimization option in modern production. Understanding the dominant sources of variability is therefore essential for designing effective adaptive dosing strategies for viscosity-modifying admixtures.

Cement Chemistry and Fineness Drift

Cement delivered to ready-mix plants commonly exhibits temporal variation in fineness, phase assemblage, and sulfate balance due to changes in clinker source, grinding conditions, and gypsum form. Small shifts in C3A content or sulfate availability significantly modify early hydration kinetics and pore solution ionic strength. These changes alter VMA adsorption equilibria and polymer conformation, leading to unpredictable viscosity development at constant dosage. The effect is particularly pronounced in low water-to-cement ratio systems, where free water is limited and rheological margins are narrow.

Variability in Supplementary Cementitious Materials

Supplementary cementitious materials contribute additional uncertainty through variable particle size distribution, glass content, and surface chemistry. Fly ash, slag, and calcined clays from different sources or production campaigns exhibit divergent water demand and reactivity profiles. Their interaction with VMAs is not purely physical, as dissolved ions and reactive surfaces influence polymer performance over time. Adaptive control becomes critical to compensate for delayed rheological changes that emerge after initial mixing, especially in transport or extended placing scenarios.

Aggregate Moisture, Absorption, and Fines Content

Aggregates represent the largest mass fraction and a major source of short-term variability. Moisture content fluctuates with weather, stockpile management, and transport, while absorption kinetics in recycled or porous aggregates introduce time-dependent water exchange. Fine content variability further modifies surface area and interparticle friction. These factors collectively cause rapid rheological drift that cannot be corrected reliably through static water adjustments, directly challenging the stability of VMA-controlled systems.

Mixing Energy and Shear History

Differences in mixer type, loading sequence, and residence time generate variability in shear exposure and dispersion quality. VMAs respond to shear through changes in polymer alignment and associative network formation. Inconsistent mixing energy can therefore produce divergent rheological states even with identical formulations. This source of variability is often underestimated but becomes critical in automated plants with high throughput and short mixing cycles.

Temperature and Environmental Conditions

Ambient and material temperature influence hydration rate, polymer mobility, and water viscosity. Elevated temperatures accelerate early reactions and increase water demand, while low temperatures delay structuration and increase risk of segregation. Temperature-induced variability interacts synergistically with moisture and cement chemistry effects, amplifying rheological instability unless actively compensated.

Process and Logistics-Related Variability

Transport duration, pumping pressure, and placement method impose additional shear and rest cycles that evolve rheology beyond the batching point. VMAs with time-dependent structuring behavior are sensitive to these histories. Variability in logistics therefore propagates backward into formulation requirements, reinforcing the need for adaptive strategies that consider the full lifecycle of fresh concrete.

Table 6.7.1: Summary of Key Variability Sources

Variability Source	Primary Rheological Impact	Relevance to Adaptive Control

Cement chemistry drift	Unstable yield stress development	High
SCM heterogeneity	Delayed viscosity changes	High
Aggregate moisture and fines	Rapid rheological drift	Very high
Mixing energy variation	Inconsistent structuration	Medium
Temperature fluctuation	Coupled hydration and viscosity effects	High
Transport and placement	Time-dependent rheology evolution	Medium

6.7.2 Adaptive Dosing Architectures for VMAs

Adaptive dosing architectures define how viscosity-modifying admixtures are translated from formulation intent into real-time rheological control actions. These architectures differ in complexity, data intensity, and degree of autonomy, yet all aim to compensate for unavoidable variability in cementitious systems without compromising robustness or safety margins. For VMAs, architecture design must account for the polymer's non-linear dose response, time-dependent structuring behavior, and sensitivity to shear and ionic environment.

In industrial practice, adaptive dosing is most effective when implemented as a layered control system. Each layer addresses variability at a different time scale, ranging from immediate batch-level correction to longer-term learning and optimization across production cycles.

Rule-Based and Threshold-Driven Architectures

Rule-based architectures represent the entry level of adaptive dosing and are widely deployed in ready-mix and precast plants. These systems rely on predefined correction rules linked to measurable variables such as aggregate moisture deviation, fines content, or mixer torque thresholds. When a parameter exceeds an acceptable range, a corresponding incremental VMA dosage adjustment is automatically applied.

For VMAs, rule-based systems require conservative correction coefficients to avoid over-structuration. They are most suitable for polymers with smooth and predictable dose–response behavior. Although limited in handling multivariate interactions, such architectures provide high transparency and are readily accepted by operators, making them effective stepping stones toward more advanced digital control.

Model-Based Adaptive Dosing Systems

Model-based architectures integrate multiple input variables into predictive algorithms that estimate rheological state rather than reacting to single thresholds. These models combine material descriptors such as cement fineness, SCM content, and aggregate absorption with process signals including mixing energy and temperature. VMA dosage is adjusted based on predicted deviation from a target rheological window.

For polymer-based VMAs, model fidelity depends strongly on the quality of underlying adsorption and structuration data. Simplified empirical models are often sufficient for industrial deployment, provided they are periodically recalibrated. Such systems significantly improve consistency in environments with frequent raw material changes and are well suited for centralized quality control across multiple plants.

Closed-Loop Feedback Control Architectures

Closed-loop architectures represent the most advanced form of adaptive dosing. In these systems, real-time sensor signals are continuously compared against predefined rheological setpoints, and VMA dosage is adjusted dynamically to minimize deviation. Feedback variables commonly include mixer torque evolution, power consumption curves, or discharge flow characteristics.

Closed-loop control demands VMAs with low hysteresis and reversible structuring under shear. Polymers that exhibit delayed or irreversible thickening introduce instability into feedback loops and increase risk of oscillatory correction. When properly matched with suitable admixtures, closed-loop systems deliver superior batch-to-batch uniformity and reduce dependence on operator intervention.

Learning and Self-Optimizing Architectures

Learning architectures extend adaptive dosing by incorporating historical data and self-optimization logic. Machine learning models identify recurring patterns in variability and refine correction strategies over time. In such systems, VMA dosage adjustments become increasingly anticipatory, accounting for seasonal trends, supplier-specific material behavior, and process drift.

For VMA suppliers, learning architectures offer a powerful pathway for technology transfer. Polymer performance characteristics are effectively embedded into digital models, allowing consistent application across plants without extensive on-site reformulation. Governance mechanisms remain essential to constrain model actions within validated safety envelopes.

Table 6.7.2: Comparative Overview of Adaptive Dosing Architectures

Architecture Type	Data Complexity	Control Speed	VMA Suitability	Industrial Maturity
Rule-based	Low	Moderate	High for robust polymers	High
Model-based	Medium	Moderate	High with calibrated response	Medium
Closed-loop	High	Fast	Requires reversible structuring	Medium
Learning systems	Very high	Anticipatory	Requires stable polymer behavior	Emerging

Implementation and Technology Transfer Considerations

Selection of an adaptive dosing architecture must align with plant digital maturity, variability profile, and admixture robustness. Overly complex systems applied to unstable polymers increase operational risk rather than reducing it. Successful implementation typically progresses from rule-based control toward closed-loop and learning architectures as confidence in data quality and polymer behavior increases.

Clear documentation of VMA response characteristics, correction limits, and failure modes is essential for safe automation. When these requirements are met, adaptive dosing architectures transform VMAs into reliable rheology actuators within digitally managed concrete production systems.

6.7.3 Polymer Design Considerations for Adaptive Dosing

Polymer architecture fundamentally determines whether a viscosity-modifying admixture can function as a controllable actuator within adaptive dosing systems. In digitally managed concrete production, the objective is not maximum thickening efficiency but predictable, proportional, and reversible rheological response. VMAs intended for adaptive environments must therefore be engineered for controllability rather than peak performance under static laboratory conditions.

A primary requirement is a smooth and continuous dose–response relationship across the relevant concentration window. Polymers that exhibit abrupt viscosity escalation near the minimum effective dosage introduce instability into automated correction loops. Such behavior is frequently associated with narrow molecular weight distributions or high density of associative sites that trigger rapid network percolation. In contrast, broader molecular weight distributions and moderated functional group density produce gradual rheological transitions that are more amenable to incremental digital correction.

Shear reversibility and low hysteresis represent equally critical design criteria. During mixing, pumping, and placement, VMAs are exposed to repeated shear and rest cycles. Polymers that undergo irreversible adsorption onto cementitious surfaces or form shear-resistant networks accumulate structuration with time, limiting the effectiveness of subsequent dosage adjustments. Associative VMAs that rely on transient hydrophobic interactions and weak physical entanglements exhibit partial network breakdown under shear and rapid rebuild at rest, enabling stable closed-loop control.

Sensitivity to pore solution chemistry must be deliberately constrained. High charge density polymers or systems strongly dependent on ionic strength display exaggerated response to fluctuations in alkali concentration, sulfate availability, and temperature. Such sensitivity undermines adaptive dosing by coupling rheological response to poorly controlled upstream variables. VMAs designed for adaptive systems typically employ balanced hydrophilic–hydrophobic architectures with limited electrostatic contribution, ensuring consistent performance across a broad range of cement chemistries and recycled material contents.

From an industrialization and technology transfer standpoint, polymer robustness simplifies system commissioning and reduces the burden on plant operators. VMAs with predictable response envelopes can be deployed across multiple plants with minimal local re-calibration, allowing digital dosing algorithms to be transferred alongside the product. This convergence of polymer design and control logic represents a decisive shift toward quality-by-design in fresh concrete rheology.

Table 6.7.3: Key Polymer Attributes for Adaptive Dosing

Polymer Attribute	Influence on Adaptive Control	Design Implication
Dose–response slope	Stability of correction loops	Avoid sharp percolation thresholds
Molecular weight distribution	Incremental viscosity control	Favor moderate and broad distributions
Associative mechanism	Shear reversibility	Use transient physical interactions
Ionic sensitivity	Robustness to cement variability	Limit charge density effects
Adsorption reversibility	Long-term controllability	Minimize irreversible binding

6.7.4 Variability Mitigation Beyond Dosage Adjustment

Adaptive dosing alone cannot fully compensate for the magnitude and frequency of variability encountered in modern concrete production, particularly when recycled constituents, alternative binders, and high-performance rheology targets are combined. Effective variability mitigation therefore extends beyond admixture dosage correction and addresses upstream material conditioning, process standardization, and data integration. These measures reduce the disturbance amplitude entering the system, allowing VMAs to operate within their optimal controllability window.

Raw Material Pre-Conditioning and Stabilization

Pre-conditioning of aggregates and supplementary cementitious materials represents a first-order mitigation strategy. Controlled pre-wetting of recycled aggregates reduces the rate and magnitude of early water absorption, stabilizing effective water availability during mixing. Fine fraction management through washing, air classification, or blending limits abrupt changes in specific surface area and adsorption demand. Such interventions directly reduce the corrective burden placed on VMA dosing systems.

For cementitious binders, blending strategies that average fineness and phase variability over time improve rheological predictability. Although these measures introduce additional handling steps, their impact on downstream rheology stability often outweighs operational complexity, particularly in high-throughput plants.

Process Standardization and Mixing Protocol Control

Variability introduced by inconsistent mixing energy, sequencing, and residence time is frequently underestimated. Standardized mixing protocols that control loading order, shear exposure, and mixing duration significantly improve reproducibility of VMA performance. Consistent dispersion of polymers and dispersants ensures that adaptive dosing corrections act on a well-defined baseline rather than compensating for process-induced artifacts.

Digital tracking of mixer power profiles and cycle times supports enforcement of standardized protocols. Deviations can be flagged in real time, enabling corrective action before rheological instability manifests at discharge or placement.

Integrated Water Management Strategies

Water management extends beyond batch water correction. Stabilizing the distribution and timing of water addition reduces transient concentration gradients that disrupt VMA structuring mechanisms. Split water addition strategies, combined with moisture-adjusted aggregate charging, improve homogeneity and limit early-stage viscosity spikes.

In plants equipped with advanced sensing, effective water evolution can be estimated dynamically and incorporated into rheology control logic. This integrated approach reduces reliance on reactive VMA adjustments and enhances overall system resilience.

Digital Material Classification and Predictive Analytics

Industry 4.0 frameworks enable mitigation of variability through anticipation rather than reaction. Digital material passports that record historical behavior of cements, aggregates, and SCMs allow predictive identification of high-risk combinations. When such materials are scheduled for production, preventive measures such as adjusted mixing protocols or pre-emptive rheology tuning can be applied.

Predictive analytics also support supplier qualification and inventory management by linking material variability to measurable rheological outcomes. Over time, this data-driven insight reduces uncertainty and strengthens process control beyond the capabilities of dosage adjustment alone.

Organizational and Technology Transfer Measures

Variability mitigation requires alignment between material suppliers, plant operators, and automation teams. Clear definition of acceptable variability ranges, standardized test methods, and shared digital dashboards promote consistent decision-making. Training programs that explain the interaction between material conditioning, process control, and VMA behavior enhance operator confidence in automated systems.

From a technology transfer perspective, embedding mitigation strategies into standard operating procedures ensures that rheological stability is not dependent on individual expertise. VMAs then function as part of an integrated control ecosystem rather than as isolated corrective additives.

Table 6.7.4: Summary of Non-Dosage Variability Mitigation Measures

Mitigation Measure	Primary Effect	Impact on VMA Control
Aggregate pre-conditioning	Stabilizes water availability	Reduces corrective dosing amplitude
Fine fraction management	Limits adsorption variability	Improves dosage predictability

Mixing protocol standardization	Ensures uniform shear history	Enhances controllability
Integrated water management	Reduces transient gradients	Improves rheology stability
Predictive analytics	Anticipates variability	Enables preventive action

6.7.5 Industrial Implementation Matrix

Translating adaptive dosing concepts for viscosity-modifying admixtures into routine industrial practice requires a structured implementation framework that aligns material behavior, process capability, and digital maturity. An industrial implementation matrix provides a practical tool to map variability sources against mitigation levers and control architectures, enabling staged deployment rather than disruptive transformation. Such matrices support decision-making across ready-mix, precast, and advanced manufacturing environments by clarifying which measures deliver the highest stability return for a given level of complexity.

At the plant level, implementation typically progresses from variability awareness toward predictive and autonomous control. Early-stage systems focus on stabilizing dominant disturbances using robust polymers and simple rules, while mature installations integrate multi-variable models, learning algorithms, and cross-plant data harmonization. VMAs must be selected and qualified in parallel with this progression to ensure compatibility with each implementation tier.

Table 6.7.5: Industrial Implementation Matrix for Adaptive VMA Control

Implementation Level	Digital Maturity	Primary Control Tools	VMA Design Requirement	Typical Application Context
Baseline stabilization	Low	Moisture probes, fixed correction rules	Broad tolerance, low sensitivity	Conventional ready-mix plants

Assisted adaptive control	Medium	Mixer torque tracking, rule-based logic	Smooth dose–response	Plants using recycled aggregates
Model-driven control	Medium–High	Multivariate predictive models	Reversible structuration, low hysteresis	Precast and performance concrete
Closed-loop rheology control	High	Real-time feedback, automated dosing	Predictable shear reversibility	SCC and pumped concrete
Learning and optimization	Very high	Machine learning, digital twins	Stable long-term polymer behavior	Advanced manufacturing, 3D printing

Alignment of Material, Process, and Control Layers

Successful implementation depends on aligning three layers that are often developed independently. The material layer includes cement chemistry, aggregate conditioning, and VMA polymer architecture. The process layer encompasses mixing protocols, water management, and logistics. The control layer integrates sensing, analytics, and dosing hardware. Misalignment between these layers leads to overreliance on dosage correction and undermines system stability.

VMAs designed for adaptive systems reduce alignment risk by providing predictable response envelopes across a wide operating space. Such polymers allow plants to progress through implementation levels without frequent reformulation, facilitating technology transfer across regions and business units.

Phased Deployment and Risk Management

A phased approach minimizes operational risk and builds organizational confidence. Initial deployment should target the dominant variability source, often aggregate moisture or fines content, using conservative correction limits. Subsequent phases can expand control scope as data quality and operator trust improve. Each phase should include validation against defined rheological performance indicators rather than subjective workability assessment.

Risk management also requires fail-safe strategies. Automated systems must revert to validated default dosages if sensor data quality degrades or model confidence thresholds are exceeded. VMAs with gradual response characteristics support such fallback modes by avoiding abrupt rheological changes.

Technology Transfer and Standardization Pathways

The implementation matrix serves as a common language between admixture suppliers, concrete producers, and automation partners. Standardized documentation of VMA behavior, correction limits, and expected response times enables consistent rollout across multiple sites. Digital templates derived from the matrix can be embedded into plant control systems, accelerating commissioning and reducing dependence on local expertise.

By structuring adaptive dosing implementation within a clear industrial matrix, VMAs evolve from auxiliary stabilizers into integral components of digitally managed concrete production systems.

6.8 Practical Compatibility Assessment Framework

A practical compatibility assessment framework for viscosity-modifying admixtures is essential to bridge laboratory rheology with full-scale concrete production. In industrial environments, VMAs must operate reliably across fluctuating cement chemistries, aggregate moisture states, and admixture packages while maintaining robustness against batching variability. The framework therefore focuses on risk reduction, formulation transferability, and predictable fresh-state performance under plant conditions rather than idealized laboratory mixes.

The assessment approach integrates materials characterization, process simulation, and data-driven feedback loops aligned with Industry 4.0 concepts. Digital capture of rheological, setting, and stability data enables rapid learning cycles, supporting consistent performance across plants and raw material sources while reducing commissioning time and rejected batches.

6.8.1 Stepwise Compatibility Evaluation Logic

Rationale for a Progressive Evaluation Strategy

A stepwise compatibility evaluation logic is fundamental for isolating the intrinsic behavior of viscosity-modifying admixtures from the complexity of full cementitious systems. VMAs interact through physicochemical mechanisms such as polymer adsorption, water structuring, and particle bridging, which are strongly influenced by cement mineralogy and ionic environment. Progressive evaluation minimizes formulation noise and allows unambiguous attribution of observed rheological responses to specific material interactions rather than system-level artifacts.

From an industrial perspective, this strategy reduces development time and raw material consumption while improving predictability during scale-up. It also aligns with technology transfer principles, where early-stage screening eliminates high-risk combinations before plant-level validation.

Stage 1: Cement and VMA Intrinsic Interaction Screening

The first evaluation stage focuses on cement paste systems without aggregates or secondary admixtures. VMAs are assessed across a controlled water-to-cement ratio to determine baseline viscosity build-up, yield stress evolution, and sensitivity to cement alkali and sulfate availability. Particular attention is paid to adsorption kinetics and early-age structuration, which often govern downstream robustness.

Isothermal calorimetry and zeta potential measurements are used to detect unintended setting acceleration or retardation induced by polymer–ion complexation. VMAs that show excessive retardation, erratic rheological build-up, or strong cement-source dependency are deprioritized at this stage.

Stage 2: Simplified Mortar-Level Validation

In the second stage, standardized fine aggregates are introduced to evaluate capillary water immobilization and particle suspension efficiency. Mortar testing reveals whether the viscosity enhancement observed in paste translates into effective bleeding control and cohesive flow under shear. Flow table spread, static bleeding columns, and rheometer-based mortar testing are combined to build a multidimensional performance profile.

This stage also identifies early signs of over-thickening, which can manifest as reduced workability or increased mixing energy. VMAs with narrow dosage windows or high sensitivity to sand grading are flagged for reformulation or restricted application use.

Stage 3: Concrete System Integration

Full concrete trials incorporate coarse aggregates, realistic grading curves, and production-relevant mixing sequences. At this level, the evaluation focuses on segregation resistance under dynamic handling, slump or slump flow retention, and robustness to minor water fluctuations. The interaction of VMAs with superplasticizers, especially polycarboxylate ethers, becomes a critical decision factor.

Concrete-level testing often includes simulated transport or pumping cycles to capture structural rebuild behavior. VMAs that maintain cohesion without compromising flow under these conditions are considered suitable for industrial deployment.

Stage 4: Process and Plant Variability Simulation

The final stage introduces controlled variability in cement source, temperature, mixing energy, and batching sequence to stress-test compatibility margins. This step reflects real-world production conditions where raw materials and environmental parameters fluctuate continuously. Performance data are collected using both laboratory instruments and plant-relevant proxies such as mixer power draw and discharge consistency.

Integration of digital data streams enables correlation of laboratory rheology with production indicators. VMAs demonstrating stable performance envelopes across simulated variability are prioritized for multi-plant technology transfer and long-term formulation stability.

Table 6.8.1: Summary of Stepwise Evaluation Stages

Evaluation Stage	System Complexity	Primary Objective	Key Acceptance Indicators
Cement paste	Low	Identify intrinsic VMA–cement interaction	Stable viscosity build-up, no adverse setting
Mortar	Medium	Validate water immobilization and cohesion	Low bleeding, consistent flow

Concrete	High	Confirm segregation resistance and workability	Stable slump, pumpability
Process simulation	Very high	Assess robustness to variability	Wide tolerance window

6.8.2 Key Test Domains and Decision Criteria

Functional Objective of Test Domain Segmentation

Key test domains are defined to translate the complex physicochemical behavior of viscosity-modifying admixtures into actionable industrial decision criteria. VMAs simultaneously influence rheology, phase stability, and admixture interactions, making single-parameter evaluation insufficient. Domain-based testing ensures that formulation decisions are driven by performance envelopes rather than isolated laboratory values, supporting reliable scale-up and cross-plant reproducibility.

From a manufacturing standpoint, each domain is linked to a specific production risk such as segregation, loss of workability, or incompatibility with superplasticizers. Decision criteria are therefore structured as acceptance windows aligned with operational tolerances and quality control limits.

Rheological Performance Domain

The rheological domain focuses on yield stress, plastic viscosity, and thixotropic rebuilding under controlled shear histories. VMAs are expected to increase low-shear viscosity and structural build-up without excessively penalizing high-shear flowability. Measurements are typically performed using rotational rheometers adapted for cement paste, mortar, or concrete, complemented by mixer torque or power consumption as a plant-relevant proxy.

Decision criteria emphasize stability over time and shear reversibility. VMAs exhibiting erratic viscosity growth, excessive hysteresis, or irreversible structuration are considered high risk for pumping, casting, and finishing operations.

Stability and Segregation Resistance Domain

This domain evaluates the ability of VMAs to immobilize free water and maintain homogeneous particle distribution under static and dynamic conditions. Bleeding columns, sieve segregation tests, and visual stability indices are combined to capture both laboratory and field-relevant behavior. Particular attention is given to delayed bleeding, which often appears after initial placement and is difficult to detect in short-term tests.

Acceptance criteria are defined as maximum allowable bleeding rates and segregation indices rather than zero-bleed targets. This pragmatic approach prevents over-thickening and preserves workability, especially in self-consolidating and highly flowable concretes.

Interaction with Other Admixtures Domain

VMAs rarely operate in isolation and must be compatible with superplasticizers, retarders, accelerators, and air-entraining agents. This domain assesses synergy or antagonism through dosage shifts, flow loss rates, and setting time modifications when VMAs are combined with full admixture packages. Polycarboxylate ether demand sensitivity is a critical indicator, as competitive adsorption or ion complexation can destabilize dispersion efficiency.

Decision criteria prioritize minimal PCE overdosage and predictable setting behavior. VMAs that require significant superplasticizer compensation or cause uncontrolled air content fluctuations are typically restricted to niche applications.

Robustness and Variability Tolerance Domain

Industrial concrete production is characterized by continuous variability in cement chemistry, aggregate moisture, and ambient temperature. The robustness domain evaluates VMA performance across intentional perturbations of these parameters. Testing includes temperature ramps, water content variation, and cement source substitution to simulate realistic operating windows.

VMAs are ranked based on the width of their acceptable performance envelope. Products demonstrating broad tolerance are favored for multi-site deployment and digital standardization initiatives.

Table 6.8.2: Consolidated Test Domains and Decision Matrix

Test Domain	Primary Measurements	Industrial Risk Addressed	Decision Threshold Type
Rheology	Yield stress, viscosity, rebuild rate	Pumping and placing failure	Stability window
Stability	Bleeding, segregation index	Surface defects, non-uniformity	Maximum allowable limit
Interactions	PCE demand, setting shift	Admixture incompatibility	Dosage neutrality
Robustness	Sensitivity to variability	Batch rejection risk	Tolerance bandwidth

Industry 4.0 Enablement and Data Use

Digitization of test domain outputs enables real-time comparison between laboratory benchmarks and plant data streams. Rheological fingerprints can be linked to mixer power curves and discharge behavior, forming predictive models for compatibility risk. Decision criteria embedded in manufacturing execution systems support automated alerts and corrective actions when materials deviate from qualified envelopes.

This data-centric approach transforms compatibility testing into a continuous assurance tool, strengthening formulation resilience and accelerating technology transfer.

6.8.3 Integration of Industry 4.0 and Digital Tools

Strategic Role of Digitalization in VMA Compatibility Management

The integration of Industry 4.0 tools into viscosity-modifying admixture compatibility assessment fundamentally shifts the process from static qualification to dynamic performance management. VMAs are highly sensitive to cement chemistry drift, moisture variability, and admixture sequencing, all of which are difficult to fully capture through conventional trial-based testing. Digital systems enable continuous data acquisition and interpretation, allowing compatibility risks to be identified before they manifest as production defects.

In industrial concrete operations, this digital layer acts as a decision-support backbone, linking formulation intent with real-time production behavior. The outcome is improved formulation robustness, reduced reliance on corrective dosing, and faster technology transfer across plants and regions.

Sensor-Based Data Acquisition and Process Fingerprinting

Modern batching and mixing systems generate large volumes of process data that are directly relevant to VMA performance. Mixer power consumption, torque evolution, mixing time, temperature, and discharge consistency serve as indirect but highly sensitive indicators of rheological state and structural build-up. When VMAs alter water immobilization or thixotropy, these changes are immediately reflected in the energy signature of the mixing process.

By establishing baseline fingerprints for VMA-compatible systems, deviations can be detected automatically. This approach allows early identification of incompatibility events caused by cement changes, overdosing, or unintended interactions with superplasticizers, without waiting for downstream quality failures.

Digital Twins and Predictive Compatibility Modeling

Digital twin concepts are increasingly applied to fresh concrete systems to simulate rheological and stability behavior under varying conditions. Laboratory-derived parameters such as yield stress evolution, viscosity index, and rebuild rate are translated into model inputs that predict plant-scale responses. VMAs are evaluated not only for nominal performance but also for their sensitivity coefficients to water, temperature, and cement variability.

Predictive models enable formulation engineers to test hypothetical scenarios virtually, such as cement source substitution or seasonal temperature shifts. This reduces physical trial requirements and supports proactive adjustment of VMA dosage strategies, particularly in large multi-plant operations.

Machine Learning and Decision Automation

Machine learning algorithms trained on historical compatibility datasets can identify non-linear relationships between raw material attributes and VMA performance. These models are particularly effective in detecting subtle incompatibilities that fall within traditional specification limits but lead to cumulative instability. Input variables typically include cement oxide composition, Blaine fineness, admixture dosages, and process conditions.

Decision thresholds derived from these models can be embedded into manufacturing execution systems. Automated alerts, dosage recommendations, or batching interlocks help operators maintain compatibility windows without requiring deep rheological expertise on the production floor.

Digital Integration Across the Technology Transfer Lifecycle

Industry 4.0 tools support compatibility management from laboratory development through commissioning and long-term production. During scale-up, digital records ensure that rheological targets and stability criteria are consistently interpreted across sites. During routine production, continuous monitoring provides feedback for incremental optimization and sustainability-driven material substitutions.

This closed-loop framework reduces formulation drift, minimizes waste, and enhances confidence in VMA deployment across diverse cementitious systems. It represents a mature approach to compatibility assessment, aligned with modern quality, productivity, and sustainability expectations.

Table 6.8.3: Digital Tools Applied to VMA Compatibility

Digital Tool	Data Source	Compatibility Insight	Industrial Benefit
Mixer energy monitoring	Torque, power curves	Rheology and build-up changes	Early incompatibility detection
Digital twin models	Lab rheology, process data	Predictive performance under variability	Reduced physical trials
Machine learning models	Historical QC datasets	Non-linear interaction detection	Robust decision-making
MES integration	Real-time production data	Automated compliance checks	Cross-plant consistency

6.8.4 Practical Assessment Matrix for VMAs

Objective and Industrial Relevance of the Assessment Matrix

A practical assessment matrix for viscosity-modifying admixtures converts complex compatibility data into structured, decision-oriented guidance for formulation engineers and plant operators. VMAs influence multiple fresh-state performance dimensions simultaneously, making isolated test interpretation insufficient for industrial deployment. The matrix approach consolidates rheological, stability, interaction, and robustness indicators into a unified framework that supports rapid, repeatable decisions under production constraints.

In technology transfer scenarios, such matrices are essential for cross-plant standardization. They enable consistent evaluation logic independent of local expertise levels, reducing formulation drift and supporting predictable performance across diverse cementitious systems.

Structure of the Practical Assessment Matrix

The matrix is organized along two principal axes: system complexity and performance criticality. Rows represent evaluation stages from paste to full concrete, while columns capture key performance domains and acceptance logic. Each matrix cell contains quantitative thresholds or qualitative ratings derived from validated test methods, ensuring traceability between laboratory data and field behavior.

Weighting factors can be applied to reflect application-specific priorities, such as segregation resistance in self-consolidating concrete or workability retention in ready-mix transport. This flexible structure allows a single matrix architecture to serve multiple product and market requirements.

Scoring Logic and Decision Thresholds

Performance within each domain is scored against predefined acceptance windows rather than absolute targets. Scores are typically normalized to a scale that reflects industrial risk, where mid-range values indicate robust behavior and extremes signal sensitivity or overdesign. Aggregated scores guide go or no-go decisions, dosage window definition, and application restriction recommendations.

Importantly, the matrix highlights trade-offs between domains. For example, high stability scores accompanied by excessive viscosity penalties are flagged for optimization rather than immediate rejection. This balanced interpretation supports iterative formulation refinement.

Digital Enablement and Continuous Updating

When integrated with digital quality systems, the assessment matrix becomes a living tool rather than a static qualification document. Real-time plant data such as mixer power, slump evolution, and temperature can be mapped back to matrix criteria, enabling continuous validation of compatibility assumptions.

Machine learning-assisted updates allow the matrix to evolve as new raw materials, cement sources, or admixture variants are introduced. This ensures long-term relevance and alignment with Industry 4.0-driven production environments.

Table 6.8.4: Example Practical Assessment Matrix for VMAs

Evaluation Stage	Rheological Control	Stability and Segregation	Admixture Interaction	Robustness to Variability	Overall Readiness
Cement paste	Controlled viscosity build-up	Not applicable	Neutral adsorption	Moderate tolerance	Screening pass
Mortar	Balanced flow and rebuild	Low bleeding	Minor PCE shift	Moderate tolerance	Conditional pass
Concrete	Stable slump flow	No visible segregation	Neutral PCE demand	High tolerance	Approved
Plant simulation	Consistent energy profile	Uniform discharge	Stable setting	Wide tolerance	Full deployment

Implementation Guidance for Industrial Use

For effective implementation, the matrix should be embedded into formulation dossiers and plant qualification protocols. Clear documentation of test methods, scoring logic, and acceptance criteria ensures transparency and reproducibility. Training plant personnel to interpret matrix outcomes fosters proactive quality control and reduces dependence on reactive troubleshooting.

By formalizing compatibility evaluation into a structured matrix, organizations can accelerate VMA adoption, enhance robustness, and support sustainable, data-driven concrete production.

6.8.5 Technology Transfer and Implementation Guidance

Translation of Laboratory Compatibility into Plant-Ready Knowledge

Effective technology transfer for viscosity-modifying admixtures requires systematic translation of laboratory compatibility outcomes into actionable production rules. Rheological curves, segregation indices, and interaction profiles must be converted into clear dosage envelopes, sequencing recommendations, and water sensitivity limits that can be reliably applied at the batching plant. Without this translation layer, even technically sound VMAs may fail during implementation due to operational misinterpretation.

Industrial experience shows that compatibility data packaged in simplified, decision-oriented formats significantly reduces commissioning time and minimizes trial-and-error adjustments during initial deployment.

Standardization Across Plants and Cement Sources

Implementation guidance must explicitly address variability in cement chemistry and aggregate characteristics across production sites. Standard operating windows for VMAs should be defined using worst-case compatibility data rather than average performance. This conservative approach ensures robustness when raw material sources change or seasonal fluctuations occur.

Cross-plant standardization benefits from harmonized test protocols and reference mixes. When identical compatibility benchmarks are applied, performance deviations can be attributed to true material changes rather than procedural differences, supporting rapid root-cause analysis.

Operational Integration and Workforce Enablement

Successful implementation depends on aligning VMA compatibility guidance with plant-level workflows. Dosage strategies, admixture addition sequences, and corrective actions should be embedded into batching instructions and quality control checklists. Visual aids and simplified decision trees are particularly effective for operators who may not have rheological training.

Training programs should emphasize cause-and-effect relationships between VMA behavior and observable production indicators such as mixer power, discharge consistency, and surface finish. This practical understanding enables early intervention before quality deviations escalate.

Digital Support and Continuous Improvement

Industry 4.0 tools play a critical role in sustaining compatibility after initial transfer. Manufacturing execution systems can store approved VMA compatibility matrices and automatically flag deviations from qualified envelopes. Real-time data streams allow continuous comparison between expected and actual performance, supporting incremental optimization rather than reactive troubleshooting.

Feedback loops between plant data and laboratory teams ensure that compatibility knowledge evolves with changing materials and market requirements. This continuous improvement model strengthens long-term formulation resilience and reduces lifecycle costs.

Sustainability and Risk Management Considerations

Well-executed technology transfer reduces waste, rework, and overdesign by preventing incompatibility-related failures. VMAs with clearly defined compatibility limits allow optimized cement and water usage, contributing to lower embodied carbon in concrete production. From a risk management perspective, documented compatibility frameworks enhance supplier accountability and support regulatory and certification audits.

These benefits position structured technology transfer as a strategic asset rather than a purely technical exercise within modern cementitious materials operations.

7. VMA Formulation and Examples

7.1 Formulation Philosophy and Industrial Boundary Conditions

VMA formulation examples are presented to illustrate how polymer chemistry, dosage strategy, and system compatibility are translated into robust, production-ready solutions. Each example reflects industrial constraints such as cement variability, admixture interoperability, and process energy efficiency rather than laboratory-optimized extremes. The formulations emphasize predictable rheological control, segregation resistance, and scalability across ready-mix, precast, and infrastructure applications.

The formulations are intentionally modular, allowing adjustment of VMA polymer architecture, molecular weight distribution, and functional group density to match local cement chemistries. This approach supports rapid customization and technology transfer without compromising qualification integrity.

7.2 Formulation Examples

Transitioning from theoretical principles and generic strategies, this section presents a series of detailed, application-specific formulation examples for viscosity-modifying admixtures. Each example translates the chemical and rheological concepts discussed earlier into practical, actionable recipes tailored to distinct concrete technologies and performance requirements. The formulations are structured to address the unique challenges of diverse applications, from conventional ready-mix and self-consolidating concrete to advanced systems like UHPC, 3D printing, and low-carbon mixes. By providing concrete ingredient profiles, dosage ranges, and processing notes, this segment serves as a practical reference for admixture producers and concrete technologists seeking to develop or optimize VMA products, while also illustrating how core formulation principles are adapted for scalability, compatibility, and integration with modern Industry 4.0 production environments.

7.2.1 Cellulose Ether–Based VMA for Conventional Ready-Mix Concrete

1. Applications

- Conventional and high-slump ready-mix concrete where enhanced stability is required without sacrificing flowability, particularly for pumped and long-haul deliveries.
- Structural elements prone to segregation such as walls, columns, deep beams, and inclined or sloped placements where paste cohesion governs surface quality.
- Concrete produced and placed under hot or windy conditions to limit rapid water evaporation, plastic shrinkage risk, and early-age rheology loss.
- Foundation piles, diaphragm walls, and tremie concrete applications where washout resistance and homogeneity under hydrostatic pressure are critical.

2. Advantages

- High Water Retention Efficiency

  Cellulose ether exhibits strong affinity for free and capillary water within the cementitious matrix, effectively suppressing bleeding and surface water migration during the plastic and early setting stages. This water immobilization promotes more uniform cement hydration, limits the formation of interconnected capillary pores, and contributes to improved long-term durability, particularly in thin sections, high surface area elements, and low water-to-cement ratio systems.

- Enhanced Mix Cohesion

  Through controlled viscosity enhancement of the cement paste phase, cellulose ether improves internal cohesion and strengthens the suspension capacity of the binder. Aggregates remain uniformly distributed in both low-slump and highly flowable concretes, significantly reducing segregation and settlement risks even under dynamic placing conditions or elevated slump values typical of modern high-workability mixes.

- Controlled and Reproducible Rheology

The rheological response of cellulose ether follows a well-defined and predictable dose–viscosity relationship within the practical formulation window. This behavior enables reliable fine-tuning of yield stress and plastic viscosity during industrial batching, supports consistent scale-up, and facilitates data-driven optimization using digital formulation tools and inline quality monitoring aligned with Industry 4.0 manufacturing practices.

- Renewable and Sustainable Raw Material Base

  Derived from chemically modified cellulose sourced from wood pulp or cotton linters, cellulose ethers offer a partially renewable alternative to fully petroleum-based rheology modifiers. Their use supports sustainability and decarbonization initiatives by reducing dependence on fossil-derived polymers, while maintaining performance levels required for demanding concrete applications.

- Broad Admixture Compatibility

  When correctly selected and dosed, cellulose ethers demonstrate acceptable compatibility with most contemporary PCE-based superplasticizers and auxiliary admixtures. A moderate increase in dispersant demand is typically observed due to competitive adsorption and viscosity effects, yet this interaction can be effectively managed through formulation balance, dosage optimization, and digitally assisted mix design protocols

3. Precautions

- Dispersion Control

  Fine powders must be pre-blended with dry constituents or introduced under sufficient shear to avoid partial hydration and gel agglomerates, commonly referred to as fish-eyes.

- Set Retardation Tendency

  Cellulose ethers intrinsically delay cement hydration kinetics by water immobilization and surface adsorption. Set-controlling agents must be balanced carefully to meet placement and finishing schedules.

- Air Entrainment Risk

Due to surface-active behavior, unintended air entrainment can occur, negatively affecting compressive strength. A compatible defoaming system is essential.

- Dosage Sensitivity at Upper Range

  At higher addition levels, viscosity and retardation effects increase disproportionately, potentially resulting in sticky, difficult-to-finish concrete. Narrow dosage control is mandatory.

4. Formula

Table 7.2.1: Representative Formulation of Cellulose Ether–Based VMA for Conventional Ready-Mix Concrete

Component	CAS No.	%	Function
HPMC (Hydroxypropyl Methyl Cellulose)	9004-65-3	85–95	Primary viscosity-modifying and water-retention polymer governing cohesion and stability.
Citric Acid or Sodium Gluconate	77-92-9 / 527-07-1	3–8	Set-modulating agent used to counterbalance or fine-tune cellulose ether–induced retardation.
Powdered Defoamer (PDMS-based)	63148-62-9	0.5–2	Air-control additive mitigating unintended air entrainment from polymer surface activity.
Inert Carrier (Milled Limestone)	1317-65-3	2–5	Diluent and flow aid ensuring uniform distribution, bulk density control, and improved handling.

5. Technical Specifications and Application Guidance

- Typical Dosage Range

  0.05 to 0.15 percent by weight of total cementitious material, with lower levels suitable for standard ready-mix and higher levels reserved for tremie or high-stability concretes.

- Primary Performance Contribution

Delivers water retention values typically exceeding 95 percent under standardized testing while significantly improving segregation resistance and surface finish quality.

- Impact on Hydration and Setting

  Induces a controlled, dosage-dependent delay of initial set, commonly in the range of 1 to 3 hours depending on cement chemistry, temperature, and formulation balance.

6. Guidance for Formulators

- Molecular Parameter Selection

  Medium molecular weight grades, typically around 120,000 to 180,000 g/mol, offer an optimal balance between viscosity efficiency and manageable retardation for ready-mix operations. Excessively high molecular weights increase stickiness and finishing difficulty.

- Degree and Type of Substitution

  HPMC generally provides superior solubility and balanced surface activity. HEC can deliver higher viscosity per unit mass but usually causes stronger retardation and is less forgiving in hot-weather concreting.

- Dosage Optimization Strategy

  Viscosity response is non-linear near the upper performance threshold. Incremental dosage steps of approximately 0.02 percent are recommended during laboratory and plant trials to avoid overshooting target rheology.

- Compatibility and Robustness Testing

  Full-scale compatibility testing with the project-specific PCE superplasticizer is mandatory. Cellulose ethers may increase dispersant demand, while citrate or gluconate regulators can interact synergistically or antagonistically with cement aluminates.

- Sustainability and Industry 4.0 Integration

  The renewable origin and low embodied carbon of cellulose ethers align with sustainability metrics. When combined with digital batching systems and real-time rheology monitoring, consistent performance can be achieved across variable raw material sources.

7.2.2 Synthetic Polymer VMA for Self-Consolidating Concrete

1. Applications

- Self-consolidating concrete in densely reinforced structures

  Used in structural elements with congested reinforcement where high deformability, excellent passing ability, and robust segregation resistance are required to coexist under minimal vibration or external energy input, ensuring uniform filling and structural integrity.

- Precast concrete with demanding surface and dimensional requirements

  Applied in precast production for elements requiring highly reproducible rheology, rapid and complete mold filling, and superior surface finish, including architectural panels, thin-wall components, and complex geometries where visual quality is critical.

- Pumped SCC under high shear and extended transport conditions

  Suitable for SCC placed by pumping over long distances or significant elevations, where the VMA must maintain viscosity stability under shear, prevent viscosity breakdown, and suppress static segregation during pauses or variable pumping regimes.

- SCC with variable aggregate grading or high powder substitution

  Effective in concrete mixes containing fluctuating aggregate grading curves or elevated powder contents, including limestone fillers, fly ash, and slag, where synthetic polymer VMAs compensate for packing variability and stabilize the paste–aggregate system.

- High-early-flow SCC for rapid placement cycles

  Employed in applications requiring immediate high flow upon discharge combined with short open times, such as fast-cycle precast lines or high-rise casting schedules, where controlled viscosity recovery is essential to prevent formwork pressure escalation and segregation.

- Digitally optimized SCC in automated production environments

  Used in concrete plants operating under Industry 4.0 frameworks, where inline rheological monitoring, pump pressure feedback, and real-time dosage adjustment enable closed-loop control of SCC performance, ensuring consistent quality despite raw material and environmental variability.

2. Advantages

- High Viscosity Efficiency at Low Dosage

  Synthetic polymer VMAs based on polyacrylamide or modified polycarboxylate backbones deliver strong yield stress enhancement at very low addition levels, minimizing side effects on workability.

- Excellent Segregation and Bleeding Control

  The polymer forms a weak three-dimensional network within the pore solution, stabilizing suspended solids without excessive water immobilization, which is critical for SCC robustness.

- Minimal Set Retardation

  Compared to cellulose ethers, most synthetic VMAs exert limited direct influence on cement hydration kinetics when properly designed, supporting predictable setting behavior.

- Consistency in Industrial Production

  Fully synthetic origin ensures narrow molecular weight distribution and batch-to-batch reproducibility, supporting automated dosing and Industry 4.0–driven quality control systems.

- Superior Compatibility with PCE Superplasticizers

  Designed to work synergistically with modern PCEs, enabling high slump flow values while maintaining viscosity and resistance to dynamic segregation.

3. Precautions

- Over-Stabilization Risk

  Excessive dosage can lead to overly high yield stress, resulting in poor passing ability, surface defects, or delayed self-leveling despite high slump flow values.

- Shear Sensitivity

  Some synthetic VMAs exhibit shear-thinning behavior that may reduce viscosity during pumping, requiring validation under realistic shear profiles.

- Water Quality Sensitivity

  Performance can be affected by high ionic strength, recycled wash water, or elevated sulfate levels, necessitating plant-specific trials.

- Entrapped Air Management

  While less air-entraining than cellulose ethers, certain polymer architectures can stabilize fine air bubbles, requiring defoamer adjustment in sensitive SCC formulations.

4. Formula

Table 7.2.2: Representative Formulation of Synthetic Polymer VMA for Self-Consolidating Concrete

Component	**CAS No.**	**%**	**Function**
Modified Polyacrylamide or Synthetic VMA Copolymer	9003-05-8	60–75	Primary viscosity-modifying agent providing yield stress enhancement and segregation resistance.
Polycarboxylate-Based Dispersant (Low MW)	9007-20-9	10–20	Rheology-balancing component improving dispersion and compatibility with SCC superplasticizers.
Powdered Defoamer (PDMS-based)	63148-62-9	0.5–2	Controls entrapped and stabilized air in high-flow SCC systems.

Inert Carrier (Precipitated or Milled Limestone)	1317-65-3	10–25	Diluent and handling aid ensuring uniform dosing and consistent bulk density.

5. Technical Specifications and Application Guidance

- Typical Dosage

  0.02 to 0.08 percent by weight of total cementitious material, depending on target slump flow, aggregate grading, and powder content.

- Primary Performance

  Delivers high static and dynamic segregation resistance while maintaining slump flow values typically between 650 and 750 mm, with improved passing ability through congested reinforcement.

- Hydration Impact

  Generally neutral to slightly retarding effect on initial set, usually less than 30 to 60 minutes, depending on cement chemistry and co-admixture interactions.

6. Guidance for Formulators

- Molecular Parameter Selection

  Medium to high molecular weight polymers with controlled anionic charge density provide optimal balance between viscosity enhancement and flow retention. Excessive molecular weight increases shear sensitivity and risk of over-stabilization.

- Dosage Strategy

  Performance response is steep near the optimal SCC window. Dosage increments as small as 0.005 to 0.01 percent are recommended during laboratory optimization and plant validation.

- Compatibility Testing

  Mandatory evaluation with the selected PCE superplasticizer, supplementary cementitious materials, and production water. Special attention should be given to slump flow retention and V-funnel time evolution.

- Sustainability Consideration

  Although synthetic in origin, low effective dosage and improved robustness enable reduced cement content and lower rejection rates. Integration with digital mix optimization tools supports reduced material variability and overall carbon footprint.

7.2.3 VMA System for Pumped Concrete with Extended Transport Time

1. Applications

- Pumped ready-mix concrete transported over long distances or subjected to extended waiting periods before placement, where rheology retention and segregation resistance must be preserved under continuous shear.
- High-rise construction and infrastructure projects involving vertical pumping over significant heights, where pressure stability and lubrication layer integrity directly influence pump efficiency.
- Concrete placed in warm climates or during peak summer conditions, where slump loss, evaporation, and premature stiffening are operational risks.
- Mixes incorporating blended cements or supplementary cementitious materials with variable reactivity, including fly ash and ground granulated blast-furnace slag, which require enhanced robustness during transport.

2. Advantages

- Shear-Stable Viscosity Control

  The VMA system maintains a stable rheological profile under prolonged pumping and recirculation, minimizing viscosity breakdown and loss of cohesion during transport.

- Reduced Slump Loss Over Time

  By moderating water migration and stabilizing the paste phase, the system extends workability retention without relying solely on superplasticizer re-dosing.

- Improved Pump Lubrication Layer

Enhanced paste cohesion supports the formation of a continuous lubrication layer along the pipeline wall, reducing friction, pressure peaks, and blockage risk.

- Balanced Hydration Moderation

  Compared with single-polymer systems, the combined VMA approach offers controlled hydration delay, allowing extended open time without excessive retardation.

- Operational Consistency with Digital Control

  The system supports real-time rheology monitoring and automated admixture dosing, aligning with Industry 4.0 production environments.

3. Precautions

- Cumulative Retardation Risk

  When combined with set-retarding superplasticizers or hot-weather admixtures, total retardation can exceed acceptable limits and must be evaluated holistically.

- Pump-Specific Sensitivity

  Performance may vary with pump type, line diameter, and pressure regime. Validation under realistic pumping conditions is essential.

- Air Stability Effects

  Although not primarily air-entraining, prolonged shear can stabilize fine air voids, necessitating careful air content monitoring.

- Water Quality Interaction

  Recycled or high-alkalinity water can modify polymer conformation and effectiveness, particularly during extended transport cycles.

4. Formula

Table 7.2.3: Representative Formulation of VMA System for Pumped Concrete with Extended Transport Time

Component	**CAS No.**	**%**	**Function**

Modified Cellulose Ether (Low Substitution HPMC or HEMC)	9004-65-3 / 9004-62-0	40–55	Primary water-retention and cohesion agent providing baseline viscosity and stability.
Synthetic Polyacrylamide-Based VMA	9003-05-8	20–30	Shear-stable viscosity modifier improving pumpability and dynamic segregation resistance.
Sodium Gluconate or Tartaric Acid	527-07-1 / 87-69-4	5–10	Set-control component moderating hydration rate during extended transport.
Powdered Defoamer (PDMS-based)	63148-62-9	0.5–2	Controls air stabilization under prolonged shear and pumping conditions.
Inert Carrier (Milled Limestone)	1317-65-3	10–20	Diluent ensuring homogeneous distribution and consistent dosing accuracy.

5. Technical Specifications and Application Guidance

- Typical Dosage

  0.08 to 0.20 percent by weight of total cementitious material, adjusted based on transport duration, pumping distance, and ambient temperature.

- Primary Performance

  Maintains stable slump or slump flow for 90 to 180 minutes with minimal segregation, while supporting smooth pumping and consistent discharge pressure.

- Hydration Impact

  Introduces a controlled delay of initial set typically ranging from 1 to 4 hours, depending on cement composition, temperature, and cumulative admixture effects.

6. Guidance for Formulators

- Molecular Parameter Selection

Moderate molecular weight cellulose ethers combined with medium molecular weight synthetic polymers provide complementary rheological behavior. Excessively high molecular weights increase stickiness and pressure demand.

- Dosage Strategy

  Optimization should be performed in narrow increments, typically 0.01 to 0.02 percent, particularly when targeting extended open time without compromising early strength development.

- Compatibility Testing

  Comprehensive testing with the selected PCE superplasticizer, cement type, SCMs, and production water is mandatory. Pump loop simulations are strongly recommended.

- Sustainability Consideration

  Hybrid VMA systems enable reduced cement content, lower pumping energy, and fewer rejected loads. Integration with digital batching, transport tracking, and rheology sensors supports overall carbon and resource efficiency.

7.2.4 VMA Strategy for Low-Carbon and SCM-Rich Concrete

1. Applications

- Low-clinker and low-carbon concrete mixtures incorporating high levels of supplementary cementitious materials such as fly ash, ground granulated blast-furnace slag, calcined clay, and limestone fillers.
- Structural and non-structural concretes designed to meet embodied carbon reduction targets while maintaining adequate robustness, cohesion, and placement reliability.
- Ready-mix and precast applications where variability in SCM reactivity and particle size distribution requires enhanced rheological control.

- Concrete produced with reduced cement content and increased powder substitution, including LC3-type systems and blended cements with high filler fractions.

2. Advantages

- Enhanced Stability in SCM-Rich Systems

  The VMA strategy compensates for reduced early-age cohesion commonly observed in low-clinker mixes, minimizing segregation and bleeding despite lower cement paste activity.

- Improved Rheology Consistency

  By stabilizing the pore solution and solid suspension, the system reduces sensitivity to SCM variability, supporting consistent slump or slump flow across production batches.

- Moderated Water Demand

  Effective viscosity enhancement allows maintenance of workability at lower total water content, supporting strength development despite reduced clinker levels.

- Compatibility with Carbon-Reduction Technologies

  The approach aligns with modern mix optimization tools, real-time batching control, and predictive models used in Industry 4.0–enabled low-carbon concrete production.

3. Precautions

- Delayed Early Strength Development

  High SCM content combined with VMA-induced water immobilization can slow early hydration, requiring careful strength profiling and curing optimization.

- Interaction with Alkali-Activated or Highly Reactive SCMs

  Certain polymers may interact unfavorably with high-alkali environments or calcined clays, necessitating targeted compatibility trials.

- Air Content Sensitivity

Fine SCM particles and VMAs can jointly stabilize air voids, making air control additives critical for strength and durability consistency.

- Over-Stabilization Risk

  Excessive viscosity can reduce finishing quality and internal compaction, particularly in mixes with very high filler or clay content.

4. Formula

Table 7.2.4: Representative Formulation of VMA Strategy for Low-Carbon and SCM-Rich Concrete

Component	CAS No.	%	Function
Modified Cellulose Ether (Low MW HPMC or HEMC)	9004-65-3 / 9004-62-0	35–50	Provides baseline water retention and cohesion in low-reactivity paste systems.
Synthetic Polyacrylate or Polyacrylamide VMA	9003-05-8	20–30	Enhances yield stress and stabilizes solid suspension without excessive retardation.
Sodium Gluconate or Sodium Citrate	527-07-1 / 68-04-2	5–10	Controls hydration kinetics and mitigates variability introduced by SCM chemistry.
Powdered Defoamer (PDMS-based)	63148-62-9	0.5–2	Regulates entrapped and stabilized air in fine-particle-rich mixes.
Inert Carrier (Calcite or Limestone Filler)	1317-65-3	10–20	Ensures homogeneous dispersion and consistent dosing accuracy.

5. Technical Specifications and Application Guidance

- Typical Dosage

  0.10 to 0.25 percent by weight of total cementitious material, depending on SCM type, replacement level, and target workability.

- Primary Performance

Delivers improved cohesion, reduced bleeding, and stable workability in mixes containing 40 to 70 percent SCM replacement, including fine fillers and calcined clays.

- Hydration Impact

  Introduces a controlled extension of setting time, typically 1 to 3 hours, which must be balanced against early-age strength requirements through curing and admixture selection.

6. Guidance for Formulators

- Molecular Parameter Selection

  Lower to medium molecular weight polymers with moderate charge density are preferred to avoid excessive retardation and allow sufficient early hydration in low-clinker systems.

- Dosage Strategy

  Optimization should proceed in small increments, typically 0.02 to 0.03 percent, with parallel monitoring of rheology, setting time, and early compressive strength.

- Compatibility Testing

  Mandatory testing with each SCM source is required, as variability in fineness, mineralogy, and alkali content can significantly affect polymer performance and adsorption behavior.

- Sustainability Consideration

  Although VMAs add marginal embodied carbon, their use enables higher SCM substitution, reduced cement content, and lower rejection rates. Integration with digital mix design platforms and real-time quality monitoring maximizes net carbon reduction benefits.

7.2.5 VMA for High-Fines Sand and Manufactured Sand Concrete

1. Applications

- Concrete produced with manufactured sand or crushed aggregates containing elevated fines content, typically below 150 μm, where water demand and rheology variability are pronounced.
- Ready-mix and precast concretes using marginal or non-traditional aggregate sources, including quarry dust-rich sands, where consistency and segregation resistance are critical.
- Pumped and slip-formed concretes in which fines-induced stickiness and pressure sensitivity must be balanced with adequate flowability.
- Concrete mixtures incorporating recycled aggregates or blended natural and manufactured sands with fluctuating particle size distribution.

2. Advantages

- Stabilization of Excess Fines

  The VMA system regulates the interaction between fine particles and the pore solution, reducing water sensitivity and mitigating excessive yield stress caused by angular fines.

- Improved Workability Robustness

  By decoupling fines content from slump variability, the system delivers more predictable fresh properties across daily and seasonal aggregate variations.

- Reduced Bleeding and Segregation

  Enhanced paste cohesion minimizes water migration and fines settlement, leading to improved surface finish and dimensional stability.

- Pumpability Optimization

  Balanced viscosity supports lubrication layer formation while preventing over-stiffening often associated with high dust content sands.

3. Precautions

- Over-Viscosification Risk

  In sands with very high microfines or clay contamination, excessive VMA dosage can lead to sticky, difficult-to-place concrete.

- Clay–Polymer Interactions

  Active clays such as montmorillonite can adsorb polymers strongly, reducing VMA efficiency and increasing dosage demand.

- Air Content Control

  Fine particles and polymers can jointly stabilize air voids, necessitating careful adjustment of defoaming systems.

- Aggregate Variability Sensitivity

  Frequent changes in fines mineralogy or gradation require ongoing monitoring and potential formulation adjustments.

4. Formula

Table 7.2.5: Representative Formulation of VMA for High-Fines Sand and Manufactured Sand Concrete

Component	CAS No.	%	Function
Modified Cellulose Ether (Low MW HPMC or HEMC)	9004-65-3 / 9004-62-0	30–45	Provides water retention and baseline cohesion to counter fines-induced water demand.
Synthetic Polyacrylamide or Polycarboxylate VMA	9003-05-8 / 9007-20-9	25–35	Enhances yield stress control and stabilizes fines suspension under shear.
Sodium Gluconate or Sodium Citrate	527-07-1 / 68-04-2	5–10	Modulates hydration and improves robustness against fines variability.
Powdered Defoamer (PDMS-based)	63148-62-9	0.5–2	Controls entrapped and stabilized air in fine-rich mixes.
Inert Carrier (Limestone or Dolomite Filler)	1317-65-3 / 16389-88-1	10–20	Ensures homogeneous dispersion and consistent bulk handling.

5. Technical Specifications and Application Guidance

- Typical Dosage

0.10 to 0.25 percent by weight of total cementitious material, depending on fines content, sand mineralogy, and target workability.

- Primary Performance

  Delivers stable slump or slump flow with reduced sensitivity to fines fluctuations, while maintaining cohesion and pumpability in mixes with up to 20 percent microfines.

- Hydration Impact

  Typically causes slight to moderate setting delay, generally within 30 to 90 minutes, depending on cement chemistry and co-admixture interactions.

6. Guidance for Formulators

- Molecular Parameter Selection

  Lower to medium molecular weight polymers with moderate surface activity are preferred to avoid excessive stickiness in fines-rich systems.

- Dosage Strategy

  Optimization should proceed in narrow increments of 0.02 to 0.03 percent, with parallel monitoring of slump, air content, and finishing behavior.

- Compatibility Testing

  Routine evaluation with site-specific sands, including clay index and methylene blue value, is essential to anticipate polymer demand and performance shifts.

- Sustainability Consideration

  Use of VMAs enables higher utilization of manufactured and recycled sands, reducing reliance on scarce natural sand resources. Integration with digital aggregate characterization and automated mix correction supports sustainable aggregate management.

7.2.6 VMA for Precast Concrete with Rapid Demolding

1. Applications

- Precast concrete elements requiring fast strength build-up and early demolding, including hollow-core slabs, beams, columns, pipes, and modular construction components.
- Factory-produced concrete with low water-to-cement ratios and high cement efficiency, where shape stability and surface quality are critical immediately after casting.
- Vibration-assisted and self-compacting precast mixes designed for short cycle times and high daily mold turnover.
- Steam-cured or thermally accelerated precast operations where rheology control must remain effective without compromising early-age hydration.

2. Advantages

- Early Shape Stability

  The VMA system increases green strength and thixotropic build-up, limiting formwork pressure and deformation prior to initial set.

- Minimal Retardation Profile

  Designed to provide viscosity enhancement with negligible delay to setting, supporting rapid demolding and early strength development.

- Improved Surface Finish

  Controlled paste cohesion reduces bugholes, laitance, and edge defects, particularly in complex molds.

- Process Consistency and Automation Readiness

  The formulation supports precise dosing and reproducible performance in automated precast plants aligned with Industry 4.0 manufacturing concepts.

3. Precautions

- Over-Stiffening Risk

  Excessive dosage may reduce mold filling ability or require higher vibration energy, potentially offsetting productivity gains.

- Interaction with Accelerators

Compatibility with calcium nitrate, calcium formate, or other accelerators must be validated to avoid unpredictable setting behavior.

- Thermal Sensitivity

  Elevated curing temperatures can amplify polymer effects on rheology and air stability, requiring temperature-adjusted formulations.

- Air Content Management

  Even low levels of stabilized air can negatively impact early strength, making air control critical.

4. Formula

Table 7.2.6: Representative Formulation of VMA for Precast Concrete with Rapid Demolding

Component	CAS No.	%	Function
Synthetic Polyacrylamide-Based VMA (Low Retardation Grade)	9003-05-8	55–70	Primary viscosity modifier providing thixotropy and green strength without significant set delay.
Low MW Polycarboxylate Polymer	9007-20-9	10–20	Rheology-balancing component improving dispersion and mold filling efficiency.
Calcium Formate or Calcium Nitrate	544-17-2 / 10124-37-5	5–10	Early-strength accelerator counteracting any residual hydration delay.
Powdered Defoamer (PDMS-based)	63148-62-9	0.5–2	Controls entrapped air to protect early and final strength.
Inert Carrier (Milled Limestone)	1317-65-3	10–20	Ensures uniform dispersion, flowability, and dosing accuracy.

5. Technical Specifications and Application Guidance

- Typical Dosage

  0.02 to 0.06 percent by weight of total cementitious material, optimized for early demolding requirements and mold geometry.

- Primary Performance

  Provides enhanced green strength, reduced formwork pressure, and stable rheology during casting, enabling demolding within 6 to 12 hours depending on curing regime.

- Hydration Impact

  Neutral to slightly accelerating overall hydration when combined with appropriate accelerators, with minimal impact on final strength.

6. Guidance for Formulators

- Molecular Parameter Selection

  Low to medium molecular weight synthetic polymers with controlled charge density are preferred to achieve viscosity enhancement without excessive water immobilization.

- Dosage Strategy

  Optimization should proceed in very small increments, typically 0.005 to 0.01 percent, to fine-tune mold filling versus green strength balance.

- Compatibility Testing

  Comprehensive trials with cement, accelerators, superplasticizers, and curing temperature profiles are mandatory for reliable plant performance.

- Sustainability Consideration

  Although synthetic in origin, the VMA enables reduced cement content, shorter curing cycles, and lower energy consumption per unit produced. Integration with digital curing control and production analytics enhances overall resource efficiency.

7.2.7 VMA for Shotcrete and Sprayed Concrete Applications

1. Applications

- Wet-mix and dry-mix shotcrete used in tunnel linings, underground mining, slope stabilization, and rock support, where adhesion and build-up control rebound and overspray.
- Sprayed concrete applied to vertical and overhead surfaces requiring high green strength and rapid structural build-up immediately after impact.
- Fiber-reinforced shotcrete systems, including steel and synthetic fibers, where uniform fiber dispersion and paste stability are critical.
- Repair and rehabilitation works involving thin layers, complex geometries, and variable substrates.

2. Advantages

- Reduced Rebound and Material Loss

  The VMA increases cohesion and impact adhesion, significantly lowering rebound rates and improving material utilization efficiency.

- Rapid Structural Build-Up

  Enhanced thixotropy supports immediate layer stability on vertical and overhead surfaces without excessive sloughing.

- Improved Fiber Suspension

  Stabilized paste rheology ensures uniform fiber distribution, reducing segregation during spraying and improving mechanical performance.

- Process Control and Digital Integration

  The system supports automated dosing and real-time monitoring of spray parameters, aligning with Industry 4.0 shotcrete operations.

3. Precautions

- Interaction with Alkali-Free Accelerators

Compatibility with aluminum sulfate or aluminum hydroxysulfate accelerators must be validated to avoid flash set or excessive stiffening.

- Nozzle Shear Sensitivity

  Some polymers may degrade under high shear at the nozzle, requiring selection of shear-resistant grades.

- Air Content Variability

  Entrapped air introduced during spraying must be controlled to avoid strength loss and surface defects.

- Environmental Exposure Sensitivity

  Wind, temperature, and humidity strongly influence fresh behavior and must be considered during formulation and application.

4. Formula

Table 7.2.7: Representative Formulation of VMA for Shotcrete and Sprayed Concrete Applications

Component	CAS No.	%	Function
Modified Cellulose Ether or Synthetic Polyacrylamide VMA	9004-65-3 / 9003-05-8	40–55	Primary viscosity modifier enhancing cohesion, thixotropy, and adhesion upon impact.
Low MW Polycarboxylate Polymer	9007-20-9	10–20	Rheology-balancing component improving pumpability and spray consistency.
Alkali-Compatible Set Regulator	527-07-1	5–10	Moderates base hydration prior to accelerator addition, improving process stability.
Powdered Defoamer (PDMS-based)	63148-62-9	0.5–2	Controls entrapped air generated during high-energy spraying.

Inert Carrier (Limestone Filler)	1317-65-3	10–20	Ensures homogeneous dispersion and accurate dosing in dry or wet systems.

5. Technical Specifications and Application Guidance

- Typical Dosage

  0.08 to 0.20 percent by weight of total cementitious material, adjusted based on spray method, fiber content, and accelerator type.

- Primary Performance

  Delivers improved adhesion, reduced rebound, and stable layer build-up, enabling application thicknesses of 50 to 100 mm per pass under controlled conditions.

- Hydration Impact

  Minimal influence prior to accelerator addition, with final setting governed primarily by the selected accelerator system.

6. Guidance for Formulators

- Molecular Parameter Selection

  Shear-resistant polymers with moderate molecular weight and controlled substitution provide optimal balance between pumpability and impact adhesion.

- Dosage Strategy

  Optimization should proceed in small increments, typically 0.01 to 0.02 percent, with direct evaluation of rebound, build-up thickness, and surface quality.

- Compatibility Testing

  Mandatory trials with accelerators, fibers, cement, and spray equipment are required to ensure robust field performance.

- Sustainability Consideration

Reduction in rebound and overspray directly lowers material consumption and waste. Integration with digital spray monitoring and automated dosing systems enhances resource efficiency and environmental performance.

7.2.8 VMA for Lightweight Aggregate Concrete

1. Applications

- Structural and non-structural lightweight aggregate concrete produced with expanded clay, shale, slate, pumice, or sintered fly ash aggregates, where density control and segregation resistance are critical.
- Pumped lightweight concrete used in high-rise buildings, bridge decks, and rehabilitation overlays, where aggregate flotation and bleeding must be controlled.
- Precast lightweight elements requiring uniform aggregate distribution and consistent surface quality.
- Insulating and semi-structural concretes incorporating high-absorption aggregates with variable moisture states.

2. Advantages

- Control of Aggregate Flotation

  The VMA system increases paste yield stress, counteracting buoyancy forces that cause lightweight aggregates to rise during placement and finishing.

- Improved Density Uniformity

  Stabilization of the fresh matrix reduces vertical density gradients, improving mechanical consistency and durability.

- Reduced Bleeding and Water Migration

  Effective water retention limits free water movement, particularly important in mixes with highly absorptive aggregates.

- Enhanced Pumping Stability

  Balanced viscosity supports stable flow and lubrication, minimizing segregation and pressure fluctuations during pumping.

3. Precautions

- Interaction with Aggregate Moisture State

  Pre-wetted versus dry lightweight aggregates significantly influence water balance and VMA efficiency, requiring moisture-specific calibration.

- Risk of Over-Viscosification

  Excessive dosage can hinder self-leveling and finishing, especially in low-density or high-porosity aggregate systems.

- Air Content Sensitivity

  Lightweight concretes are sensitive to air stabilization, making careful defoamer selection and dosage essential.

- Cumulative Retardation Effects

  When combined with water-reducing admixtures or set retarders, overall setting time must be evaluated to avoid delays.

4. Formula

Table 7.2.8: Representative Formulation of VMA for Lightweight Aggregate Concrete

Component	CAS No.	%	Function
Modified Cellulose Ether (Low to Medium MW HPMC or HEMC)	9004-65-3 / 9004-62-0	35–50	Provides water retention and viscosity to stabilize lightweight aggregate suspension.
Synthetic Polyacrylamide or Polyacrylate VMA	9003-05-8	20–30	Enhances yield stress and cohesion without excessive water immobilization.
Sodium Gluconate or Sodium Citrate	527-07-1 / 68-04-2	5–10	Regulates hydration kinetics under variable internal curing conditions.

Powdered Defoamer (PDMS-based)	63148-62-9	0.5–2	Controls entrapped and stabilized air in low-density systems.
Inert Carrier (Limestone Filler)	1317-65-3	10–20	Ensures homogeneous dispersion and consistent dosing accuracy.

5. Technical Specifications and Application Guidance

- Typical Dosage

  0.10 to 0.30 percent by weight of total cementitious material, depending on aggregate type, density, and absorption capacity.

- Primary Performance

  Maintains uniform aggregate distribution, limits bleeding, and stabilizes workability in concretes with fresh densities ranging from 1,600 to 2,000 kg/m³.

- Hydration Impact

  Typically introduces slight to moderate set delay, generally within 1 to 2 hours, influenced by aggregate moisture and admixture synergy.

6. Guidance for Formulators

- Molecular Parameter Selection

  Medium molecular weight polymers with balanced hydrophilicity are preferred to provide sufficient viscosity without excessive stickiness or retardation.

- Dosage Strategy

  Optimization should be conducted in small increments of 0.02 to 0.03 percent, with simultaneous monitoring of density profile, slump retention, and finishing behavior.

- Compatibility Testing

  Testing must include aggregate moisture conditioning scenarios, pump simulations, and evaluation of density uniformity in hardened concrete.

- Sustainability Consideration

  VMAs facilitate the use of lightweight aggregates that reduce structural dead load and embodied carbon. Integration with digital moisture sensing and automated mix correction enhances material efficiency and sustainability outcomes.

7.2.9 VMA for Ultra-High Performance Concrete

1. Applications

- Ultra-high performance concrete used in structural elements demanding exceptional strength, durability, and homogeneity, including bridges, façade panels, thin-shell components, and protective structures.
- Precast UHPC with very low water-to-binder ratios and extremely high powder content, where rheology control is essential for mold filling and fiber dispersion.
- Steel and hybrid fiber-reinforced UHPC systems requiring uniform fiber orientation and minimized sedimentation during placement.
- UHPC produced for 3D casting, thin overlays, and complex geometries where segregation or bleeding is unacceptable.

2. Advantages

- Stabilization of Extremely Dense Particle Systems

  The VMA system enhances yield stress without compromising flow, enabling stable suspension of fine powders and fibers in mixes with particle packing densities approaching theoretical limits.

- Improved Fiber Dispersion and Orientation

  Controlled viscosity minimizes fiber balling and sedimentation, supporting consistent mechanical performance and crack-bridging efficiency.

- Reduction of Casting Defects

  Enhanced cohesion limits segregation, surface laitance, and bleeding, producing dense, defect-free surfaces.

- Compatibility with Advanced Production Technologies

The formulation supports automated dosing, rheology monitoring, and closed-loop control systems commonly applied in Industry 4.0–enabled UHPC production.

3. Precautions

- Extremely Narrow Processing Window

  Small deviations in VMA dosage can cause disproportionate changes in flowability, making precise dosing essential.

- Interaction with High-Dosage PCEs

  Strong adsorption competition between VMAs and high-performance PCE superplasticizers can affect dispersion efficiency and setting behavior.

- Air Entrapment Sensitivity

  Even minimal stabilized air content can significantly reduce compressive strength, necessitating rigorous air control.

- Thermal and Curing Sensitivity

  Heat treatment or steam curing can amplify polymer effects on rheology and hydration, requiring temperature-specific validation.

4. Formula

Table 7.2.9: Representative Formulation of VMA for Ultra-High Performance Concrete

Component	CAS No.	%	Function
Synthetic Polyacrylamide or Polyvinyl Alcohol-Based VMA	9003-05-8 / 9002-89-5	55–70	Primary viscosity modifier providing cohesion in ultra-dense powder systems.
Low MW Polycarboxylate Polymer	9007-20-9	10–20	Rheology-balancing component ensuring adequate dispersion at very low water content.
Sodium Gluconate or Sodium Citrate	527-07-1 / 68-04-2	3–8	Fine-tunes hydration kinetics and improves robustness

			against adsorption competition.
Powdered Defoamer (PDMS-based)	63148-62-9	0.5–1.5	Eliminates micro-air voids detrimental to UHPC strength.
Inert Carrier (Ultrafine Limestone or Silica)	1317-65-3 / 14808-60-7	10–20	Ensures uniform dispersion and precise micro-dosing.

5. Technical Specifications and Application Guidance

- Typical Dosage

  0.01 to 0.05 percent by weight of total binder, reflecting the extreme sensitivity of UHPC rheology to polymer addition.

- Primary Performance

  Maintains stable flow, homogeneous fiber distribution, and zero bleeding in systems with water-to-binder ratios typically below 0.20.

- Hydration Impact

  Generally neutral to slightly retarding, with set delay usually below 60 minutes when properly balanced with dispersants and curing regime.

6. Guidance for Formulators

- Molecular Parameter Selection

  Low to medium molecular weight polymers with narrowly controlled distribution and moderate charge density are essential to avoid excessive viscosity spikes.

- Dosage Strategy

  Optimization must proceed in very fine increments, typically 0.002 to 0.005 percent, with simultaneous monitoring of flow spread, fiber distribution, and air content.

- Compatibility Testing

Extensive compatibility testing with cement, silica fume, quartz flour, fibers, and PCE superplasticizers is mandatory. Heat-curing simulations are strongly recommended.

- Sustainability Consideration

  Although UHPC has high binder content, VMAs enable reduced defect rates, longer service life, and material minimization through thinner sections. Digital mix optimization and performance-based design amplify life-cycle sustainability benefits.

7.2.10 VMA for Hot-Weather and High-Temperature Concreting

1. Applications

- Ready-mix concrete produced, transported, and placed in hot climates where ambient temperatures frequently exceed 30 °C, accelerating slump loss and evaporation.
- Large pours, pavements, and infrastructure works executed under direct solar exposure and low relative humidity.
- Pumped concrete and long-haul deliveries where elevated temperature amplifies rheology loss and segregation risk.
- Concrete incorporating rapid-reacting cements or finely ground SCMs that further intensify early-age hydration in hot conditions.

2. Advantages

- Workability Retention at Elevated Temperature

  The VMA stabilizes the paste phase, slowing rheology loss and maintaining cohesion despite accelerated evaporation and hydration kinetics.

- Reduction of Plastic Shrinkage Risk

  Improved water retention limits surface drying and capillary stress development during the critical early hours after placement.

- Segregation Resistance Under Thermal Stress

Enhanced viscosity counters temperature-induced reduction in paste yield stress, supporting aggregate suspension in high-slump mixes.

- Compatibility with Digital Thermal Control

  The formulation supports integration with temperature monitoring, predictive batching adjustments, and automated admixture dosing typical of Industry 4.0 concrete plants.

3. Precautions

- Cumulative Retardation Effects

  When combined with conventional set retarders or extended-slump PCEs, total setting delay can exceed project limits and must be managed holistically.

- Temperature-Dependent Polymer Response

  Some VMAs show reduced effectiveness or altered hydration interaction at very high temperatures, requiring validation across the expected temperature range.

- Air Stability Concerns

  Elevated temperature can destabilize entrained air, while polymers may stabilize microbubbles, making air control additives essential.

- Increased Water Demand Misinterpretation

  Apparent stiffening caused by evaporation should not be compensated solely by water addition, as this undermines VMA efficiency and durability.

4. Formula

Table 7.2.10: Representative of Formulation of VMA for Hot-Weather and High-Temperature Concreting

Component	CAS No.	%	Function
Modified Cellulose Ether (Thermally Stable HPMC or HEMC)	9004-65-3 / 9004-62-0	40–55	Provides water retention and viscosity stability under elevated temperature.

Synthetic Polyacrylamide or Polyacrylate VMA	9003-05-8	20–30	Enhances yield stress and cohesion during extended hot-weather handling.
Sodium Gluconate or Tartaric Acid	527-07-1 / 87-69-4	5–10	Moderates accelerated hydration and extends open time.
Powdered Defoamer (PDMS-based)	63148-62-9	0.5–2	Controls air stability in high-temperature mixing and placement.
Inert Carrier (Milled Limestone)	1317-65-3	10–20	Ensures uniform dispersion and dosing consistency.

5. Technical Specifications and Application Guidance

- Typical Dosage

  0.10 to 0.25 percent by weight of total cementitious material, adjusted according to temperature, haul time, and evaporation rate.

- Primary Performance

  Maintains cohesive, pumpable concrete with controlled slump loss for 60 to 120 minutes under hot-weather conditions.

- Hydration Impact

  Introduces a controlled delay of initial set, typically 1 to 3 hours, balancing workability retention with finishing and curing schedules.

6. Guidance for Formulators

- Molecular Parameter Selection

  Medium molecular weight polymers with stable substitution patterns are preferred to ensure viscosity retention without excessive thermal sensitivity.

- Dosage Strategy

  Optimization should be performed in increments of 0.02 to 0.03 percent, with evaluation at the highest expected concrete temperature.

- Compatibility Testing

Mandatory testing with cement, SCMs, PCE superplasticizers, and temperature-conditioned materials is required. Field trials under realistic hot-weather scenarios are strongly recommended.

- Sustainability Consideration

  VMAs reduce re-tempering, rejected loads, and excessive water use in hot weather. Integration with digital temperature tracking and adaptive batching improves resource efficiency and lowers overall project carbon footprint.

7.2.11 VMA for 3D Printed and Extrudable Cementitious Materials

1. Applications

- Additive manufacturing of cement-based materials using extrusion-based 3D printing for structural and non-structural components, including walls, façade elements, and customized architectural features.
- Robotic and gantry-based printing systems requiring precise control of extrudability, shape retention, and interlayer bonding.
- Printable mortars and micro-concretes with high binder and fine aggregate content, often incorporating fibers or functional fillers.
- On-site and off-site digital construction platforms integrating automated mixing, pumping, and deposition.

2. Advantages

- Controlled Extrudability and Buildability

  The VMA system balances low dynamic viscosity during extrusion with rapid static structural build-up, enabling stable layer stacking without collapse.

- Enhanced Shape Retention

  Increased thixotropy and yield stress prevent deformation, edge rounding, and slump immediately after deposition.

- Improved Interlayer Cohesion

Stabilized moisture distribution and rheology support consistent bonding between successive layers, reducing cold-joint formation.

- Compatibility with Digital Manufacturing

  The formulation supports real-time rheology feedback, closed-loop dosing, and adaptive process control aligned with Industry 4.0 construction.

3. Precautions

- Narrow Rheological Processing Window

  Small formulation deviations can cause either nozzle blockage or loss of buildability, requiring precise control of VMA dosage.

- Shear and Pressure Sensitivity

  High shear during pumping and extrusion can degrade certain polymers, necessitating selection of shear-stable grades.

- Interaction with Accelerators and Set Modifiers

  Accelerated setting strategies must be carefully balanced to avoid premature stiffening within hoses or print heads.

- Air Entrapment Risks

  Entrapped air disrupts filament continuity and reduces mechanical performance, requiring effective defoaming strategies.

4. Formula

Table 7.2.11: Representative Formulation of VMA for 3D Printed and Extrudable Cementitious Materials

Component	CAS No.	%	Function
Modified Cellulose Ether or Synthetic Polyacrylamide VMA	9004-65-3 / 9003-05-8	45–60	Primary viscosity modifier providing thixotropy and shape stability after extrusion.
Low MW Polycarboxylate Polymer	9007-20-9	10–20	Controls dispersion and flow during pumping and extrusion.

Sodium Gluconate or Tartaric Acid	527-07-1 / 87-69-4	3–8	Fine-tunes hydration rate to match printing speed and layer time.
Powdered Defoamer (PDMS-based)	63148-62-9	0.5–2	Eliminates entrapped air and improves filament continuity.
Inert Carrier (Ultrafine Limestone)	1317-65-3	10–20	Ensures homogeneous blending and accurate micro-dosing.

5. Technical Specifications and Application Guidance

- Typical Dosage

  0.10 to 0.30 percent by weight of total binder, depending on print speed, nozzle diameter, and layer height.

- Primary Performance

  Delivers stable extrusion, rapid shape retention, and consistent layer geometry with minimal deformation or collapse.

- Hydration Impact

  Provides controlled setting delay, typically 30 to 90 minutes, allowing sufficient open time for continuous printing without hose blockage.

6. Guidance for Formulators

- Molecular Parameter Selection

  Medium molecular weight polymers with pronounced thixotropic response and shear stability are preferred for printable systems.

- Dosage Strategy

  Optimization should proceed in very small increments, typically 0.01 to 0.02 percent, with simultaneous evaluation of extrudability, build height, and interlayer adhesion.

- Compatibility Testing

Extensive trials with cement, fillers, fibers, superplasticizers, and printing equipment are mandatory. Rheological characterization under printing-relevant shear rates is strongly recommended.

- Sustainability Consideration

  VMAs enable material-efficient, formwork-free construction and reduced waste. Integration with digital design, mix optimization, and automated quality control enhances sustainability across the construction lifecycle.

7.2.12 VMA for Mass Concrete with Thermal Control Requirements

1. Applications

- Mass concrete placements such as dams, raft foundations, thick mat slabs, bridge piers, and large footings where heat of hydration and thermal gradients govern cracking risk.
- Low-cement or blended cement concretes incorporating fly ash, slag, or limestone fillers to reduce peak temperature rise in massive sections.
- Concrete placed over extended durations with staged pours, where workability retention and internal stability must be maintained without accelerating hydration.
- Infrastructure projects requiring compatibility with embedded cooling pipes and temperature monitoring systems.

2. Advantages

- Stabilized Rheology at Low Heat Release

  The VMA system enhances cohesion and segregation resistance while allowing reduction of cement content and heat-generating clinker phases.

- Controlled Water Distribution

  Improved water retention minimizes bleeding and differential moisture movement, supporting uniform hydration and thermal development across large cross-sections.

- Reduced Risk of Thermal Cracking

By enabling lower water and cement contents while maintaining workability, the system indirectly lowers peak temperature and thermal gradients.

- Integration with Digital Thermal Management

  The formulation is compatible with real-time temperature sensing, predictive modeling, and automated batching strategies consistent with Industry 4.0 construction practices.

3. Precautions

- Cumulative Set Delay

  Combination with low-heat cements, SCMs, and chemical retarders can significantly extend setting time, requiring careful scheduling and curing planning.

- Early-Age Strength Development

  Excessive viscosity or water immobilization may further slow strength gain in already slow-reacting mass concrete systems.

- Air Content Stability

  VMAs can stabilize entrapped air in large pours, potentially affecting density and thermal conductivity if not properly controlled.

- Thermal Sensitivity of Polymers

  Performance must be validated across the full temperature range expected during hydration, including peak internal temperatures.

4. Formula

Table 7.2.12: Representative Formulation of VMA for Mass Concrete with Thermal Control Requirements

Component	**CAS No.**	**%**	**Function**
Modified Cellulose Ether (Low Substitution HPMC or HEMC)	9004-65-3 / 9004-62-0	35–50	Provides water retention and baseline cohesion with limited impact on hydration rate.

Synthetic Polyacrylamide or Polyacrylate VMA	9003-05-8	20–30	Enhances yield stress and segregation resistance under low-cement conditions.
Sodium Gluconate or Sodium Citrate	527-07-1 / 68-04-2	5–10	Moderates hydration kinetics and complements thermal control strategy.
Powdered Defoamer (PDMS-based)	63148-62-9	0.5–2	Controls entrapped air in large-volume placements.
Inert Carrier (Milled Limestone)	1317-65-3	10–20	Ensures homogeneous dispersion and accurate dosing.

5. Technical Specifications and Application Guidance

- Typical Dosage

  0.08 to 0.20 percent by weight of total cementitious material, adjusted according to cement type, SCM level, and placement duration.

- Primary Performance

  Provides stable workability, low bleeding, and improved homogeneity in massive pours while supporting reduced cement and heat generation.

- Hydration Impact

  Introduces a controlled extension of setting time, typically 1 to 4 hours, which must be coordinated with thermal control and curing plans.

6. Guidance for Formulators

- Molecular Parameter Selection

  Low to medium molecular weight polymers with moderate substitution levels are preferred to minimize excessive retardation while ensuring adequate cohesion.

- Dosage Strategy

  Optimization should proceed in increments of 0.02 percent, with concurrent monitoring of temperature rise, setting time, and early-age strength.

- Compatibility Testing

  Comprehensive testing with low-heat cements, SCMs, superplasticizers, and temperature-controlled curing simulations is mandatory.

- Sustainability Consideration

  VMAs enable reduced clinker content, lower peak temperatures, and improved durability, extending service life of massive structures. Integration with digital thermal modeling and automated batching enhances material efficiency and sustainability.

7.2.13 VMA for Recycled Aggregate Concrete

1. Applications

- Structural and non-structural concrete incorporating recycled coarse and fine aggregates derived from construction and demolition waste, where high water absorption and variable quality affect fresh performance.
- Ready-mix concrete produced with partial or full replacement of natural aggregates, including pavements, foundations, and general construction elements.
- Pumped concrete using recycled aggregates, where segregation, bleeding, and pressure fluctuations are amplified by aggregate heterogeneity.
- Concrete designed for circular economy projects and green building certifications requiring high recycled content.

2. Advantages

- Compensation for High Aggregate Absorption

  The VMA stabilizes the paste phase, reducing sensitivity to rapid water uptake by recycled aggregates and limiting early slump loss.

- Improved Cohesion and Homogeneity

  Enhanced viscosity counteracts the irregular shape and rough surface texture of recycled aggregates, reducing segregation and settlement.

- More Predictable Fresh Properties

By buffering variability in recycled aggregate grading and porosity, the system improves batch-to-batch consistency.

- Alignment with Digital Quality Control

  The formulation supports integration with moisture sensors, aggregate tracking, and automated mix adjustments in Industry 4.0–enabled plants.

3. Precautions

- Variability of Recycled Aggregate Sources

  Significant differences in absorption, residual mortar content, and fines require continuous monitoring and frequent recalibration.

- Clay and Contaminant Interaction

  Residual clays, gypsum, or organic matter can adsorb polymers and reduce VMA efficiency.

- Air Content Sensitivity

  Increased fines and polymer presence may stabilize air voids, affecting strength and durability if not controlled.

- Potential Set Delay

  Combined effects of VMAs, water reducers, and SCMs may slow early hydration, requiring careful balance.

4. Formula

Table 7.2.13: Representative Formulation of VMA for Recycled Aggregate Concrete

Component	CAS No.	%	Function
Modified Cellulose Ether (Low MW HPMC or HEMC)	9004-65-3 / 9004-62-0	30–45	Provides water retention and cohesion to counter high absorption and fines content.
Synthetic Polyacrylamide or Polyacrylate VMA	9003-05-8	25–35	Enhances yield stress and stabilizes heterogeneous aggregate systems.

Sodium Gluconate or Sodium Citrate	527-07-1 / 68-04-2	5–10	Moderates hydration kinetics and improves robustness against material variability.
Powdered Defoamer (PDMS-based)	63148-62-9	0.5–2	Controls air stabilization caused by fines and polymer interactions.
Inert Carrier (Limestone Filler)	1317-65-3	10–20	Ensures homogeneous dispersion and accurate dosing.

5. Technical Specifications and Application Guidance

- Typical Dosage

  0.10 to 0.30 percent by weight of total cementitious material, adjusted based on recycled aggregate content, absorption, and target workability.

- Primary Performance

  Delivers improved slump retention, reduced bleeding, and stable cohesion in mixes containing up to 50 percent recycled aggregates.

- Hydration Impact

  Typically introduces a slight to moderate set delay, usually within 30 to 120 minutes, depending on aggregate condition and admixture synergy.

6. Guidance for Formulators

- Molecular Parameter Selection

  Low to medium molecular weight polymers with moderate hydrophilicity are preferred to manage water balance without excessive retardation.

- Dosage Strategy

  Optimization should proceed in increments of 0.02 to 0.03 percent, with parallel monitoring of slump retention, air content, and early strength.

- Compatibility Testing

Routine testing with each recycled aggregate source is mandatory. Evaluation should include moisture state variation, fines content, and contaminant screening.

- Sustainability Consideration

  VMAs enable higher recycled aggregate utilization, reducing natural resource depletion and landfill waste. Integration with digital aggregate characterization and automated mix correction enhances environmental performance and circularity.

7.2.14 VMA for Fiber-Reinforced Concrete with High Fiber Loading

1. Applications

- Fiber-reinforced concrete containing high volumes of steel, synthetic, basalt, or glass fibers, where uniform fiber dispersion and segregation resistance are critical.
- Structural elements subjected to impact, fatigue, or crack control requirements, including industrial floors, tunnel linings, precast segments, and shotcrete.
- Self-consolidating and pumped fiber-reinforced concretes with fiber dosages exceeding conventional limits.
- Hybrid fiber systems combining macro and micro fibers requiring stable rheology during mixing and placement.

2. Advantages

- Uniform Fiber Suspension

  The VMA system increases paste yield stress, preventing fiber settlement or flotation and ensuring homogeneous distribution throughout the matrix.

- Reduced Fiber Balling

  Improved cohesion and controlled flow minimize fiber agglomeration during mixing, pumping, and placement.

- Enhanced Workability Robustness

Balanced viscosity enables placement of high-fiber mixes without excessive water or superplasticizer demand.

- Process Consistency with Digital Monitoring

  The formulation supports integration with automated fiber dosing, rheology monitoring, and quality control systems aligned with Industry 4.0 production.

3. Precautions

- Over-Viscosification and Placement Difficulty

  Excessive VMA dosage can restrict flow and increase pumping pressure, particularly in high fiber volume systems.

- Interaction with Fiber Surface Treatments

  Sizing agents or surface coatings on synthetic fibers may interact with polymers, affecting dispersion efficiency.

- Air Entrainment Sensitivity

  Fibers and VMAs can jointly stabilize air voids, requiring careful defoamer management.

- Equipment Wear and Shear Effects

  High fiber content combined with viscous mixes increases shear demand and mechanical stress on mixing and pumping equipment.

4. Formula

Table 7.2.14: Representative Formulation of VMA for Fiber-Reinforced Concrete with High Fiber Loading

Component	CAS No.	%	Function
Synthetic Polyacrylamide or Polycarboxylate VMA	9003-05-8 / 9007-20-9	50–65	Primary viscosity modifier enhancing fiber suspension and cohesion.

Modified Cellulose Ether (Low MW HPMC or HEMC)	9004-65-3 / 9004-62-0	15–25	Provides supplemental water retention and rheological stability.
Sodium Gluconate or Sodium Citrate	527-07-1 / 68-04-2	5–10	Adjusts hydration kinetics to accommodate fiber-induced rheology changes.
Powdered Defoamer (PDMS-based)	63148-62-9	0.5–2	Controls air stabilization associated with fibers and polymers.
Inert Carrier (Limestone Filler)	1317-65-3	10–20	Ensures uniform dispersion and accurate dosing.

5. Technical Specifications and Application Guidance

- Typical Dosage

  0.10 to 0.30 percent by weight of total cementitious material, depending on fiber type, aspect ratio, and volume fraction.

- Primary Performance

  Maintains uniform fiber distribution, stable workability, and reduced segregation in concretes with fiber volumes up to 2.5 percent by volume.

- Hydration Impact

  Typically results in slight set delay, usually within 30 to 90 minutes, depending on cement chemistry and admixture synergy.

6. Guidance for Formulators

- Molecular Parameter Selection

  Medium molecular weight polymers with controlled charge density provide optimal balance between fiber suspension and workable flow.

- Dosage Strategy

  Optimization should proceed in increments of 0.02 to 0.03 percent, with direct evaluation of fiber dispersion, slump flow, and pumping pressure.

- Compatibility Testing

Mandatory testing with fiber type, surface treatment, superplasticizers, and mixing sequence is required to ensure reproducible performance.

- Sustainability Consideration

  VMAs enable efficient use of fibers to replace traditional reinforcement, reducing steel consumption and extending service life. Integration with digital mix optimization and automated fiber dosing enhances resource efficiency.

7.2.15 VMA for High-Alkali and Rapid-Hardening Cement Systems

1. Applications

- Concrete produced with high-alkali Portland cements, rapid-hardening cements, or calcium aluminate cements where very fast hydration and stiffening challenge fresh stability.
- Precast and repair mortars requiring rapid strength development combined with controlled rheology for placement and finishing.
- Shotcrete and sprayed concrete using accelerators or alkali-rich binders, where segregation and rebound must be minimized.
- Cold-weather concreting where high-alkali or fast-reacting binders are selected to compensate for low ambient temperatures.

2. Advantages

- Rapid Rheology Stabilization

  The VMA system develops viscosity quickly, counterbalancing accelerated hydration and early stiffening without excessive water addition.

- Improved Fresh-State Robustness

  Enhanced cohesion limits bleeding and segregation during the narrow placement window typical of rapid-hardening systems.

- Compatibility with Accelerated Processes

Formulated to withstand high pH environments, maintaining performance in alkali-rich pore solutions.

- Process Control in Automated Production

  Supports integration with high-speed batching, inline rheology measurement, and digital feedback control in Industry 4.0 precast operations.

3. Precautions

- Polymer Alkaline Stability

  Certain cellulose ethers and polyacrylamides can degrade or lose efficiency at very high pH and temperature, requiring careful grade selection.

- Over-Retardation Risk

  Although designed for fast systems, excessive dosage can counteract desired rapid strength gain.

- Interaction with Accelerators

  Alkali-free and alkali-based accelerators may compete or interfere with VMA adsorption and performance.

- Air Entrainment Control

  High shear mixing combined with polymers may stabilize entrapped air, impacting early strength.

4. Formula

Table 7.2.15: Representative Formulation of VMA for High-Alkali and Rapid-Hardening Cement Systems

Component	CAS No.	%	Function
Alkali-Stable Synthetic Polyacrylate or Polyacrylamide VMA	9003-05-8 / 9007-20-9	55–70	Primary viscosity modifier resistant to high pH and rapid hydration conditions.

Modified Cellulose Ether (Low Substitution HEMC)	9004-62-0	10–20	Provides supplemental cohesion with improved alkali tolerance.
Sodium Gluconate or Sodium Citrate (Low Level)	527-07-1 / 68-04-2	2–5	Fine-tunes setting without suppressing rapid hardening.
Powdered Defoamer (PDMS-based)	63148-62-9	0.5–2	Controls air entrainment under high shear and rapid setting.
Inert Carrier (Limestone Filler)	1317-65-3	10–15	Ensures uniform dispersion and accurate dosing.

5. Technical Specifications and Application Guidance

- Typical Dosage

  0.05 to 0.20 percent by weight of total cementitious material, depending on alkali content, cement type, and accelerator use.

- Primary Performance

  Provides immediate cohesion, reduced bleeding, and stable workability during the short handling time typical of rapid-hardening concretes.

- Hydration Impact

  Minimal extension of initial set, typically limited to 15 to 60 minutes, preserving early strength development.

6. Guidance for Formulators

- Molecular Parameter Selection

  Low to medium molecular weight polymers with proven alkaline stability and limited calcium complexation are preferred.

- Dosage Strategy

  Optimization should proceed in small increments of 0.01 to 0.02 percent, with emphasis on early-age strength and set time monitoring.

- Compatibility Testing

Mandatory evaluation with specific cement alkali levels, accelerators, and curing temperatures is required to ensure reliable performance.

- Sustainability Consideration

  VMAs reduce material waste and rework in fast-track construction and precast production. Improved process control and digital optimization lower energy use and enhance productivity.

7.2.16 VMA for Marine and High-Salinity Environments

1. Applications

- Concrete structures exposed to seawater, tidal zones, and splash zones, including piers, wharves, offshore platforms, and breakwaters.
- Onshore infrastructure subjected to saline groundwater or deicing salt exposure, such as coastal foundations, tunnels, and bridge substructures.
- Ready-mix and precast concretes formulated with seawater or saline aggregates where freshwater availability is limited.
- Repair mortars and grouts for marine rehabilitation requiring stable rheology under high ionic strength conditions.

2. Advantages

- Salt-Tolerant Rheological Control

  The VMA system maintains viscosity and cohesion in high ionic strength pore solutions where conventional polymers lose efficiency.

- Improved Paste Stability and Water Retention

  Enhanced control of free water reduces bleeding and washout risk during underwater or tidal placement.

- Robust Performance with Chlorides

  Formulated polymers exhibit reduced sensitivity to chloride ions, preserving fresh-state stability and placement quality.

- Compatibility with Digitally Controlled Marine Construction

Supports integration with automated batching, salinity monitoring, and adaptive dosing systems consistent with Industry 4.0 marine projects.

3. Precautions

- Polymer Electrolyte Sensitivity

  Certain anionic polymers can collapse or precipitate in high-salinity environments, requiring careful selection.

- Interaction with Corrosion Inhibitors

  VMAs may interact with nitrite-based or organic corrosion inhibitors commonly used in marine concrete.

- Air Void Stability

  Salts can alter surface tension and stabilize air bubbles in the presence of polymers, impacting density and durability.

- Long-Term Durability Validation

  Although VMAs act primarily in the fresh state, indirect effects on permeability and microstructure must be verified.

4. Formula

Table 7.2.16: Representative Formulation of VMA for Marine and High-Salinity Environments

Component	CAS No.	%	Function
Salt-Tolerant Synthetic Polyacrylate or Polyacrylamide VMA	9003-05-8 / 9007-20-9	50–65	Primary viscosity modifier stable in high ionic strength and chloride-rich systems.
Modified Cellulose Ether (HEMC or HPMC, Low Ionic Sensitivity)	9004-62-0 / 9004-65-3	15–25	Provides water retention and supplementary cohesion under saline conditions.
Sodium Gluconate (Low Level)	527-07-1	2–5	Moderates hydration without increasing chloride mobility.

Powdered Defoamer (PDMS-based)	63148-62-9	0.5–2	Controls air entrainment influenced by salts and polymers.
Inert Carrier (Limestone Filler)	1317-65-3	10–20	Ensures homogeneous dispersion and accurate dosing.

5. Technical Specifications and Application Guidance

- Typical Dosage

  0.08 to 0.25 percent by weight of total cementitious material, adjusted according to salinity level and placement method.

- Primary Performance

  Delivers stable cohesion, reduced bleeding, and improved washout resistance in saline and marine exposure conditions.

- Hydration Impact

  Typically introduces minimal set delay, generally within 30 to 90 minutes, depending on cement chemistry and salt concentration.

6. Guidance for Formulators

- Molecular Parameter Selection

  Polymers with low charge density and proven electrolyte tolerance are preferred to prevent viscosity loss in saline environments.

- Dosage Strategy

  Optimization should proceed in increments of 0.02 percent, with evaluation under simulated seawater conditions rather than freshwater.

- Compatibility Testing

  Testing with seawater, saline aggregates, corrosion inhibitors, and supplementary cementitious materials is mandatory.

- Sustainability Consideration

VMAs enable the use of seawater and local aggregates, reducing freshwater consumption and material transport. Digital salinity monitoring and automated admixture adjustment improve resource efficiency and environmental performance.

7.2.17 VMA for Nuclear and Radiation-Shielding Concrete

1. Applications
   - Radiation-shielding concrete for nuclear power plants, research reactors, medical radiology facilities, and radioactive waste storage structures.
   - High-density concretes incorporating barite, magnetite, hematite, or steel aggregates, where segregation control is critical to shielding efficiency.
   - Massive containment walls and biological shields requiring uniform density and homogeneity over large cross-sections.
   - Precast nuclear components produced under strict quality assurance, traceability, and regulatory compliance frameworks.
2. Advantages
   - Density Preservation and Aggregate Suspension

     The VMA system prevents settlement of heavyweight aggregates, ensuring uniform density and consistent radiation attenuation performance.
   - Enhanced Rheological Stability

     Improved cohesion enables placement of very high-density mixes without excessive water or loss of workability.
   - Process Reliability in Safety-Critical Construction

     Predictable rheology supports reproducible batching and placement, aligning with stringent nuclear construction standards.
   - Compatibility with Digital QA Systems

     The formulation supports integration with real-time density monitoring, automated batching, and data logging consistent with Industry 4.0–driven nuclear quality systems.

3. Precautions

- Over-Viscosification Risk

  Excessive VMA dosage can hinder compaction and vibration, increasing entrapped air in dense mixes.

- Interaction with Heavy Mineral Fines

  High surface area fines from barite or magnetite can adsorb polymers, reducing effective VMA performance.

- Air Content Control

  Stabilized air voids negatively affect density and shielding efficiency, requiring stringent defoamer control.

- Regulatory Material Restrictions

  All components must comply with nuclear industry specifications regarding purity, traceability, and long-term stability.

4. Formula

Table 7.2.17: Representative Formulation of VMA for Nuclear and Radiation-Shielding Concrete

Component	CAS No.	%	Function
High-Performance Synthetic Polyacrylamide or Polyacrylate VMA	9003-05-8 / 9007-20-9	55–70	Primary viscosity modifier ensuring suspension of heavyweight aggregates.
Modified Cellulose Ether (Low MW HEMC)	9004-62-0	10–20	Provides supplemental cohesion and water retention without excessive retardation.
Sodium Gluconate or Sodium Citrate (Low Level)	527-07-1 / 68-04-2	2–5	Fine-tunes hydration kinetics in massive, dense sections.
Powdered Defoamer (PDMS-based)	63148-62-9	0.5–2	Minimizes entrapped air to preserve density and shielding capacity.

Inert Carrier (High-Purity Limestone)	1317-65-3	10–15	Ensures uniform dispersion and controlled dosing under QA conditions.

5. Technical Specifications and Application Guidance

- Typical Dosage

  0.08 to 0.25 percent by weight of total cementitious material, depending on aggregate density and grading.

- Primary Performance

  Delivers stable cohesion, minimal segregation, and uniform density in concretes exceeding 3.5 t/m^3 fresh density.

- Hydration Impact

  Introduces a slight and predictable extension of setting time, typically within 30 to 120 minutes, compatible with mass placements.

6. Guidance for Formulators

- Molecular Parameter Selection

  Medium molecular weight polymers with low charge density are preferred to balance viscosity development and compaction efficiency.

- Dosage Strategy

  Optimization should proceed in increments of 0.01 to 0.02 percent, with continuous monitoring of density, air content, and workability.

- Compatibility Testing

  Comprehensive testing with heavyweight aggregates, low-alkali cements, and superplasticizers is mandatory under simulated placement conditions.

- Sustainability Consideration

  VMAs reduce rework, material rejection, and variability in safety-critical structures. Digital process control and data traceability improve material efficiency and long-term reliability of nuclear infrastructure.

7.2.18 VMA for Underwater and Tremie Concrete

1. Applications

- Underwater concrete placement for bridge piers, marine foundations, caissons, and offshore structures using tremie pipes or pumping systems.
- Concrete for diaphragm walls, bored piles, and secant piles placed below groundwater level where washout resistance is essential.
- Repair and grouting operations in submerged or partially submerged environments requiring stable flow and minimal cement loss.
- Mass underwater placements where extended discharge times and low turbulence stability are required.

2. Advantages

- Exceptional Washout Resistance

  The VMA system significantly increases cohesion, preventing cement paste loss and segregation when concrete is placed directly into water.

- Stable Flow Through Tremie Systems

  Controlled viscosity ensures continuous, plug-like flow, reducing turbulence and maintaining concrete integrity during discharge.

- Improved Surface Quality and Strength Uniformity

  Minimized paste dilution leads to consistent strength development and reduced laitance formation.

- Integration with Digital Placement Control

  The formulation supports real-time monitoring of flow rate, discharge pressure, and concrete quality consistent with Industry 4.0 marine construction practices.

3. Precautions

- Over-Viscosity and Flow Interruption

  Excessive VMA dosage can hinder flow through tremie pipes and increase pumping pressure.

- Interaction with Anti-Washout Admixtures

  When used alongside dedicated anti-washout admixtures, synergy or redundancy must be carefully evaluated.

- Air Content Management

  High cohesion systems can trap air, potentially reducing density and strength.

- Temperature and Salinity Effects

  Polymer performance may vary with water temperature and salinity, requiring site-specific validation.

4. Formula

Table 7.2.18: Representative Formulation of VMA for Underwater and Tremie Concrete

Component	CAS No.	%	Function
High-Performance Synthetic Polyacrylamide or Polyacrylate VMA	9003-05-8 / 9007-20-9	55–70	Primary viscosity modifier providing washout resistance and cohesion.
Modified Cellulose Ether (HEMC or HPMC)	9004-62-0 / 9004-65-3	15–25	Enhances water retention and stabilizes paste under submerged conditions.
Sodium Gluconate or Sodium Citrate	527-07-1 / 68-04-2	3–6	Moderates hydration during extended placement times.
Powdered Defoamer (PDMS-based)	63148-62-9	0.5–2	Controls entrapped air in highly cohesive mixes.
Inert Carrier (Limestone Filler)	1317-65-3	10–15	Ensures uniform dispersion and accurate dosing.

5. Technical Specifications and Application Guidance

- Typical Dosage

0.15 to 0.40 percent by weight of total cementitious material, depending on water depth, flow conditions, and placement duration.

- Primary Performance

  Provides high washout resistance, stable tremie flow, and excellent segregation control under submerged placement.

- Hydration Impact

  Typically results in a moderate extension of initial set, generally 1 to 3 hours, adjustable through set control additives.

6. Guidance for Formulators

- Molecular Parameter Selection

  High molecular weight polymers with low charge density are preferred to maximize cohesion and washout resistance.

- Dosage Strategy

  Optimization should proceed in increments of 0.03 percent, with evaluation under simulated underwater placement conditions.

- Compatibility Testing

  Mandatory testing with superplasticizers, anti-washout admixtures, seawater, and varying temperatures is required.

- Sustainability Consideration

  VMAs reduce cement loss to the environment and improve placement efficiency, lowering material waste and environmental impact. Digital monitoring of placement parameters enhances resource efficiency and construction quality.

7.2.19 VMA for High-Pressure Pumped Concrete in Long Pipelines

1. Applications

- High-pressure pumped concrete used in long-distance horizontal pipelines exceeding 300 m and vertical pumping heights above 100 m in high-rise buildings and infrastructure megaprojects.
- Concrete pumping for tunnels, dams, wind turbine foundations, and large-scale industrial floors where stable rheology under sustained shear is mandatory.
- Low-clinker, blended cement concretes and high-SCM mixes where lubrication layer stability governs pumpability.
- Digitally controlled pumping operations integrating pressure sensors, flow meters, and real-time rheology feedback consistent with Industry 4.0 construction sites.

2. Advantages

- Lubrication Layer Stability Under Extreme Shear

  The VMA promotes formation and persistence of a cohesive paste-rich lubrication layer, reducing friction losses and preventing pressure spikes along extended pipelines.

- Segregation and Bleeding Suppression During Pumping

  Controlled viscosity minimizes paste filtration and water migration under high pressure, preserving uniform composition from pump inlet to discharge.

- Reduced Pump Wear and Energy Demand

  Stable rheology lowers pulsation and peak pumping pressures, extending equipment lifetime and improving energy efficiency in continuous pumping cycles.

- Digital Process Compatibility

  Predictable pressure-flow relationships support closed-loop pumping control, enabling adaptive pressure management and predictive maintenance algorithms.

3. Precautions

- Excessive Viscosity and Pressure Build-Up

Overdosage increases yield stress, leading to elevated pumping pressures, risk of line blockage, and mechanical stress on couplings.

- Interaction with High-Range Water Reducers

  Certain polycarboxylate ethers may amplify viscosity development, requiring joint optimization rather than independent admixture adjustment.

- Temperature Sensitivity

  Low temperatures increase apparent viscosity and thixotropic rebuild, necessitating recalibration of dosage in cold-weather pumping.

- Air Entrainment Control

  High shear stability may trap microbubbles, requiring compatible defoaming strategies to avoid compressibility effects during pumping.

4. Formula

Table 7.2.19: Representative VMA Formulation for High-Pressure Pumped Concrete

Component	CAS No.	%	Function
Synthetic Polyacrylamide or Polyacrylate VMA (Medium–High MW)	9003-05-8 / 9007-20-9	50–65	Primary viscosity modifier ensuring lubrication layer stability under high pressure.
Modified Cellulose Ether (Low to Medium MW HPMC or HEMC)	9004-65-3 / 9004-62-0	15–25	Enhances shear stability and water retention without excessive yield stress increase.
Sodium Gluconate or Sodium Citrate	527-07-1 / 68-04-2	2–5	Controls hydration kinetics during prolonged pumping and delays pressure rise.
Powdered Defoamer (PDMS-based)	63148-62-9	0.5–2	Limits entrapped air and compressibility during high-pressure pumping.
Inert Carrier (Ground Limestone or Dolomite)	1317-65-3 / 16389-88-1	10–15	Ensures homogeneous dispersion and accurate dosing

	in automated batching systems.

5. Technical Specifications and Application Guidance

- Typical Dosage:

  0.05 to 0.20 percent by weight of total cementitious material, adjusted according to pipeline length, pressure regime, and fines content.

- Primary Performance:

  Maintains continuous, low-friction pumping with stable discharge consistency over long distances and extended pumping durations.

- Hydration Impact:

  Produces a mild and controllable delay in initial set, typically 20 to 90 minutes, with minimal impact on ultimate strength development.

6. Guidance for Formulators

- Molecular Parameter Selection:

  Medium to high molecular weight polymers with low to moderate charge density are optimal to balance viscosity, shear thinning, and pump restart capability.

- Dosage Strategy:

  Incremental optimization in steps of 0.01 percent is recommended, combined with real-time monitoring of pumping pressure and flow stability.

- Compatibility Testing:

  Full-scale pumping trials with selected superplasticizers, SCM combinations, and temperature scenarios are essential to validate pressure profiles and restart behavior.

- Sustainability Consideration:

  Improved pumpability reduces rejected concrete, energy consumption, and equipment wear. Integration with digital pumping analytics enhances material efficiency and aligns with low-carbon, Industry 4.0 construction workflows.

7.2.20 VMA for Ultra-Low Water-to-Binder Ratio Concrete

1. Applications

- Ultra-low water-to-binder ratio concrete with $w/b \leq 0.25$ used in ultra-high-performance concrete, reactive powder concrete, and high-strength precast elements.
- Structural elements requiring very high compressive strength, low permeability, and enhanced durability such as bridge girders, façade panels, offshore elements, and military or protective infrastructure.
- Concrete incorporating high volumes of silica fume, metakaolin, or finely ground slag where particle packing density and rheological control are critical.
- Digitally optimized batching and mixing lines using rheology sensors, torque monitoring, and AI-assisted mix adjustment aligned with Industry 4.0 production environments.

2. Advantages

- Rheology Control at Extreme Low Water Contents

  The VMA system stabilizes paste flow and prevents shear-induced segregation in mixes where free water is nearly absent.

- Enhanced Homogeneity and Fiber Dispersion

  Improved cohesion supports uniform dispersion of steel, glass, or polymer fibers commonly used in ultra-high-performance systems.

- Improved Casting Reliability

  Controlled viscosity enables mold filling and surface quality without relying on excessive superplasticizer dosage or additional water.

- Compatibility with Digital Mix Optimization

Predictable viscosity response facilitates closed-loop control using mixer power consumption, slump flow imaging, and rheology-based feedback systems.

3. Precautions

- Risk of Over-Stiffening

  Excessive VMA dosage can rapidly increase yield stress, making placement difficult even with high-range water reducers.

- Strong Interaction with Silica Fume and Nano-Fillers

  High specific surface materials can adsorb polymers, reducing effective VMA availability and requiring formulation recalibration.

- Sensitivity to Mixing Energy

  Insufficient mixing intensity may prevent full polymer activation, while overmixing can lead to thixotropic breakdown.

- Air Entrainment Susceptibility

  Highly cohesive systems may stabilize entrapped air, negatively affecting density and compressive strength.

4. Formula

Table 7.2.20: Representative VMA Formulation for Ultra-Low w/b Concrete

Component	CAS No.	%	Function
Synthetic Polyacrylamide or Polyacrylate VMA (Medium MW, Low Charge)	9003-05-8 / 9007-20-9	45–60	Primary viscosity modifier providing cohesion at minimal water availability.
Modified Cellulose Ether (Low MW HPMC or HEMC)	9004-65-3 / 9004-62-0	15–25	Enhances water retention and stabilizes micro-scale paste structure.
Sodium Gluconate or Sodium Citrate	527-07-1 / 68-04-2	1–4	Fine control of early hydration and workability window.

Powdered Defoamer (PDMS-based)	63148-62-9	0.5–2	Limits air stabilization in highly cohesive matrices.
Inert Carrier (High-Purity Limestone or Silica Flour)	1317-65-3 / 14808-60-7	15–25	Ensures accurate dosing and homogeneous dispersion in dry blends.

5. Technical Specifications & Application Guidance

- Typical Dosage:

  0.03 to 0.15 percent by weight of total binder, strongly dependent on silica fume content and target flow class.

- Primary Performance:

  Delivers stable, non-segregating rheology with controlled flowability, enabling casting and compaction of ultra-dense mixes.

- Hydration Impact:

  Induces a slight delay in initial hydration, typically 15 to 60 minutes, with negligible effect on long-term strength development.

6. Guidance for Formulators

- Molecular Parameter Selection:

  Medium molecular weight polymers with very low charge density are preferred to avoid excessive yield stress while maintaining cohesion.

- Dosage Strategy:

  Optimization should proceed in increments of 0.005 to 0.01 percent, supported by rheometer data and mixer torque monitoring.

- Compatibility Testing:

  Mandatory evaluation with selected polycarboxylate superplasticizers, silica fume sources, fiber types, and curing regimes under realistic mixing energy.

- Sustainability Consideration:

  Ultra-low w/b concretes enable material-efficient, long-life structures. VMAs reduce rejected batches and reprocessing. Integration with digital quality control improves resource efficiency and supports low-carbon, high-durability construction strategies.

7.2.21 VMA for Geopolymer and Alkali-Activated Binder Systems

1. Applications

- Geopolymer concretes and mortars based on fly ash, ground granulated blast furnace slag, metakaolin, or blended aluminosilicate precursors activated with alkaline solutions.
- Structural and precast elements requiring low Portland cement content, reduced CO_2 footprint, and high chemical resistance such as sewer infrastructure, marine structures, and industrial floors.
- Cast-in-place and precast alkali-activated concretes where segregation control and workability retention are challenged by high alkalinity and rapid reaction kinetics.
- Digitally managed production lines employing real-time viscosity tracking, activator dosing control, and AI-assisted mix optimization consistent with Industry 4.0 low-carbon construction strategies.

2. Advantages

- Rheological Stabilization in High-Alkalinity Media

  The VMA system enhances cohesion and yield stress control in mixes containing concentrated alkali activators where conventional rheology modifiers fail.

- Segregation Resistance with Diverse Precursors

  Improved suspension of unreacted particles and aggregates ensures homogeneity across fly ash–rich and slag-dominant systems.

- Improved Casting and Surface Quality

  Controlled viscosity reduces bleeding and surface defects while supporting self-compacting or highly flowable geopolymer formulations.

- Compatibility with Digital Process Control

Predictable rheological response enables integration with sensor-driven batching, activator concentration monitoring, and adaptive mixing protocols.

3. Precautions

- Chemical Stability in Alkaline Environments

  Certain cellulose ethers and anionic polymers may undergo degradation or viscosity loss at pH > 13, requiring careful polymer selection.

- Interaction with Alkali Activators

  High sodium or potassium concentrations can alter polymer conformation, reducing thickening efficiency or causing delayed viscosity development.

- Rapid Setting and Structural Build-Up

  Overdosage may exacerbate early stiffening in slag-rich systems with fast geopolymerization kinetics.

- Air Entrainment Sensitivity

  Highly cohesive geopolymer pastes can stabilize entrapped air, adversely affecting density and mechanical performance.

4. Formula

Table 7.2.21: Representative VMA Formulation for Geopolymer and Alkali-Activated Systems

Component	CAS No.	%	Function
Alkali-Stable Synthetic Polyacrylamide or Modified Polyacrylate (Low Charge)	9003-05-8 / 9007-20-9	45–60	Primary viscosity modifier providing cohesion under high pH conditions.
Alkali-Resistant Modified Cellulose Ether (Specialty HEMC)	9004-62-0	10–20	Enhances water retention and stabilizes paste without rapid degradation.
Sodium Gluconate or Sodium Citrate	527-07-1 / 68-04-2	1–4	Moderates early reaction kinetics and workability window.

Powdered Defoamer (PDMS-based, Alkali-Stable)	63148-62-9	0.5–2	Controls entrapped air in highly cohesive geopolymer matrices.
Inert Carrier (Calcined or Low-Reactivity Limestone)	1317-65-3	15–25	Enables uniform dispersion and accurate dosing in dry admixture systems.

5. Technical Specifications & Application Guidance

- Typical Dosage:

  0.05 to 0.25 percent by weight of total binder, depending on precursor fineness, activator concentration, and target flow class.

- Primary Performance:

  Provides stable rheology, high segregation resistance, and consistent flow behavior across a wide range of alkali-activated binder chemistries.

- Hydration Impact:

  Slight and controllable delay in early geopolymerization or slag activation, typically 20 to 120 minutes, without compromising final strength.

6. Guidance for Formulators

- Molecular Parameter Selection:

  Medium molecular weight polymers with very low anionic charge density and proven alkali stability are essential for sustained performance.

- Dosage Strategy:

  Optimization should proceed in increments of 0.01 percent, supported by rheological measurements and early-age stiffness monitoring.

- Compatibility Testing:

  Extensive testing with specific alkali activators, precursor sources, curing temperatures, and mixing sequences is mandatory due to system sensitivity.

- Sustainability Consideration:

Geopolymer systems significantly reduce embodied carbon. VMAs improve robustness, reduce rejected batches, and enable digital quality control, reinforcing the environmental and economic benefits of alkali-activated construction.

7.3 Implementation and Adaptation Guidance

The implementation and adaptation guidance presented here provides a structured pathway for transferring laboratory-validated VMA formulations into reliable, large-scale industrial production. Effective commercialization requires careful management of scale-up effects, raw material variability, and process-induced rheological shifts, while preserving the performance envelope established during development. Critical focus areas include faithful translation of laboratory mixing and hydration conditions to plant operations, systematic adaptation to regional variations in cement chemistry and aggregate characteristics, and verification of compatibility within complex, multi-admixture systems.

In parallel, successful deployment increasingly depends on the integration of digital process control, inline quality monitoring, and data-driven adjustment strategies consistent with Industry 4.0 manufacturing paradigms. These tools enable real-time detection of deviations, predictive correction of viscosity and stability trends, and continuous improvement across diverse climatic conditions and application scenarios. By embedding sustainability criteria, durability targets, and lifecycle performance metrics into implementation decisions, this guidance supports the delivery of robust, adaptable, and commercially deployable VMA solutions across global concrete markets.

7.3.1 Translation from Laboratory Formulation to Industrial Production

Successful transfer of VMA concepts from laboratory-scale validation to full-scale concrete production requires structured upscaling that preserves rheological intent while accommodating industrial variability. Laboratory rheometer data and mini-slump tests provide directional guidance but must be correlated with mixer torque, power consumption, and real batching behavior at plant scale. Differences in shear history, mixing energy density, and residence time often shift the effective viscosity window, particularly for polymer-based VMAs interacting with fines-rich or SCM-heavy systems. Pilot-scale trials under realistic mixing sequences are therefore mandatory to recalibrate dosage and dispersion protocols before commercial deployment.

7.3.2 Adaptation to Cement, SCM, and Aggregate Variability

VMAs are highly sensitive to binder mineralogy, alkali content, and particle size distribution. Changes in clinker chemistry, fly ash source, slag reactivity, or limestone filler fineness can alter polymer adsorption and hydration kinetics. Aggregate grading and shape influence shear-induced filtration and lubrication layer formation, especially in pumped and self-compacting concretes. Formulators should establish adaptable dosage envelopes rather than fixed recipes, supported by statistical process control charts linking raw material variability to rheological response. This adaptive approach improves robustness across multiple supply chains and project locations.

7.3.3 Integration with Admixture Systems

VMAs must be designed and implemented as part of a complete admixture package rather than as isolated additives. Interactions with polycarboxylate superplasticizers, set regulators, air entrainers, and defoamers determine the final flow profile and stability. Sequential addition strategies, including delayed VMA dosing or split superplasticizer addition, often enhance performance in demanding applications such as ultra-low w/b or geopolymer systems. Compatibility matrices generated through systematic screening reduce the risk of over-viscosification, loss of flow, or unpredictable setting behavior.

7.3.4 Process Control and Industry 4.0 Integration

Modern VMA implementation benefits significantly from Industry 4.0 tools that convert rheology from an empirical attribute into a controlled process variable. Online monitoring of mixer power, pumping pressure, slump flow imaging, and discharge consistency enables real-time adjustment of VMA dosage within predefined limits. Data-driven models and machine learning algorithms can predict rheological drift caused by temperature changes, material variability, or extended transport times. Such digital integration enhances repeatability, supports traceability, and aligns VMA usage with performance-based specifications rather than prescriptive limits.

7.3.5 Quality Assurance and Performance Verification

Quality assurance protocols for VMA-containing concretes should extend beyond conventional slump or flow tests. Key indicators include segregation resistance indices, bleeding rate, air content stability, and time-dependent rheology. Application-specific tests such as column segregation, washout resistance, pumping loop trials, or early-age stiffness monitoring provide higher confidence in field performance. Documentation of VMA batch consistency, polymer molecular parameters, and carrier quality is essential, particularly in safety-critical or regulated applications.

7.3.6 Adaptation Across Applications and Climates

VMAs must be tuned to local climatic conditions and construction practices. High temperatures accelerate hydration and polymer hydration, while low temperatures increase viscosity and thixotropic rebuild. Seasonal adjustment of dosage and molecular weight distribution improves placement reliability. In regions with long transport distances or intermittent placing, VMAs with predictable time-dependent behavior reduce risk of rejection and retempering. Cross-application learning, for example between pumped concrete and underwater concrete, can accelerate optimization when supported by shared rheological datasets.

7.3.7 Sustainability and Lifecycle Perspective

From a lifecycle standpoint, VMAs contribute to sustainability by enabling lower cement contents, higher SCM utilization, and reduced waste due to segregation or placement failure. Improved rheological control minimizes energy consumption during pumping and vibration and extends equipment service life. Digital monitoring further reduces overdesign and conservatism in mix design. When evaluated across the full construction cycle, VMA implementation supports durable, resource-efficient concrete aligned with low-carbon construction objectives.

8. Formulation Processes and Procedures

Formulation of concrete viscosity-modifying admixtures represents the critical interface between polymer chemistry, manufacturing discipline, and field performance. The processes described in this chapter translate rheological design intent into reproducible industrial products, addressing raw material selection, polymer architecture, blending methodology, and quality control under real manufacturing constraints. Emphasis is placed on dry-blend production routes, polymer compatibility, and dispersion control, while accounting for application-driven requirements ranging from conventional ready-mix to ultra-high-performance, geopolymer, and digitally controlled construction systems. By linking formulation architecture with processing procedures and Industry 4.0-enabled manufacturing practices, the chapter provides a structured framework for robust, scalable, and performance-consistent VMA production across diverse concrete technologies.

8.1 Group 1: Dry-Blend Powder Manufacturing (Standard Physical Blending)

Dry-blend powder manufacturing represents the most established and industrially accessible route for producing viscosity-modifying admixtures based on physical blending of solid constituents. The process relies on controlled mechanical mixing of base polymers, most commonly cellulose ethers, with auxiliary functional components such as dispersants, setting modifiers, or flow aids to achieve a homogeneous, free-flowing powder. Its technical attractiveness lies in operational simplicity, low capital intensity, and ease of scale-up, making it well suited for formulations where chemical compatibility in the solid state is assured and rapid dissolution or dispersion is expected during concrete mixing.

Despite its apparent simplicity, consistent product performance requires tight control of critical material and process variables, including particle size distribution, bulk density matching, moisture content, and blending energy. Inadequate control can lead to segregation, poor dosing accuracy, or delayed dissolution, directly affecting rheological efficiency in concrete. When integrated with Industry 4.0 concepts such as automated weighing, inline moisture monitoring, and digital batch traceability, dry-blend manufacturing can achieve high reproducibility and cost efficiency. As a result, this manufacturing group remains a foundational production pathway for powder-based VMAs in high-volume, cost-sensitive markets where robustness and logistical simplicity are paramount.

8.1.1 Manufacturing Method and Process Control

Dry-blend powder manufacturing relies on purely physical mixing of solid constituents to produce a free-flowing, dust-controlled VMA with consistent performance at very low dosage. Polymers and auxiliaries are incorporated in powder or granular form and homogeneously distributed onto an inert mineral carrier, most commonly finely milled limestone or calcite with controlled particle size distribution. Blending is typically performed in ribbon, paddle, or ploughshare mixers under controlled residence time and shear intensity. Particular attention is given to dispersion efficiency to prevent agglomeration and delayed dissolution, commonly referred to as fish-eye formation, by ensuring adequate pre-dilution of the active polymer with carrier and controlled addition sequences during batching at the concrete plant.

8.1.2 Typical Component Architecture

- Cellulose ether, predominantly HPMC or HEMC, acting as the primary
- Viscosity-modifying and water-retention polymer.
- Set modulators such as sodium gluconate or citric acid used at low levels to stabilize workability and mitigate temperature-driven hydration variability.
- Powdered PDMS-based defoamer to limit air stabilization associated with cellulose-driven cohesion.
- Inert mineral carrier, usually milled limestone or calcite, serving as a dilution medium to ensure accurate dosing, flowability, and consistent bulk density.

8.1.3 Applicable Formulations

- 7.2.1 Cellulose Ether–Based VMA for Conventional Ready-Mix Concrete
- 7.2.3 VMA System for Pumped Concrete with Extended Transport Time
- 7.2.4 VMA Strategy for Low-Carbon and SCM-Rich Concrete
- 7.2.5 VMA for High-Fines Sand and Manufactured Sand Concrete
- 7.2.8 VMA for Lightweight Aggregate Concrete
- 7.2.10 VMA for Hot-Weather and High-Temperature Concreting
- 7.2.12 VMA for Mass Concrete with Thermal Control Requirements
- 7.2.13 VMA for Recycled Aggregate Concrete
- 7.2.16 VMA for Marine and High-Salinity Environments
- 7.2.17 VMA for Nuclear and Radiation-Shielding Concrete
- 7.2.18 VMA for Underwater and Tremie Concrete
- 7.2.21 VMA for Geopolymer and Alkali-Activated Binder Systems

8.1.4 Performance Characteristics and Limitations

Formulations in this group deliver moderate viscosity enhancement, effective water retention, and improved segregation resistance under conventional mixing and placement conditions. They are robust, tolerant to raw material variability, and well suited to decentralized ready-mix operations where simplicity and handling safety are prioritized. However, cellulose ether–dominant systems exhibit sensitivity to temperature, mixing energy, and high shear environments, which limits their effectiveness in ultra-low water-to-binder ratios, fiber-rich mixes, and extreme pumping regimes.

8.1.5 Quality Assurance and Manufacturing Considerations

Quality control focuses on carrier moisture content, bulk density, blend homogeneity, and dissolution time in water. Polymer substitution degree and particle size distribution are critical parameters influencing field performance.

From a manufacturing standpoint, these systems are well suited to standardized dry blending and automated packaging lines, with minimal need for complex process controls. Integration with Industry 4.0 platforms typically centers on batch traceability, gravimetric dosing accuracy, and statistical monitoring of blend uniformity rather than real-time rheology prediction.

8.1.6 Common Traits and Industrial Positioning

Formulations in this group are characterized by reliance on cellulose ether chemistry, inclusion of mild set retarders and defoamers, and use of limestone carriers to support dry handling and accurate dosing. They represent the most established and widely deployed VMA manufacturing route, forming the baseline against which more advanced synthetic and hybrid systems are developed and evaluated.

8.2 Group 2: Synthetic Polymer–Based Dry Blends with Dispersant Integration

Group 2 covers engineered dry blends based on synthetic polymers where dispersant functionality is incorporated directly into the powdered admixture architecture. These systems combine viscosity-modifying polymers, typically polyacrylamides or polyacrylates, with solid-state dispersants such as polycarboxylate ethers, enabling concurrent regulation of rheology, water retention, and particle dispersion within a single additive package. The design philosophy targets multifunctionality at the formulation level, reducing dependency on multiple liquid admixtures while improving robustness during handling and storage.

From a manufacturing perspective, these dry blends rely on controlled co-drying, spray granulation, or precision co-blending under Industry 4.0-enabled process monitoring to achieve uniform component distribution and minimize segregation risks. Process parameters are optimized to preserve polymer integrity and dispersant reactivity, ensuring predictable performance upon rehydration. Such admixtures are tailored for high-performance mortars, self-consolidating concrete, and binder-rich systems with elevated supplementary cementitious material content, where simultaneous water management and dispersion efficiency are critical. The integrated dry-blend approach streamlines batching operations, enhances dosing accuracy, and delivers reproducible performance in complex cementitious formulations.

8.2.1 Manufacturing Method and Formulation Strategy

Group 2 formulations are built around synthetic, water-soluble polymers as the primary rheology-controlling agents and are manufactured exclusively through high-precision dry blending. Polyacrylamides, polyacrylates, or tailored copolymers are selected with controlled molecular weight distribution and low to moderate charge density to deliver shear-thinning behavior, pressure stability, and predictable time-dependent rheology. These primary VMAs are deliberately combined with low–molecular weight polycarboxylate ether fractions acting as secondary dispersants to fine-tune yield stress, improve cement compatibility, and stabilize flow in highly optimized mixes. Blending is performed under tightly controlled shear and residence time to avoid polymer degradation and to ensure uniform distribution at very low dosage levels.

8.2.2 Typical Component Architecture

- Synthetic VMA based on polyacrylamide, polyacrylate, or modified acrylic copolymer, providing the dominant viscosity and cohesion contribution.
- Low–molecular weight PCE polymer functioning as a rheology-balancing and dispersion aid, particularly critical in high-flow and fiber-reinforced systems.
- Optional accelerator or set regulator, commonly calcium formate or formate-based blends, to support rapid strength development in precast or rapid-cycle production.
- PDMS-based powdered defoamer to control air stabilization under high shear and pumping conditions, combined with an inert mineral carrier to ensure dosing accuracy and bulk density control.

8.2.3 Applicable Formulations

- 7.2.2 Synthetic Polymer VMA for Self-Consolidating Concrete

- 7.2.6 VMA for Precast Concrete with Rapid Demolding
- 7.2.9 VMA for Ultra-High Performance Concrete
- 7.2.14 VMA for Fiber-Reinforced Concrete with High Fiber Loading
- 7.2.15 VMA for High-Alkali and Rapid-Hardening Cement Systems
- 7.2.19 VMA for High-Pressure Pumped Concrete in Long Pipelines
- 7.2.20 VMA for Ultra-Low Water-to-Binder Ratio Concrete

8.2.4 Performance Characteristics and Application Scope

These formulations deliver precise control of both yield stress and plastic viscosity in concretes operating near rheological limits. They are particularly effective in self-consolidating, ultra-low water-to-binder, and fiber-rich systems where small deviations in viscosity can lead to segregation, fiber balling, or pumping instability. The absence or minimal use of cellulose ethers reduces temperature sensitivity and thixotropic overbuild, supporting stable performance under high shear, extended pumping, and rapid casting cycles.

8.2.5 Quality Control and Manufacturing Considerations

Quality assurance emphasizes polymer molecular weight consistency, charge density control, and blend homogeneity at low active content. Flowability, bulk density, and dust behavior of the finished blend are critical for automated dosing systems. Industry 4.0 integration typically includes gravimetric control, inline particle size monitoring of carriers, and digital batch records linking polymer parameters to downstream rheological performance in concrete.

8.2.6 Common Traits and Industrial Positioning

Formulations in this group are characterized by dominance of synthetic polymer chemistry, frequent inclusion of low–molecular weight PCE co-polymers, and minimal reliance on cellulose ethers. They are engineered for advanced concrete technologies, including UHPC, SCC, precast, and high-performance pumped systems, where reproducibility, rheological precision, and compatibility with modern admixture packages are paramount.

8.3 Group 3: Hybrid Cellulose + Synthetic Polymer Blends (Dual-Polymer Systems)

Group 3 comprises dual-polymer dry blends that synergistically combine cellulose ethers with synthetic polymers to deliver balanced control over water retention, rheology, and dispersion in cementitious systems. Cellulose derivatives such as HPMC or HEC provide robust water management and open time control, while complementary synthetic polymers including polyacrylates, vinyl-based copolymers, or redispersible polymer powders contribute flow optimization, strength development, and substrate adhesion. The hybridization strategy exploits the distinct hydration and interaction mechanisms of each polymer class to achieve performance profiles that are not attainable with single-polymer systems.

Manufacturing of these hybrid blends requires tight control over particle size distribution, bulk density, and inter-polymer compatibility to prevent phase separation during storage and mixing. Advanced blending protocols supported by Industry 4.0 tools such as in-line NIR spectroscopy and digital batch traceability are increasingly applied to ensure formulation consistency and rapid scale-up. These dual-polymer systems are widely specified in tile adhesives, repair mortars, renders, and high-performance dry-mix products where tolerance to raw material variability and application robustness are critical. The hybrid approach enables formulators to fine-tune fresh-state behavior and hardened properties while maintaining operational simplicity at the job site.

8.3.1 Manufacturing Method and Formulation Strategy

Group 3 systems are engineered as dry blends integrating cellulose ethers with synthetic, water-soluble polymers to exploit synergistic rheological mechanisms that cannot be achieved by single-polymer approaches. Modified cellulose ethers deliver robust water retention, bleed control, and early-age cohesion, while synthetic VMAs contribute shear stability, pressure resistance, and pumpability under dynamic flow. Formulation design focuses on balancing molecular hydration kinetics of cellulose with the shear-responsive behavior of synthetic polymers to ensure consistent rheology across variable raw materials, transport durations, and ambient conditions. Manufacturing relies on staged dry blending with controlled energy input to prevent cellulose particle damage and to preserve synthetic polymer molecular integrity at low active dosages.

8.3.2 Typical Component Architecture

- Modified cellulose ether, typically HPMC, HEMC, or blended ether grades at 30–55%, acting as the primary water retention and cohesion provider under static and low-shear conditions.
- Synthetic VMA, commonly polyacrylamide, polyacrylate, or associative copolymers at 20–35%, delivering shear-thinning behavior, pumping stability, and resistance to segregation under high shear.
- Organic set regulators such as gluconates, citrates, or tartaric acid derivatives to moderate hydration kinetics, particularly important in SCM-rich, hot-weather, or mass concrete applications.
- PDMS-based defoamer pre-adsorbed on an inert mineral carrier to manage air stability and ensure accurate dosing, flowability, and bulk density consistency in automated systems.

8.3.3 Applicable Formulations

- 7.2.3 Pumped Concrete with Extended Transport Time
- 7.2.4 Low-Carbon and SCM-Rich Concrete
- 7.2.5 High-Fines Sand Concrete
- 7.2.8 Lightweight Aggregate Concrete
- 7.2.10 Hot-Weather Concreting
- 7.2.12 Mass Concrete
- 7.2.13 Recycled Aggregate Concrete
- 7.2.16 Marine and High-Salinity Exposure
- 7.2.17 Nuclear Shielding Concrete

- 7.2.18 Underwater and Tremie Concrete
- 7.2.21 Geopolymer Binder Systems

These formulations frequently overlap with cellulose-only solutions but are deliberately positioned in this group due to the intentional inclusion of synthetic polymers to manage variability, durability risk, and rheological drift.

8.3.4 Performance Characteristics and Application Scope

Dual-polymer blends provide a broader operational window than single-polymer systems, particularly in concretes exposed to fluctuating moisture demand, fines content, or chemical aggressiveness. Cellulose ethers stabilize fresh-state water distribution and surface quality, while synthetic polymers mitigate shear-induced viscosity loss, pressure filtration, and pumping instability. The hybrid approach is especially valuable in recycled aggregate, marine, underwater, and geopolymer concretes where ionic strength, alkali content, or binder heterogeneity can destabilize conventional VMA solutions.

8.3.5 Quality Control and Manufacturing Considerations

Quality management emphasizes tight control of cellulose substitution degree, particle size distribution, and hydration rate, alongside synthetic polymer molecular weight and charge density. Blend homogeneity is critical due to the wide difference in component bulk densities and hydration kinetics. Industry 4.0 practices increasingly include digital formulation fingerprints, inline bulk density monitoring, and batch-to-batch traceability linking polymer ratios to rheological performance indicators such as slump retention, yield stress evolution, and pumping pressure profiles.

8.3.6 Common Traits and Industrial Positioning

Group 3 formulations are defined by intentional dual-polymer design rather than additive redundancy. They supersede pure cellulose systems wherever both polymer families are present, ensuring clear technical differentiation and avoiding formulation duplication. Industrially, these blends occupy the space between robustness and precision, supporting infrastructure, energy, marine, and specialty concreting applications where environmental exposure, material variability, and operational risk demand multi-functional rheology control.

Table 8.3.6: Refinement of Polymer Group Classification

Polymer Group	Defining Criterion	Qualified Formulations
Group 1: Pure Cellulose Ether Systems	Cellulose ether as the sole rheology modifier	7.2.1
Group 3: Hybrid Dual-Polymer Systems	Mandatory inclusion of cellulose ether and synthetic polymer	7.2.3, 7.2.4, 7.2.5, 7.2.8, 7.2.10, 7.2.12, 7.2.13, 7.2.16, 7.2.17, 7.2.18, 7.2.21

8.4 Group 4: Application-Specific Printable or Spray-Optimized Blends

Group 4 addresses polymer-engineered dry blends tailored for advanced placement technologies including 3D concrete printing, robotic spraying, and wet or dry shotcrete application. These systems rely on highly customized polymer architectures that deliver tightly controlled rheological profiles, combining pronounced shear-thinning during extrusion or spraying with rapid structural rebuild immediately after deposition. The objective is to secure shape stability, interlayer adhesion, and dimensional accuracy under continuous, automated processing conditions.

From a formulation and manufacturing standpoint, these blends incorporate associative thickeners, selected cellulose derivatives, and reactive or latent modifiers that activate upon hydration. Production typically involves precision granulation, surface-functionalized powders, or hybrid spray-drying routes to ensure uniform particle morphology, predictable dissolution kinetics, and uninterrupted flow through pumps and nozzles. Industry 4.0 process control tools, including real-time moisture monitoring, digital rheology fingerprints, and closed-loop quality analytics, are increasingly applied to guarantee batch-to-batch consistency. Such materials are engineered to meet the stringent requirements of automated construction, where dispersion kinetics, moisture sensitivity, and particle integrity directly govern print fidelity, spray efficiency, and long-term structural performance.

8.4.1 Manufacturing Method and Formulation Strategy

Group 4 formulations are engineered specifically for placement methods dominated by extreme shear, rapid deformation, and impact-driven deposition, including shotcrete, sprayed concrete, and extrusion-based 3D printing. Dry blending remains the preferred manufacturing route, but with significantly tighter control of polymer molecular architecture, particle size distribution, and dosage resolution compared to conventional VMAs. Formulation strategy prioritizes rapid thixotropic rebuild after shear release, controlled structural build-up during rest, and resilience against viscosity collapse under nozzle shear and high-energy mixing. Both cellulose ethers and synthetic polymers may be employed, either individually or in hybrid form, provided they meet the stringent rheological response profile required for additive manufacturing and spray applications.

8.4.2 Typical Component Architecture

- High-purity cellulose ether or shear-responsive synthetic polymer selected for fast viscosity recovery and minimal hysteresis between shear and rest states.
- Thixotropy-modifying synthetic VMA, often associative or lightly crosslinked, designed to maintain extrusion pressure stability and filament integrity.
- Rapid-acting set or stiffening regulator, frequently aluminate-compatible accelerators or tailored organic modifiers, enabling early shape retention without premature nozzle blockage.
- PDMS-based or advanced polymeric defoamer supported on a fine mineral carrier to manage air entrainment under dynamic mixing, high turbulence, and impact spraying conditions.

8.4.3 Applicable Formulations

- 7.2.7 VMA for Shotcrete and Sprayed Concrete
- 7.2.11 VMA for 3D Printed and Extrudable Cementitious Materials

These systems are classified independently of polymer family due to their dominant dependence on placement mechanics and rheological response rather than solely on chemical composition.

8.4.4 Performance Characteristics and Application Scope

Group 4 blends deliver a highly tuned balance between pumpability, extrudability, and immediate structural stability after placement. In shotcrete, they support adhesion to substrates, reduced rebound, and uniform layer build-up under high-velocity impact. In 3D printing and extrusion, they enable filament continuity, high green strength, and dimensional accuracy across successive layers. The emphasis on thixotropic control and shear resilience distinguishes these formulations from conventional VMAs used in cast or pumped concrete.

8.4.5 Quality Control and Manufacturing Considerations

Quality assurance focuses on thixotropic index reproducibility, shear recovery time, and air content stability under simulated spraying or extrusion conditions. Minor deviations in polymer molecular weight or substitution pattern can lead to nozzle pressure spikes, filament tearing, or layer collapse. Industry 4.0 practices increasingly include rheological fingerprinting using rotational rheometry, digital tracking of extrusion pressure profiles, and closed-loop feedback between blend composition and in-process performance metrics.

8.4.6 Common Traits and Industrial Positioning

Formulations in this group are defined by functional placement requirements rather than polymer taxonomy. Although they may chemically resemble Group 2 or

Group 3 systems, their design intent centers on thixotropy, rapid structural build-up, and resistance to extreme shear environments. Industrially, Group 4 occupies a strategic role in automated construction, robotic spraying, and digital fabrication, supporting the transition of cementitious materials toward Industry 4.0-enabled manufacturing and placement technologies.

8.5 Group 5: Classification by Manufacturing Method and Processing Strategy

Group 5 consolidates viscosity-modifying admixtures according to manufacturing method and processing strategy, translating formulation concepts into an actionable industrial framework. This grouping aligns each admixture architecture with its most suitable production route by correlating chemical composition, targeted performance attributes, and scalability constraints. By focusing on manufacturing logic rather than solely on chemistry, the classification supports informed selection of process technologies that ensure reproducible quality and economic viability at commercial scale.

The framework organizes products based on shared processing pathways, including conventional dry blending, co-processed dispersant integration, hybrid polymer compounding, and application-conditioned systems for printing or spraying. Such alignment highlights process-sensitive quality attributes such as particle size distribution, bulk density control, moisture tolerance, and segregation resistance. With increasing adoption of Industry 4.0 tools such as digital batch genealogy, in-line process analytics, and predictive maintenance, this manufacturing-centered grouping strengthens consistency, shelf stability, and field performance. The approach facilitates efficient technology transfer, structured scale-up, and resilient supply chain management across the viscosity-modifying admixture value chain.

8.5.1 Rationale for Manufacturing-Oriented Classification

The final grouping consolidates all VMA formulations based on shared manufacturing logic, raw material handling constraints, and formulation architecture rather than end-use alone. Dry blending is the common production platform across all groups, yet polymer chemistry, sensitivity to shear and residence time, and interaction with auxiliary components impose distinct processing windows. Grouping by manufacturing method enables standardization of blending protocols, carrier selection, quality control metrics, and digital production tracking, while still preserving application-specific performance differentiation. This approach supports scalable production, minimizes formulation redundancy, and aligns formulation design with Industry 4.0-enabled manufacturing control.

8.5.2 Overview of Manufacturing Groups and Formulation Allocation

Table 8.5.2: Overview of Manufacturing Groups and Formulation Allocation

Group	Manufacturing Method	Defining Formulation Architecture	Formulation IDs
A. Pure Cellulose Ether Dry Blends	Single-polymer dry blend using HPMC or HEMC with set regulator, defoamer, and mineral carrier	Cellulose ether as the sole rheology modifier	7.2.1
B. Synthetic Polymer + PCE Dispersant Blends	Dry blends dominated by synthetic VMAs combined with low–molecular weight PCE fractions	Synthetic polymer-driven viscosity with dispersant-assisted flow control	7.2.2, 7.2.6, 7.2.9, 7.2.14, 7.2.15, 7.2.19, 7.2.20
C. Hybrid Dual-Polymer Blends	Dry blends integrating cellulose ethers and synthetic polymers	Multi-functional stability through complementary polymer mechanisms	7.2.3, 7.2.4, 7.2.5, 7.2.8, 7.2.10, 7.2.12, 7.2.13, 7.2.16, 7.2.17, 7.2.18, 7.2.21
D. Thixotropic	Dry blends engineered for	Placement-driven rheology for	7.2.7, 7.2.11

and Shear-Optimized Blends	extreme shear and rapid structural rebuild	spraying and extrusion

8.5.3 Manufacturing and Quality Control Implications

Each group imposes distinct requirements on blending energy, sequence of component addition, and carrier optimization. Pure cellulose systems demand strict control of particle integrity and hydration rate, while synthetic polymer-dominated blends require molecular weight preservation and dispersion uniformity at very low dosage. Hybrid systems introduce additional complexity due to differing bulk densities and hydration kinetics, necessitating advanced homogeneity control. Thixotropic and shear-optimized blends require the narrowest tolerance windows, as small deviations can translate directly into nozzle pressure instability or buildability failure. Across all groups, Industry 4.0 tools such as gravimetric dosing, digital batch records, and rheological fingerprint databases enable consistent production and rapid troubleshooting.

8.5.4 Industrial Value and Standardization Benefits

This manufacturing-centric classification provides a practical framework for portfolio rationalization, plant layout optimization, and technology transfer between production sites. By aligning formulation families with defined processing envelopes, manufacturers can reduce changeover risk, simplify quality audits, and accelerate scale-up of new products. The structure also facilitates data-driven formulation management, allowing performance feedback from field applications to be directly linked to manufacturing parameters and polymer specifications, supporting continuous improvement in high-performance and specialty concrete technologies.

9. Ready-to-Use Formulation Checklists for Viscosity-Modifying Admixtures

9.1 Purpose and Scope of Formulation Checklists

The formulation checklists provided here translate polymer selection principles and manufacturing constraints into actionable tools for industrial development, scale-up, and routine production of VMAs. Each checklist is structured to support rapid validation of raw material compatibility, blending robustness, and application performance prior to pilot or full-scale manufacturing. The intent is to reduce formulation risk, accelerate technology transfer, and ensure reproducibility across plants, raw material sources, and operating conditions. Integration of digital quality metrics enables alignment with Industry 4.0 production environments.

9.2 Checklist A: Pure Cellulose Ether Dry Blend VMAs

Polymer Selection

☐ HPMC or HEMC grade with controlled substitution pattern and narrow viscosity tolerance
☐ Verified compatibility with cement alkali content and sulfate balance
☐ Low sensitivity to temperature-driven hydration variability

Auxiliary Components

☐ Organic set regulator matched to desired open time and early strength development
☐ PDMS-based powdered defoamer with proven air-release efficiency
☐ Limestone or inert mineral carrier with consistent particle size and moisture content

Manufacturing Controls

☐ Low-shear dry blending to preserve cellulose particle integrity

☐ Controlled blending sequence to avoid segregation or demixing

☐ Bulk density and flowability validated for automated dosing

Performance Verification

☐ Water retention and bleed resistance under static conditions

☐ Slump and surface finish consistency

☐ Absence of excessive thixotropic overbuild

9.3 Checklist B: Synthetic Polymer + PCE Dispersant Blends

Polymer Architecture

☐ Synthetic VMA with defined molecular weight distribution and charge density

☐ Low–molecular weight PCE fraction optimized for rheology balancing

☐ Proven compatibility with modern superplasticizer systems

Formulation Balance

☐ Viscosity control achieved without yield stress overdevelopment

☐ Shear-thinning behavior maintained under pumping and mixing

☐ Minimal temperature sensitivity

Manufacturing Controls

☐ Tight control of blending energy and residence time

☐ Homogeneous distribution at very low active polymer dosage

☐ Dust suppression and flowability suitable for gravimetric feeders

Performance Verification

☐ Pumping pressure stability over extended distances

☐ Segregation resistance in ultra-low water-to-binder systems

☐ Reproducible rheological profile across batches

9.4 Checklist C: Hybrid Cellulose + Synthetic Polymer VMAs

Polymer Synergy

☐ Cellulose ether selected for water retention and cohesion

☐ Synthetic polymer tuned for shear stability and pressure resistance

☐ Balanced hydration kinetics between both polymer families

Formulation Robustness

☐ Stable performance across variable fines content and aggregate moisture

☐ Resistance to ionic strength and alkali fluctuations

☐ Controlled interaction with SCMs and recycled materials

Manufacturing Controls

☐ Multi-stage dry blending to manage density and particle size differences

☐ Segregation risk assessment during transport and storage

☐ Digital batch tracking of polymer ratios and specifications

Performance Verification

☐ Slump retention and yield stress evolution over time

☐ Pumpability and placement stability under variable conditions

☐ Consistent air content and surface quality

9.5 Checklist D: Thixotropic and Shear-Optimized VMAs

Rheological Design Targets

☐ Rapid viscosity recovery after high shear

☐ Controlled structural build-up during rest

☐ Minimal hysteresis between shear and rebuild cycles

Polymer and Additive Selection

☐ Cellulose or synthetic polymer with fast thixotropic response

☐ Optional accelerator or stiffening agent compatible with nozzle systems

☐ Advanced defoamer effective under turbulent and impact placement

Manufacturing Controls

☐ Narrow tolerance windows for polymer dosage and particle size

☐ Verification of blend uniformity at micro-dosage levels

☐ Storage stability under vibration and handling stress

Performance Verification

☐ Nozzle pressure stability during spraying or extrusion

☐ Layer buildability and green strength development

☐ Reduced rebound in shotcrete or filament collapse in printing

9.6 Digital and Industry 4.0 Integration Checklist

☐ Gravimetric dosing and automated blending control

☐ Digital formulation fingerprints linked to rheological data

☐ Inline bulk density and flow monitoring

☐ Traceability from polymer batch to field performance feedback

☐ Closed-loop optimization using production and application data

10. Ready-Mix Plant Implementation and Process Control

Successful deployment of viscosity-modifying admixtures in ready-mix concrete plants depends less on formulation chemistry alone and more on disciplined process control at the plant level. VMAs interact dynamically with mixing energy, water availability, superplasticizer action, and time-dependent hydration, making their performance highly sensitive to dosing strategy and operational variability. Effective implementation therefore requires clear rules for addition sequence, defined rheological targets, and feedback-driven adjustment mechanisms that align plant practice with formulation design intent. When embedded into digitally assisted plant operations, VMA process control becomes a stabilizing tool rather than a corrective measure.

10.1 Dosing Strategies and Addition Sequences

The performance of viscosity-modifying admixtures is governed not only by molecular design but also by the precision of dosing strategy and the sequence of addition during mixing. Controlled introduction of VMAs directly influences polymer hydration, dispersion kinetics, and interaction with other chemical admixtures, most notably polycarboxylate ether superplasticizers. Improper timing or dosing often results in incomplete activation, excessive viscosity buildup, or competitive adsorption effects that compromise rheological stability and workability retention.

From an operational standpoint, dosing protocols must be aligned with mixing energy, water availability, and equipment configuration. Central mixer addition typically favors early VMA incorporation under high shear to promote uniform dispersion, whereas in-truck or on-site dosing requires delayed or pre-dissolved addition to avoid agglomeration. The relative timing of VMA addition with respect to cement, aggregates, and mixing water is critical, particularly in systems with high fines content or elevated supplementary cementitious materials, where adsorption dynamics are intensified. Structured sequencing minimizes antagonistic interactions and ensures predictable yield stress and plastic viscosity development.

Advanced strategies increasingly employ split dosing, staged activation, or on-demand correction to accommodate raw material variability and transport-induced changes. Integration of Industry 4.0-enabled dosing systems, incorporating real-time rheology sensing, automated feedback loops, and digital batch traceability, allows dynamic optimization of VMA dosage under fluctuating production conditions. Such approaches enhance robustness, reduce overdosage risk, and support consistent fresh-state performance in both conventional and high-performance concrete applications.

10.1.1 Fundamental Dosing Principles

VMA dosing in ready-mix production must account for the low active content of polymeric rheology modifiers and their non-linear impact on yield stress and plastic viscosity. Small absolute dosing errors can result in disproportionate rheological shifts, particularly in low water-to-binder and high-performance concretes. Dosing systems must therefore prioritize accuracy, repeatability, and compatibility with plant automation. Gravimetric dosing is preferred over volumetric systems, especially for synthetic and hybrid VMAs, to minimize drift caused by bulk density variation or moisture uptake. Where manual dosing is unavoidable, formulations should be selected with wider tolerance windows and lower sensitivity to over-dosage.

10.1.2 Central Mixer versus Truck Addition

Addition of VMAs in a central mixer provides the highest level of process control and reproducibility. Central mixing allows defined shear input, controlled residence time, and uniform dispersion of polymeric components before discharge, which is critical for achieving predictable rheological baselines. This approach is particularly advantageous for synthetic polymer and hybrid VMA systems that require precise dispersion to develop shear stability and pumping resistance. Truck addition offers operational flexibility and retrofit capability but introduces variability linked to drum efficiency, loading sequence, and residual material. It is most suitable for cellulose-dominated or thixotropy-tolerant formulations specifically designed to accommodate dispersion uncertainty.

10.1.3 Timing Relative to Water Addition

The point at which VMAs encounter free water governs their hydration kinetics and initial rheological impact. Early addition with batch water promotes uniform polymer wetting and interaction with fines, enhancing water retention and segregation resistance. This strategy is commonly used for cellulose-based systems but increases the risk of early viscosity build-up under high temperatures or extended mixing. Delayed addition, after partial wetting of cement and aggregates, moderates initial thickening and supports better control of slump development. Selection of early or delayed water contact should be formulation-specific and validated against time-dependent rheological evolution.

10.1.4 Interaction with PCE Superplasticizers

The sequence of VMA and PCE addition critically influences dispersion efficiency and viscosity development. When VMAs are added before or simultaneously with PCE, polymer adsorption competition can reduce dispersant effectiveness, leading to higher yield stress at equivalent PCE dosage. Adding VMAs after PCE establishment allows particle dispersion to occur first, followed by controlled viscosity build-up, which is preferred for synthetic polymer systems in high-flow concretes. Split-dosing strategies, where a portion of VMA is added early and the remainder after PCE, are sometimes employed to balance stability and workability retention in long-haul or high-temperature scenarios.

10.1.5 Split Dosing and On-Demand Adjustment

Split dosing offers a practical method to manage variability in haul time, temperature, and placement schedule. An initial base dosage establishes segregation resistance and cohesion, while a secondary adjustment at the plant or jobsite fine-tunes rheology based on real-time observations. This approach requires VMAs with predictable incremental response and low risk of sudden viscosity spikes. On-demand adjustment protocols should be standardized, documented, and supported by operator training to avoid ad hoc corrections that compromise quality consistency.

10.1.6 Integration with Plant Automation and Industry 4.0

Advanced ready-mix plants increasingly integrate VMA dosing into automated control systems linked to predefined rheological targets. Manufacturing Execution Systems can store formulation-specific dosing windows, preferred addition sequences, and allowable adjustment ranges. Real-time data from mixer power consumption, batch weight verification, and slump flow measurements can be used to validate correct execution and trigger alerts when deviations occur. Such digital integration transforms VMA dosing from a manual intervention into a controlled process variable, enhancing consistency, traceability, and scalability across production sites.

10.2 Robustness to Operational Variability

Robust VMA performance in ready-mix operations is ultimately defined by the ability of the polymer system to tolerate unavoidable fluctuations in production and delivery conditions without inducing excessive corrective intervention. Variability arises from ambient temperature swings, inconsistent haul durations, changes in mixer efficiency, and batch-to-batch differences in fines and moisture. From a technology transfer perspective, VMAs intended for industrial deployment must therefore be evaluated not only under ideal laboratory conditions but also across realistic operating envelopes that reflect plant-level uncertainty. Integration of robustness criteria into formulation selection reduces dependence on operator judgment and supports scalable, multi-plant implementation aligned with digital production control.

10.2.1 Temperature, Haul Time, and Mixer Energy Effects

Thermal Sensitivity and Polymer Hydration Kinetics

Ambient and material temperature variations exert a dominant influence on the field performance of viscosity-modifying admixtures by directly controlling polymer hydration rate, molecular mobility, and adsorption kinetics at cement and fines surfaces. Elevated temperatures accelerate water diffusion into polymer domains and intensify early-stage cement reactions, which can drive rapid increases in yield stress and apparent viscosity shortly after mixing. Cellulose ethers and strongly hydrophilic synthetic VMAs are especially prone to this behavior, often exhibiting steep viscosity ramps that challenge workability retention during hot-weather production. More robust formulations rely on associative or hybrid polymer architectures with moderated hydration rates and distributed hydrophobic domains, allowing viscosity to develop progressively rather than abruptly. From an industrial standpoint, such controlled thermal sensitivity reduces the need for seasonal re-optimization and supports consistent performance across geographically dispersed plants.

Haul Time Effects and Shear History Accumulation

Extended haul times expose fresh concrete to prolonged low-intensity shear while hydration and flocculation processes continue to evolve. Under these conditions, VMAs that rely on irreversible adsorption or strong network formation may show cumulative viscosity drift, leading to gradual slump loss that cannot be easily corrected without overdosing superplasticizer or water. Robust systems instead exhibit reversible shear thinning combined with partial structural rebuilding, enabling the material to recover cohesion without excessive stiffening. Polymer features that promote this behavior include balanced molecular weight distributions, flexible backbones, and side-chain architectures that favor transient associations over permanent network formation. Such designs are particularly advantageous in urban delivery scenarios with unpredictable traffic patterns, where rheological stability over time is a primary quality requirement.

Mixer Energy as a Rheological Activation Variable

Mixer energy defines the effectiveness with which VMA chemistry is translated into macroscopic rheological control. High-energy central mixers provide uniform dispersion and sufficient shear to fully activate both soluble and associative polymer segments, resulting in predictable viscosity development and reduced sensitivity to batch-to-batch variability. In contrast, truck mixers operate across a wide and often lower energy spectrum, which can limit activation of certain synthetic or hybrid VMAs and delay their rheological contribution. Industrially robust VMAs are therefore designed with low activation thresholds and rapid dispersion characteristics, ensuring functional performance even under suboptimal mixing conditions. When mixer power consumption is digitally monitored and correlated with rheological outcomes, this parameter becomes a controllable variable within Industry 4.0 frameworks, enabling early detection of deviation and reinforcing consistency without reliance on manual adjustment.

10.2.2 MES-Linked Rheology Targets

Digital Definition of Rheological Acceptance Windows

Manufacturing Execution Systems enable rheological control to be formalized as a set of digitally defined acceptance windows rather than fixed, recipe-based dosages. For VMAs, this approach recognizes that equivalent performance can be achieved through slightly different dosage levels depending on temperature, raw material variability, and mixing history. Rheological targets such as slump, slump flow diameter, or mixer power draw ranges can be encoded as permissible operating bands linked to specific mix designs. Once these windows are established through validation trials, the MES acts as a gatekeeper, confirming compliance at the batch level and flagging deviations before concrete leaves the plant. This structured definition of acceptability reduces subjective decision-making and supports consistent quality across shifts and production sites.

Translation of Formulation Intent into Executable Control Rules

MES platforms provide a mechanism for embedding formulation knowledge directly into plant operations. VMA suppliers and concrete producers can jointly define rules governing addition sequence, dosage adjustment limits, and temperature or haul-time compensation factors, all of which are stored as formulation-specific logic. For example, associative synthetic VMAs may be assigned narrower adjustment increments due to their sharper dose–response behavior, while cellulose-based systems may be allowed wider correction bands. By converting laboratory-derived rheological behavior into executable digital rules, MES-linked control preserves formulation intent during scale-up and routine production. This approach is central to effective technology transfer, as it ensures that polymer design assumptions remain valid under industrial conditions.

Feedback Loops and Continuous Performance Refinement

When MES platforms are connected to real-time or near-real-time rheological indicators, they enable closed feedback loops that continuously refine VMA application strategies. Data from slump measurements, mixer power consumption, and batch weight verification can be captured and correlated with final placement performance. Over time, statistical analysis of this data supports recalibration of rheology targets and adjustment rules, improving robustness without increasing admixture consumption. Such data-driven refinement aligns with Industry 4.0 principles by transforming operational variability into a source of process intelligence. In mature implementations, MES-linked rheology control shifts VMAs from corrective additives to proactive stabilizers of fresh concrete behavior, reinforcing consistency and scalability in ready-mix operations.

10.2.3 Real-Time Feedback from Slump Flow and Mixer Power Sensors

Sensor-Based Rheology Indicators in Fresh Concrete Production

Real-time feedback from slump flow measurements and mixer power sensors provides a practical bridge between laboratory rheology and plant-level process control. Slump flow reflects the combined effects of yield stress, plastic viscosity, and thixotropic rebuilding, making it a robust field indicator of fresh concrete consistency. Mixer power consumption, torque, and rotational resistance offer continuous, indirect measurement of internal resistance to shear, correlating closely with viscosity evolution and structural build-up when properly calibrated. Together, these indicators allow rheological behavior to be monitored dynamically rather than inferred post hoc, enabling earlier detection of deviations linked to VMA performance, moisture variation, or dosing error. Their combined use is increasingly viewed as a foundational element of digitally assisted ready-mix production.

Closed-Loop Control and Controlled On-Demand Adjustment

When integrated with automated dosing and MES platforms, sensor-derived rheological signals support closed-loop control strategies for VMA application. Predefined decision rules translate measured slump flow or mixer power values into limited, incremental dosage adjustments that remain within validated safety margins. Such control logic prevents over-correction and avoids abrupt viscosity changes that compromise pumpability or finishing. For these systems to function reliably, VMAs must exhibit smooth, monotonic dose–response behavior with minimal hysteresis under shear. From a formulation perspective, polymers with gradual associative mechanisms and reversible structural interactions are therefore better suited to sensor-driven adjustment than highly reactive or sharply thickening systems.

Data Integrity, Calibration, and Industrial Scalability

The industrial value of real-time rheology feedback depends on sensor reliability, calibration discipline, and data integrity. Mixer power signals must be normalized for mixer type, batch size, and rotational speed, while slump flow measurements require standardized procedures to ensure comparability across plants. Once validated, these data streams can be archived and analyzed to refine rheology targets, dosage rules, and robustness criteria over time. In an Industry 4.0 context, accumulated sensor data becomes a strategic asset, supporting predictive analytics and continuous improvement of VMA deployment. Such data-driven scalability reduces dependence on individual operator judgment and reinforces consistent fresh concrete performance under variable operating conditions.

11. Measurement, Validation, and Quality Control Strategies

Measurement, validation, and quality control form the operational backbone of viscosity-modifying admixture technology, translating polymer design intent into predictable field performance. Because VMAs function at low dosages yet govern yield stress, plastic viscosity, and thixotropic evolution, even minor deviations in polymer architecture, blending quality, or cement chemistry can propagate into significant placement risk. A rigorous measurement framework is therefore required to connect raw material specifications, dry-blend manufacturing control, and rheological performance under realistic processing and placement conditions, enabling reproducibility, scalability, and safe technology transfer.

Beyond compliance testing, modern quality control strategies increasingly serve as predictive tools within data-driven production environments. Integration of rheological characterization, process analytics, and field feedback allows manufacturers to detect early performance drift, manage raw material variability, and continuously optimize formulations. When embedded into Industry 4.0 systems, measurement and validation evolve from passive checkpoints into active control mechanisms that safeguard performance in high-shear, low-margin, and application-critical concrete technologies.

11.1 Strategic Role of Measurement in VMA Technology

Measurement and validation define the reliability boundary of viscosity-modifying admixtures from formulation development through industrial deployment. VMAs operate at low dosage yet exert disproportionate influence on yield stress, plastic viscosity, and thixotropic recovery, making them highly sensitive to small variations in polymer architecture, blending quality, and cement chemistry. Robust quality control frameworks must therefore link polymer-level parameters to rheological performance under realistic processing and placement conditions. Data-driven measurement strategies enable predictive control, reduce batch-to-batch variability, and support continuous improvement within Industry 4.0 manufacturing environments.

11.2 Raw Material Characterization and Incoming Quality Control

Polymer-Level Metrics

- Molecular weight distribution by GPC or SEC, with tight control of polydispersity
- Degree of substitution and substitution pattern for cellulose ethers via NMR or titration
- Charge density and functional group content for synthetic polymers
- Moisture content and thermal stability of polymer powders

Auxiliary Component Verification

- Particle size distribution and surface area of mineral carriers
- Active content and adsorption efficiency of powdered defoamers
- Purity and ionic contribution of set regulators

Incoming raw material data should be digitized and linked to downstream performance records to enable traceability and root-cause analysis.

11.3 In-Process Manufacturing Control

Dry Blending Validation

- Blend homogeneity assessment using tracer-based or spectroscopic methods
- Bulk density and flowability monitoring for automated dosing reliability
- Control of blending energy, sequence, and residence time

Process Stability Indicators

- Statistical process control on key blend parameters
- Alarm thresholds for density drift and segregation risk
- Batch record integration with polymer specification databases

Advanced plants increasingly employ inline sensors and digital twins of the blending process to predict performance deviations before release.

11.4 Rheological Performance Measurement

Fresh-State Concrete Testing

- Yield stress and plastic viscosity determination using rotational rheometers
- Slump flow, V-funnel, and L-box tests adapted to VMA-sensitive systems
- Thixotropic rebuild measurement through shear–rest–shear protocols

Dynamic Placement Simulation

- Pumping pressure loss measurements in loop systems
- Nozzle shear and extrusion force monitoring for spray and printing applications
- Segregation resistance under pressure filtration

Correlation between laboratory rheometry and field-relevant tests is essential for meaningful validation.

11.5 Time-Dependent and Environmental Stability Assessment

Open Time and Slump Retention

- Rheological evolution over extended rest and agitation periods
- Sensitivity to temperature and humidity variation

Chemical Robustness

- Performance under high-alkali, high-sulfate, or saline conditions
- Compatibility with SCMs, recycled aggregates, and alternative binders

These evaluations support reliable performance in variable and aggressive environments.

11.6 Field Validation and Feedback Integration

Application-Level Verification

- Pumping stability, surface finish, and placement quality

- Air content stability during transport and placement
- Rebound, buildability, or filament integrity in dynamic applications

Digital Feedback Loops

- Linking field data to formulation and manufacturing parameters
- Continuous refinement of specification limits
- Predictive maintenance and formulation optimization using historical data

11.7 Release Criteria and Continuous Improvement

Product Release Framework

- Defined acceptance windows for rheological and physical parameters
- Deviation handling protocols and corrective action pathways
- Periodic revalidation following raw material or process changes

Continuous Improvement Tools

- Multivariate analysis linking polymer chemistry to performance outcomes
- Machine learning models for anomaly detection and optimization
- Cross-plant benchmarking supported by standardized data structures

12. VMA Performance in Concrete Applications

The functional contribution of viscosity-modifying admixtures varies significantly across concrete applications, as each system imposes distinct rheological, placement, and stability requirements. VMAs do not act as universal thickeners but as application-specific rheology regulators that balance flowability, cohesion, and segregation resistance under defined shear and boundary conditions. Effective industrial use therefore requires aligning polymer chemistry, molecular architecture, and dose–response behavior with the dominant deformation mode of each application, whether gravity-driven flow, pressure-induced pumping, or extrusion-controlled shaping. From a technology transfer standpoint, performance must be evaluated in application-relevant test configurations rather than relying solely on generic laboratory metrics.

12.1 Self-Consolidating Concrete (SCC)

Rheological Stabilization of Highly Flowable Systems

Self-consolidating concrete relies on extremely low yield stress to achieve self-leveling and complete form filling under gravity, which inherently increases susceptibility to segregation and bleeding. VMAs are therefore applied as stabilizing agents that raise plastic viscosity without materially increasing yield stress, preserving flow while controlling phase separation. In SCC formulations with high paste volume and elevated fines content, even small variations in VMA dosage can shift the balance between stability and passing ability. Industrially suitable VMAs exhibit smooth dose–response behavior and limited sensitivity to shear history, allowing rheological targets to be maintained within narrow acceptance windows defined by slump flow and visual stability indices.

Interaction with Superplasticizers and Fines

SCC typically operates at high polycarboxylate ether dosage, making VMA–PCE interaction a critical design variable. Early or simultaneous addition of VMAs may compete with PCE adsorption on cement and filler surfaces, reducing dispersing efficiency and increasing apparent yield stress. For this reason, synthetic and hybrid VMAs are commonly introduced after initial dispersion has been established, enabling controlled viscosity build-up without sacrificing flow spread. The nature and quantity of fines, including limestone fillers and supplementary cementitious materials, further influence VMA effectiveness by modifying available surface area and water demand. Robust SCC systems are formulated to tolerate fines variability without requiring frequent dosage adjustment.

Robustness, Placement Performance, and Digital Control

Field performance of SCC is judged by its ability to maintain uniform flow, pass through congested reinforcement, and achieve defect-free surface finish without mechanical vibration. VMAs contribute by dampening segregation during transport and placement while supporting stable rheology over time. Polymers with associative thickening mechanisms are increasingly favored due to their reversible shear thinning and limited irreversible build-up under extended mixing. When integrated into MES-driven plant control, SCC VMA dosing can be linked to slump flow targets, mixer power ranges, and temperature compensation rules. This Industry 4.0–aligned approach reduces reliance on visual assessment and enhances reproducibility across production sites.

12.2 Pumped Concrete

Rheology Control under High Shear and Pressure

Pumped concrete is subjected to sustained high shear rates and elevated pressure gradients that fundamentally alter paste structure during transport through pipelines. VMAs in this application function to stabilize the lubricating paste layer that forms along the pipe wall while preventing filtration of water into the aggregate skeleton. Adequate plastic viscosity is required to maintain a continuous lubrication layer, yet excessive thickening increases pumping pressure and energy demand. Industrially effective VMAs therefore exhibit pronounced shear-thinning behavior, allowing viscosity to decrease under pumping conditions and recover rapidly after pressure release at discharge.

Lubrication Layer Formation and Filtration Resistance

The efficiency of concrete pumping is strongly dependent on the quality and stability of the lubrication layer composed of fine particles, cement paste, and polymer-modified water. VMAs enhance cohesion within this layer, reducing water migration and minimizing the risk of blockage or pressure spikes. Polymers with reversible associative interactions are particularly effective, as they reinforce the paste network under low shear while disengaging under high shear to lower flow resistance. Overly strong network-forming VMAs or high molecular weight systems can disrupt this balance, leading to elevated line pressure and reduced pumping reliability, especially over long distances or vertical rises.

Operational Robustness and Digital Monitoring

Pumping performance is sensitive to variations in mix temperature, moisture content, and shear history, making robustness a critical selection criterion for VMAs. Formulations intended for pumped concrete should tolerate moderate variability without sharp changes in rheology that force on-site correction. Increasingly, mixer power consumption and pump pressure data are monitored and correlated with rheological behavior to support predictive control. When integrated into digital plant and site monitoring systems, these data streams enable early identification of abnormal viscosity development and facilitate proactive adjustment within validated limits. Such Industry 4.0–aligned strategies position VMAs as enablers of reliable, energy-efficient pumping rather than reactive problem-solving additives.

12.3 Shotcrete (Sprayed Concrete)

Cohesion Control during Projection and Impact

Shotcrete is characterized by extreme shear conditions during pumping and spraying, followed by an abrupt transition to near-zero shear upon impact with the substrate. VMAs play a critical role in maintaining sufficient cohesion during transport while enabling rapid structural build-up immediately after placement. During projection, the admixture must support stable flow through hoses and nozzles without excessive pressure loss or pulsation. Upon impact, rapid viscosity recovery and yield stress increase are required to minimize rebound, reduce material loss, and ensure adhesion to vertical or overhead surfaces. Industrially effective VMAs therefore exhibit pronounced thixotropic behavior with fast reversible structuring rather than slow, irreversible thickening.

Interaction with Accelerators and Early Stiffening Systems

In most wet-mix shotcrete systems, VMAs operate in combination with alkali-free or alkaline accelerators that trigger rapid setting at the nozzle. The polymer chemistry must be compatible with these accelerator systems to avoid premature thickening in the hose or loss of accelerator efficiency at impact. Cellulose ethers are widely used due to their strong water retention and thickening capacity, which improve adhesion and reduce dust generation. Synthetic and hybrid VMAs with tailored associative mechanisms are increasingly applied to achieve finer control over pumpability and early build-up, particularly in high-performance shotcrete. Compatibility testing under realistic spraying conditions is essential, as laboratory rheology alone does not capture the complex shear and reaction environment at the nozzle.

Rebound Reduction, Layer Build-Up, and Digital Support

Effective VMA selection directly influences rebound rate, achievable layer thickness, and surface quality in sprayed concrete. Excessively low viscosity leads to high rebound and poor build-up, while overly strong thickening increases pumping pressure and risks hose blockage. Robust shotcrete VMAs strike a balance by providing controlled cohesion during spraying and rapid structural recovery after deposition. Emerging Industry 4.0 practices integrate pump pressure monitoring, nozzle velocity data, and rebound measurement into digital feedback systems that support consistent application quality. In such frameworks, VMAs function as predictable rheology regulators that enable stable, repeatable spraying performance across variable site conditions.

12.4 Grouts for Post-Tensioning and Repair

Stability Control in Low-Solids Cementitious Systems

Post-tensioning and repair grouts operate at relatively low solids content and high fluidity, making them inherently prone to bleeding, sedimentation, and particle separation during placement. VMAs are introduced primarily to stabilize the cementitious suspension while preserving sufficient flow for injection through narrow ducts, tendons, or cracks. Effective grout VMAs increase plastic viscosity without inducing excessive yield stress, allowing uniform filling over long distances and complex geometries. Industrial formulations favor polymers that provide strong anti-bleeding performance at very low dosage levels, as even modest overdosing can sharply increase injection pressure and compromise filling completeness.

Injectability, Filtration Resistance, and Compatibility

Injectability is a defining performance criterion for grouts, requiring predictable rheology under low to moderate shear and sustained pressure. VMAs enhance filtration resistance by limiting water migration ahead of the grout front, reducing the risk of voids and incomplete encapsulation of steel elements. Synthetic VMAs with narrow molecular weight distributions and controlled hydration kinetics are particularly suited to this application, as they offer consistent performance and reduced sensitivity to temperature and mixing variability. Compatibility with expansive agents, corrosion inhibitors, and shrinkage-compensating systems must be carefully validated, as adverse interactions can negate the stabilizing effect of the VMA or alter setting behavior.

Field Robustness, Quality Assurance, and Digital Oversight

Grout placement often occurs under constrained site conditions with limited opportunities for real-time correction, elevating the importance of formulation robustness. VMAs selected for post-tensioning and repair must tolerate variability in water quality, temperature, and mixing energy without sudden rheological shifts. Increasingly, injection pressure, flow rate, and volume balance data are digitally logged and evaluated as indirect indicators of grout rheology and stability. When aligned with predefined acceptance criteria, these data streams support Industry 4.0–oriented quality assurance, ensuring that VMA performance contributes to durable corrosion protection and long-term structural integrity rather than short-term workability alone.

12.5 3D Printing and Extrudable Systems

Rheological Window for Extrusion and Shape Retention

3D printing and other extrudable cementitious systems impose a narrow and highly specific rheological window that differs fundamentally from conventional cast concrete. The material must exhibit low dynamic viscosity under shear to enable continuous extrusion, combined with high static yield stress and rapid structural build-up immediately after deposition to preserve geometric fidelity. VMAs are central to achieving this balance, as they regulate both flow during extrusion and stiffness recovery between successive layers. Polymers with pronounced thixotropic behavior and reversible network formation are therefore preferred, as they allow repeated shear-induced breakdown followed by fast rebuild without irreversible viscosity drift.

Buildability, Interlayer Bonding, and Process Stability

Beyond initial extrusion, VMAs influence buildability, defined as the ability of printed layers to support subsequent layers without collapse or excessive deformation. Adequate viscosity and yield stress development must occur within seconds to minutes, depending on print speed and layer height. At the same time, excessive early stiffening can impair interlayer bonding, leading to anisotropic mechanical performance. VMA chemistry must therefore be co-optimized with hydration control, particle packing, and accelerator systems to synchronize structural build-up with printing kinematics. Industrially viable formulations prioritize predictable time-dependent rheology over maximum thickening efficiency, ensuring repeatable performance across extended print runs.

Digital Integration and Industry 4.0 Alignment

3D printing inherently operates within a digitally controlled environment, making predictable VMA response a prerequisite for automation. Extrusion pressure, screw torque, print speed, and layer geometry are continuously monitored and adjusted in real time. VMAs used in these systems must exhibit consistent dose–response behavior that can be reliably embedded into control algorithms without risk of sudden rheological transitions. Sensor data collected during printing can be analyzed to refine rheology models and dosage rules, enabling closed-loop control of material behavior. Within an Industry 4.0 framework, VMAs thus function not only as material modifiers but as enablers of digitally driven construction processes that link formulation design directly to manufacturing execution.

12.6 Ultra-High-Performance Concrete (UHPC)

Rheology Management in Extremely Dense Particle Systems

Ultra-high-performance concrete is characterized by very low water-to-binder ratios, exceptionally high fines content, and intensive use of high-range water reducers. These features create a rheological system that is highly sensitive to minor compositional or processing variations. VMAs in UHPC function primarily as stabilizers of the dense particle suspension, preventing segregation of fine constituents and controlling flow behavior during casting, vibration, or self-placement. Because UHPC operates near the limit of workable rheology, even marginal VMA overdosing can induce sharp increases in yield stress or loss of flow, making dosage precision and polymer efficiency critical industrial considerations.

Interaction with PCEs, Fines, and Fiber Distribution

UHPC formulations typically rely on high polycarboxylate ether dosages to achieve dispersion of cement, silica fume, and other ultrafine components. VMAs must therefore be selected for minimal interference with PCE adsorption and dispersing efficiency. Synthetic VMAs with controlled molecular architecture and low effective dosage are preferred, as they allow viscosity modulation without compromising particle dispersion. In fiber-reinforced UHPC, VMAs also contribute indirectly to uniform fiber distribution by stabilizing the paste matrix during flow and vibration. Excessive viscosity, however, can hinder fiber orientation and increase entrapped air, underscoring the need for narrowly defined rheological targets validated under realistic mixing and placement conditions.

Process Sensitivity, Robustness, and Digital Qualification

UHPC production is highly sensitive to mixing sequence, shear intensity, and temperature, amplifying the importance of VMA robustness. Effective systems tolerate reasonable variability without abrupt rheological transitions that disrupt casting or surface quality. Industrial qualification increasingly combines laboratory rheometry with mixer power monitoring and controlled pilot-scale trials to establish safe operating windows. When integrated into MES-based control systems, UHPC VMA dosing can be linked to power draw thresholds and temperature compensation rules, supporting Industry 4.0-aligned production. In this context, VMAs serve as precision rheology regulators that enable reproducible UHPC performance rather than as corrective additives applied at the plant or site.

12.7 High-Performance Fiber-Reinforced Concrete (HPFRC)

Matrix Cohesion and Fiber Dispersion Control

High-performance fiber-reinforced concrete combines high paste quality with elevated fiber volume fractions, creating a rheological system in which matrix cohesion directly governs fiber distribution and orientation. VMAs are applied to increase plastic viscosity and suspension stability, preventing fiber settlement, flotation, or clustering during mixing and placement. The admixture must enhance matrix cohesiveness without excessively increasing yield stress, which would impair flow, self-leveling behavior, and form filling. Associative synthetic VMAs are particularly effective in this context, as they selectively raise viscosity under low shear while maintaining sufficient deformability around fibers under applied stress.

Influence on Fiber Orientation and Mechanical Performance

The rheology of the fresh matrix influences not only fiber dispersion but also fiber alignment, which in turn affects tensile strength, crack control, and post-cracking ductility. VMAs contribute by damping segregation-driven flow patterns that promote preferential fiber orientation or accumulation. Overly strong thickening can restrict fiber mobility, leading to non-uniform orientation and increased entrapped air, while insufficient viscosity allows fibers to migrate under gravity or shear. Industrial optimization therefore requires balancing VMA dosage with fiber geometry, surface treatment, and volume fraction to achieve reproducible mechanical performance rather than maximum fresh stability alone.

Robustness, Mixing Sensitivity, and Digital Process Support

HPFRC mixtures are sensitive to mixing sequence, shear intensity, and temperature, particularly at high fiber contents. VMAs selected for industrial use must tolerate reasonable variability in these parameters without sudden rheological shifts that complicate casting or finishing. Increasingly, mixer power consumption and workability indicators are monitored to infer matrix–fiber interaction during production. When integrated into MES-driven control systems, these data enable formulation-specific rheological targets and adjustment limits, aligning HPFRC production with Industry 4.0 principles. In this role, VMAs function as enablers of consistent fiber performance and scalable manufacturing rather than as corrective measures applied at the jobsite.

12.8 Lightweight Concrete

Stabilization of Low-Density Aggregate Systems

Lightweight concrete incorporates aggregates with low particle density and high internal porosity, creating a strong tendency toward segregation, floating, and non-uniform distribution during mixing, transport, and placement. VMAs are applied to increase paste cohesion and stabilize the aggregate skeleton, counteracting buoyancy-driven movement and differential settlement. The admixture must raise plastic viscosity sufficiently to immobilize lightweight particles while preserving workable flow for casting or pumping. Industrially robust VMAs provide consistent stabilizing efficiency across a range of aggregate types, including expanded clay, shale, and pumice, each of which exhibits distinct absorption behavior.

Interaction with Aggregate Absorption and Moisture Variability

The high and often variable water absorption of lightweight aggregates introduces additional complexity to rheology control. Rapid water uptake can locally increase viscosity and alter effective VMA concentration, while delayed absorption during mixing or haul time may release water back into the paste. VMAs selected for lightweight concrete must therefore tolerate fluctuations in free water content without sharp rheological transitions. Synthetic and hybrid VMAs with moderated hydration kinetics are generally preferred over highly hydrophilic cellulose systems in mixes with poorly controlled aggregate moisture. Formulation strategies often combine VMA use with pre-wetting protocols and digital moisture correction to maintain stable workability.

Robustness, Placement Performance, and Digital Monitoring

Field performance of lightweight concrete depends on maintaining uniform aggregate distribution through transport, pumping, and placement, particularly in vertical elements and long pours. VMAs contribute by reducing bleeding and segregation while supporting cohesive flow around reinforcement. Over-thickening increases the risk of incomplete consolidation and surface defects, making dosage control critical. Increasingly, mixer power draw, slump flow, and density measurements are digitally monitored to detect early signs of segregation or rheological drift. Within an Industry 4.0 framework, these data enable formulation-specific control rules that reinforce consistency and scalability of lightweight concrete production across multiple plants.

12.9 Mass Concrete

Rheological Requirements for Large-Volume Placement

Mass concrete is defined by large pour volumes, extended placement durations, and limited opportunities for corrective intervention once casting begins. VMAs are applied conservatively in these systems, with the primary objective of controlling bleeding and modestly improving cohesion rather than significantly modifying flow behavior. The admixture must support uniform paste distribution over long discharge times while maintaining low yield stress to facilitate placement and internal consolidation. Industrially suitable VMAs exhibit stable rheological performance under prolonged low-shear mixing and do not promote premature stiffening that could interfere with finishing or joint formation.

Interaction with Heat Development and Time-Dependent Behavior

Thermal evolution plays a dominant role in mass concrete, as heat generated by cement hydration accumulates and gradually raises internal temperature. VMAs must remain rheologically stable across this slow temperature increase without accelerating viscosity build-up or interfering with heat management strategies. Highly reactive or strongly hydrophilic polymers may amplify temperature sensitivity, leading to unintended stiffening during extended pours. Low-reactivity synthetic VMAs with gradual hydration kinetics are therefore preferred, as they maintain consistent behavior over hours rather than minutes. From a formulation standpoint, compatibility with low-heat cements and supplementary cementitious materials is a key selection criterion.

Robustness, Quality Control, and Digital Oversight

Operational robustness is critical in mass concrete due to the scale and continuity of placement. VMAs selected for this application must tolerate variability in mixing energy, aggregate moisture, and haul time without requiring on-site adjustment. Increasingly, slump, temperature, and discharge rate data are digitally recorded and evaluated to verify compliance with predefined acceptance ranges. When integrated into MES-based systems, these data support early identification of rheological drift and reinforce consistency across large pours. In an Industry 4.0 context, VMAs function as low-risk stabilizers that enhance quality assurance and repeatability in mass concrete construction.

12.10 Underwater and Anti-Washout Concrete

Cohesion and Washout Resistance in Direct Water Contact

Underwater and anti-washout concretes are placed in environments where fresh cementitious material is exposed directly to surrounding water, creating extreme risk of cement and fines loss. VMAs serve as the primary functional component that imparts high cohesion and water retention, enabling the concrete to resist dilution, segregation, and paste erosion during placement. Effective systems substantially increase plastic viscosity and yield stress to immobilize fine particles while still allowing controlled flow under gravity or pumping. Industrial formulations must strike a precise balance, as insufficient VMA dosage leads to rapid washout, while excessive thickening compromises placement efficiency and homogeneity.

Polymer Selection, Dosage Sensitivity, and Flow Control

Cellulose ethers remain the dominant VMA class for anti-washout applications due to their strong thickening efficiency and exceptional water retention capacity. These polymers form highly cohesive networks that bind free water and limit dispersion of cement particles into the surrounding medium. Synthetic anti-washout polymers are increasingly adopted where improved pumpability, reduced temperature sensitivity, or smoother flow profiles are required. Regardless of chemistry, dosage sensitivity is high, and narrow operating windows must be respected to avoid excessive injection pressure or incomplete filling. Formulation development therefore emphasizes controlled hydration kinetics and predictable dose–response behavior validated under realistic underwater placement conditions.

Qualification, Monitoring, and Digital Process Integration

Performance verification for underwater concrete relies on standardized washout resistance tests combined with flowability and segregation assessments. In industrial practice, placement rate, discharge method, and hydrostatic head further influence effective rheology, making robustness a key selection criterion for VMAs. Increasingly, digital logging of placement parameters, including flow rate, tremie depth, and batch sequence, is used to support quality assurance. When integrated into MES-driven documentation systems, these data reinforce traceability and compliance with project specifications. Within an Industry 4.0 framework, VMAs in underwater concrete function as engineered rheology controls that enable predictable, auditable performance in one of the most demanding placement environments.

12.11 Concrete for Extreme Weather Conditions

Rheological Stability under Thermal Extremes

Concrete placed under extreme weather conditions, including hot, cold, and rapidly fluctuating climates, experiences accelerated or retarded hydration coupled with significant changes in water mobility. VMAs in these environments must maintain rheological stability despite sharp temperature gradients that can occur between batching, transport, and placement. At high temperatures, increased polymer hydration rates and cement reactivity can trigger premature viscosity build-up and rapid slump loss, while at low temperatures delayed polymer activation may reduce cohesion and increase segregation risk. Industrially suitable VMAs therefore exhibit moderated temperature sensitivity and predictable rheological response across a wide thermal range rather than maximum thickening efficiency at standard laboratory conditions.

Interaction with Setting Control and Environmental Countermeasures

Extreme weather concreting often employs auxiliary measures such as chilled or heated mixing water, ice, accelerators, retarders, and air-entraining agents to control setting and durability. VMAs must be compatible with these interventions to avoid unintended interactions that destabilize fresh concrete behavior. In hot climates, synthetic and hybrid VMAs with controlled hydration kinetics are favored to prevent early stiffening when combined with retarders or high PCE dosages. In cold conditions, VMAs must not excessively delay flow development or interfere with early strength gain promoted by accelerators. Formulation robustness is therefore evaluated under combined chemical and thermal stresses representative of field practice.

Operational Robustness and Digital Weather-Adaptive Control

Production under extreme weather magnifies operational variability, increasing dependence on robust process control rather than on-the-fly correction. VMAs selected for such conditions must tolerate fluctuations in haul time, mixing energy, and raw material temperature without sharp rheological transitions. Increasingly, weather data, concrete temperature, and rheological indicators such as slump and mixer power draw are integrated into digital plant control systems. Within an Industry 4.0 framework, these inputs enable weather-adaptive dosing rules and acceptance windows that preserve consistent performance across seasonal extremes. In this role, VMAs function as stabilizing elements that support reliable construction schedules and quality assurance under challenging environmental conditions.

12.12 Decorative and Architectural Concrete

Surface Quality, Uniformity, and Rheological Control

Decorative and architectural concretes place a high premium on surface appearance, color uniformity, and defect-free finishes, often under demanding casting geometries and exposed surfaces. VMAs contribute by stabilizing the fresh mix to prevent segregation, bleeding, and laitance formation, all of which directly affect surface texture and visual consistency. Adequate control of plastic viscosity is essential to ensure uniform paste distribution against formwork and around reinforcement without entrapping air or creating flow marks. Industrially effective VMAs provide smooth, predictable rheology that supports consistent surface reproduction across repeated pours.

Compatibility with Pigments, Fillers, and Formwork Systems

Architectural concretes frequently incorporate pigments, fine fillers, and specialty aggregates to achieve specific aesthetic outcomes. VMAs must be chemically compatible with these components to avoid color variation, streaking, or uneven dispersion. Strongly adsorbing or highly reactive polymers may interact with pigment surfaces, altering effective dosage or causing localized viscosity changes. Synthetic and hybrid VMAs with controlled adsorption behavior are therefore preferred, as they minimize interference with color development and maintain stable rheology. Compatibility with form-release agents and complex formwork materials is also critical, as surface defects often originate from subtle rheological imbalances during casting.

Reproducibility, Quality Assurance, and Digital Support

Architectural projects demand high reproducibility across multiple casting cycles, making robustness and process consistency more important than maximum rheological modification. VMAs must tolerate reasonable variability in temperature, mixing energy, and raw materials without compromising surface quality. Increasingly, slump flow, air content, and mixer power data are digitally monitored to detect early deviations that could affect appearance. When integrated into MES-driven quality systems, these parameters support standardized execution and traceability, aligning decorative concrete production with Industry 4.0 principles. In this context, VMAs act as precision tools that translate formulation design into repeatable aesthetic performance rather than as corrective measures applied after defects appear.

13. Troubleshooting and Failure Analysis

Failure analysis of viscosity-modifying admixtures in concrete systems requires a formulation-centric and process-aware approach, integrating rheological science with field feedback and digital diagnostics. VMAs operate at low dosage windows with nonlinear effects on yield stress and plastic viscosity, making them particularly sensitive to raw material variability, mixing energy, temperature, and admixture sequencing.

From an industrial technology transfer perspective, troubleshooting must bridge laboratory rheometry, pilot batching, and real-time jobsite data. Industry 4.0 tools such as sensor-enabled mixers, MES-linked dosage control, and data-driven root cause analytics significantly reduce trial-and-error cycles and improve corrective action reliability.

13.1 Over-Viscosification and Workability Loss

Over-viscosification represents one of the most frequent and costly failure modes associated with viscosity-modifying admixtures in cementitious systems. It is characterized by excessive plastic viscosity increase, impaired flowability, and reduced placement efficiency, often without proportional gains in segregation resistance. From an industrial formulation standpoint, this failure mode reflects a misalignment between polymer architecture, cement chemistry, and process conditions rather than simple overdosing alone.

13.1.1 Root cause analysis

The primary root cause of over-viscosification is excessive polymer-induced water immobilization within the cement paste pore solution. High molecular weight or highly associative VMAs form extended polymer networks that trap mixing water through hydrogen bonding and physical entanglement, sharply increasing plastic viscosity under low to moderate shear. This effect is amplified when the polymer concentration exceeds the percolation threshold of the paste system.

A second dominant contributor is unfavorable interaction with fine particles, including cement fines, limestone fillers, and clay-contaminated aggregates. VMAs adsorb onto high surface area solids, promoting particle bridging and flocculation when dispersant efficiency is insufficient. In mixes with elevated fines or high clay activity index, even nominal VMA dosages can trigger pseudo-gel formation and rapid workability loss.

Cement chemistry plays a critical role, particularly high C3A content, elevated alkali levels, and sulfate balance. These factors accelerate early hydration and modify pore solution ionic strength, inducing premature polymer contraction and enhanced adsorption. The result is an early-stage viscosity spike that cannot be corrected through additional water or superplasticizer without compromising performance.

Process-related root causes are frequently underestimated in field diagnostics. Inadequate mixing energy, short wet mixing time, or poor admixture dispersion create localized VMA over-concentration zones. Similarly, inaccurate dosing systems and delayed calibration drift contribute to batch-to-batch variability that manifests as intermittent over-viscosification rather than systematic failure.

From an Industry 4.0 perspective, over-viscosification correlates strongly with abnormal digital fingerprints such as elevated mixer torque, increased power draw, and non-linear viscosity rise detected by inline rheology sensors. When integrated into MES or SCADA platforms, these signals enable early-stage root cause attribution by correlating formulation variables, raw material changes, and process deviations across production cycles.

13.1.2 Corrective formulation actions

Corrective formulation actions for over-viscosification must prioritize polymer architecture optimization rather than simplistic dosage reduction. Selecting VMAs with lower effective molecular weight, reduced hydrodynamic volume, or weaker associative groups directly limits excessive network formation while preserving shear-thinning behavior. In industrial practice, blending two VMA grades with complementary rheological profiles often delivers a wider processing window than single-component systems.

Rebalancing the dispersant system is a critical secondary measure. Increasing superplasticizer efficiency through tailored polycarboxylate ether side-chain length and charge density restores particle dispersion, reducing polymer-induced flocculation. In highly fines-rich systems, partial replacement of adsorbing VMAs with non-adsorbing, water-structuring polymers minimizes competitive surface interactions.

Admixture sequencing and water management provide rapid corrective leverage without reformulating the full mix design. Delaying VMA addition until after complete cement wetting and superplasticizer adsorption prevents premature polymer network locking. Split water addition, with a portion reserved for post-dispersion adjustment, improves polymer distribution and mitigates localized thickening.

Fine content and filler optimization are frequently overlooked corrective pathways. Reducing ultra-fine mineral additions or substituting inert fillers with lower surface area grades decreases polymer demand and lowers the risk of percolation-driven viscosity spikes. Where clay contamination is unavoidable, incorporation of clay-tolerant VMAs or dedicated clay-blocking additives stabilizes rheology.

From an Industry 4.0 implementation perspective, closed-loop control systems linking inline rheological measurements to automated dosing units enable real-time corrective action. Machine learning algorithms trained on historical batch performance can recommend immediate VMA and superplasticizer adjustments, converting corrective formulation actions from reactive field interventions into predictive, plant-level control strategies.

13.2 Incompatibility with Superplasticizers

Incompatibility between viscosity-modifying admixtures and superplasticizers represents a critical formulation risk, particularly in high-performance and low water-to-cement ratio concretes. The failure manifests as unexpected slump loss, poor flow development, delayed dispersion, or excessive viscosity despite adequate superplasticizer dosage. From a technology transfer perspective, this incompatibility is fundamentally governed by surface chemistry and competitive adsorption phenomena rather than bulk polymer concentration alone.

13.2.1 Competitive Adsorption Mechanisms

Competitive adsorption arises when VMAs and superplasticizers contend for the same active sites on cement and supplementary cementitious material surfaces. Anionically charged VMAs, especially polysaccharide-based and sulfonated polymers, can preferentially adsorb onto positively charged aluminate and ferrite phases, reducing the availability of anchoring sites required for polycarboxylate ether adsorption. This limits electrosteric repulsion and suppresses particle dispersion efficiency.

The severity of competitive adsorption is strongly influenced by cement phase composition and sulfate balance. High C3A content and elevated alkali sulfate levels accelerate early surface charge development, promoting rapid VMA uptake before superplasticizer molecules can establish stable adsorption layers. Under these conditions, even delayed superplasticizer addition may fail to fully recover workability.

Polymer architecture further modulates adsorption hierarchy. VMAs with high charge density or multiple functional groups exhibit strong multi-point attachment, leading to quasi-irreversible adsorption. In contrast, superplasticizers rely on a limited number of anchoring groups combined with long side chains, making them more susceptible to displacement or steric hindrance once VMA adsorption has occurred.

Pore solution chemistry introduces an additional competitive dimension. Increasing ionic strength during early hydration compresses the electrical double layer, enhancing polymer-surface interactions and favoring strongly adsorbing VMAs. Simultaneously, calcium ion complexation alters superplasticizer conformation, reducing its effective steric stabilization capability.

Advanced analytical approaches such as adsorption isotherms, zeta potential evolution, and time-resolved total organic carbon depletion provide quantitative insight into these mechanisms. When coupled with digital twins of cement–admixture interactions, these datasets enable predictive screening of incompatibility risks prior to field application, aligning with Industry 4.0-driven formulation development.

13.2.2 Field Mitigation Strategies

Field mitigation of incompatibility between viscosity-modifying admixtures and superplasticizers requires a pragmatic balance between formulation robustness and operational flexibility. The first line of mitigation is admixture selection, favoring VMAs with low anionic charge density or predominantly non-adsorbing, water-structuring mechanisms. Such polymers reduce occupation of cement surface sites and preserve superplasticizer anchoring efficiency under variable cement chemistries.

Admixture sequencing remains one of the most effective on-site control levers. Introducing the superplasticizer during initial wetting ensures early particle dispersion, followed by delayed VMA addition once electrosteric stabilization is established. In critical placements, split dosing of the superplasticizer further stabilizes flow development and minimizes sensitivity to minor adsorption disturbances.

Operational adjustments at plant and jobsite level provide rapid mitigation without altering approved mix designs. Extending wet mixing time enhances polymer distribution and reduces localized adsorption dominance. Maintaining tight control of water temperature and avoiding prolonged truck agitation limit time-dependent adsorption and polymer rearrangement that contribute to incompatibility-related slump loss.

For high-risk systems, dual-admixture strategies are increasingly adopted. Combining a fast-adsorbing polycarboxylate ether with a delayed-release or low-affinity dispersant improves tolerance to VMA presence. Compatibility screening using mini-slump flow, rheometry, and rapid adsorption tests should be institutionalized for each cement source and SCM blend.

Industry 4.0 implementation elevates field mitigation from reactive to predictive. MES-linked compatibility matrices, batch-level data logging, and AI-assisted decision rules enable automatic sequencing and dosage adjustments when cement chemistry, ambient temperature, or raw material sources change. These digital interventions significantly reduce field variability and enhance confidence during technology transfer to new plants or regions.

13.3 Temperature and Time-Dependent Instability

Temperature and time-dependent instability constitutes a major risk factor in the field performance of viscosity-modifying admixtures, as ambient and process conditions continuously influence polymer activation and cement hydration kinetics. Elevated temperatures accelerate polymer swelling, adsorption, and cement reactions, frequently resulting in rapid viscosity build-up, accelerated slump loss, and shortened open time. Under hot-weather conditions, these effects are further intensified by increased water demand and faster depletion of dispersing capacity, particularly in systems containing polycarboxylate ether superplasticizers.

In contrast, low-temperature environments retard polymer hydration and delay microstructural development, increasing susceptibility to segregation, bleeding, and delayed structural build-up. During extended mixing, transport, or placement delays, thermal history interacts with shear exposure and ongoing hydration, producing non-linear changes in yield stress, plastic viscosity, and thixotropic recovery. Such time-dependent evolution often manifests as delayed thickening followed by abrupt stiffening, complicating placement control and finishability in ready-mix and specialty concrete applications.

Mitigation of these instabilities requires an integrated formulation and process strategy. Seasonal optimization of polymer chemistry, including molecular weight distribution and substitution patterns, improves thermal robustness and activation predictability. Process-side measures such as controlled addition sequences, split dosing, and temperature-compensated water management further stabilize rheological response. Increasingly, Industry 4.0-enabled production systems incorporating real-time temperature tracking, rheology monitoring, and predictive analytics support proactive adjustment of VMA dosage and admixture sequencing, ensuring consistent fresh-state performance across diverse climatic conditions and extended project timelines.

13.3.1 Delayed Thickening and Slump Loss

Temperature and time-dependent instability represents a complex failure mode in VMA-containing concrete systems, often misdiagnosed as cement inconsistency or improper water control. In reality, it reflects the kinetic response of polymeric admixtures to evolving hydration chemistry, thermal conditions, and shear history. From an industrial formulation viewpoint, these instabilities become critical in ready-mix operations, long haul times, and hot climate placements.

13.3.2 Seasonal Adjustment Protocols

Delayed thickening is primarily driven by time-dependent changes in polymer conformation and adsorption behavior as cement hydration progresses. VMAs that rely on associative hydrophobic interactions or hydrogen-bonded network formation may remain relatively inactive during initial mixing, then undergo rapid structuring as ionic strength increases and calcium ion concentration rises. This results in a sudden increase in plastic viscosity after an apparent induction period.

Elevated temperatures accelerate both cement hydration and polymer dynamics, intensifying delayed thickening phenomena. Higher thermal energy promotes polymer chain mobility and enhances adsorption onto newly formed hydration products such as C–S–H and ettringite. The combined effect is accelerated water immobilization and progressive slump loss, even in mixes that initially meet workability specifications.

Time-dependent slump loss is further compounded by gradual superplasticizer depletion. Competitive adsorption, calcium complexation, and partial polymer entrapment within hydration products reduce dispersant availability over time. As electrosteric repulsion weakens, particle flocculation increases, amplifying the viscosity contribution of the VMA network.

From a process diagnostics perspective, delayed thickening is identifiable through non-linear slump decay curves and late-stage increases in mixer torque or truck drum resistance. Smart rheology tracking tools and maturity-linked prediction models enable differentiation between temperature-driven hydration effects and admixture-induced instability, supporting targeted corrective actions aligned with Industry 4.0-driven quality control frameworks.

14. Testing, Characterization, and Digital Analytics

Ensuring consistent and reliable performance of viscosity-modifying admixtures requires a deliberately integrated testing strategy that spans molecular-scale interactions, fresh-state rheology, and placement-relevant concrete behavior. Empirical field tests such as slump, slump flow, and visual stability indices remain indispensable for compliance verification and rapid screening. However, their limited sensitivity often masks early-stage incompatibilities, progressive rheological drift, and subtle interactions with cement phases, supplementary cementitious materials, and co-admixtures.

Modern industrial practice therefore extends beyond empirical testing toward multi-scale characterization. Physicochemical analysis of polymer structure, adsorption behavior, and pore solution interactions is combined with advanced rheometry to capture time, temperature, and shear-dependent effects. This integrated framework enables differentiation between formulation-driven instability and variability originating from raw materials or process conditions. As a result, testing evolves from pass–fail evaluation into a predictive performance management tool.

From a technology transfer perspective, robustness and reproducibility of test protocols are paramount. Methods developed at R&D scale must translate reliably to pilot batching and full-scale production, despite variations in cement chemistry, SCM composition, aggregate grading, ambient temperature, and mixing energy. Harmonized testing methodologies, supported by statistical process control and standardized data structures, are essential to ensure consistent interpretation across laboratories and production sites.

Industry 4.0 digital frameworks fundamentally reshape the role of testing and characterization. Real-time sensors measuring torque, temperature, moisture, and power consumption embed rheological diagnostics directly into mixers, batching plants, and transport units. Cloud-based data platforms consolidate laboratory, plant, and field datasets, while machine learning models identify emerging incompatibility patterns and predict performance drift before it becomes operationally critical. In this way, static test results are transformed into dynamic, actionable intelligence that strengthens formulation resilience and accelerates reliable technology deployment.

14.1 Laboratory Rheological Methods

Laboratory rheological methods form the quantitative backbone for evaluating viscosity-modifying admixture performance and diagnosing formulation-related instability. Unlike empirical workability tests, rheological measurements resolve yield stress, plastic viscosity, and shear-dependent flow behavior, enabling direct correlation with polymer structure and adsorption dynamics. For industrial formulation and technology transfer, these methods provide reproducible, material-sensitive data that support mix design optimization across varying cement chemistries and environmental conditions.

Rheological testing is particularly valuable for detecting early-stage over-viscosification, competitive adsorption effects, and delayed thickening phenomena that are not apparent during initial slump measurements. Time-resolved protocols, combined with controlled temperature conditioning, allow systematic separation of hydration-driven effects from polymer-induced structuring.

14.1.1 Rotational Rheometry and Vane Geometry Testing

Rotational rheometry is the preferred laboratory technique for quantifying the flow behavior of cement pastes and mortars containing VMAs. By applying controlled shear rates or stresses, rotational rheometers generate flow curves that accurately describe yield stress development and plastic viscosity evolution. These parameters directly reflect polymer network formation, water immobilization, and particle dispersion efficiency.

Vane geometry testing offers distinct advantages for heterogeneous, particle-rich systems typical of cementitious materials. The vane-in-cup configuration minimizes wall slip and structural breakdown during loading, preserving the native microstructure of the paste. This is particularly critical when evaluating VMAs that induce flocculated or weakly gelled structures, where conventional concentric cylinder geometries often underestimate yield stress.

Time-dependent rheological protocols, including stepped shear and rest–shear cycles, are essential for assessing structural build-up and breakdown behavior. VMAs exhibiting delayed association or temperature-sensitive interactions reveal characteristic hysteresis and thixotropic recovery patterns under these test conditions. Such signatures are highly relevant for predicting slump retention and pumpability under field-relevant shear histories.

From an industrial quality and scale-up perspective, standardization of rheometer geometry, shear protocols, and sample preparation is mandatory. Variations in pre-shear history, temperature control, and measurement timing can overshadow formulation effects. Integration of rheometer outputs into digital databases enables trend analysis and supports Industry 4.0-driven model calibration linking laboratory rheology to plant and jobsite performance.

14.1.2 Correlation Between Lab Rheology and Field Flow Behavior

Establishing a robust correlation between laboratory rheological parameters and field-observed flow behavior is essential for effective formulation design and reliable technology transfer of viscosity-modifying admixtures. Laboratory-derived yield stress and plastic viscosity provide mechanistic insight into fresh concrete performance, yet their practical value depends on consistent translation to placement, pumping, and finishing conditions. Misalignment between lab and field data often arises from differences in shear history, scale, and environmental exposure rather than from rheological methodology itself.

Yield stress measured in controlled laboratory environments correlates strongly with initial flow initiation in the field, including slump, slump flow, and the ability of concrete to self-level. Excessive yield stress identified by rotational rheometry typically manifests as poor discharge, incomplete mold filling, or delayed flow under gravity. Plastic viscosity, in contrast, governs flow maintenance and energy demand during pumping, correlating with observed resistance in pipelines and increased mixer or truck drum torque.

Time-dependent rheological measurements provide the most reliable bridge between laboratory and field behavior. Rest–shear cycles and structural build-up indices closely mirror slump retention and delayed stiffening observed during transport. VMAs exhibiting low structural recovery in rheometry tend to maintain stable workability under prolonged agitation, whereas systems showing rapid rebuild are prone to late-stage thickening and placement difficulties.

Temperature-controlled rheological testing further enhances field relevance. Laboratory simulations incorporating realistic thermal profiles reproduce seasonal flow behavior with high fidelity, allowing differentiation between hydration-driven slump loss and polymer-induced instability. These correlations are particularly valuable in hot climates, where delayed thickening often emerges beyond the timeframe of standard laboratory tests.

Industry 4.0 integration strengthens lab–field correlation by embedding rheological data into digital performance models. Laboratory rheology datasets are used to calibrate predictive algorithms that ingest real-time plant signals such as mixing energy, torque, and temperature. This data fusion enables near real-time estimation of field flow behavior from laboratory-derived rheological fingerprints, reducing reliance on empirical adjustment and enhancing formulation robustness across production sites.

14.2 Plant and Field Test Protocols

Plant- and field-level testing translates laboratory-validated viscosity-modifying admixture behavior into operational reality, where variability in materials, equipment, and environment becomes dominant. Unlike controlled rheometry, these protocols operate under mixed shear regimes, fluctuating temperatures, and non-ideal batching sequences. Their primary value lies in validating robustness, detecting gross instability, and generating performance data under true manufacturing and placement conditions.

At industrial scale, testing must accommodate production throughput and decision speed. Methods are therefore selected not only for technical relevance but also for simplicity, repeatability, and operator independence. When properly harmonized with laboratory benchmarks, plant and field tests provide an indispensable feedback loop that closes the gap between formulation intent and delivered performance.

14.2.1 Stability column, sieve segregation, and visual indices

Plant- and field-oriented stability tests provide direct evidence of whether viscosity-modifying admixtures deliver the intended balance between flowability and particle suspension under practical conditions. These methods deliberately expose fresh concrete to gravity-driven separation, dynamic flow, and operator observation, thereby stressing the polymer-mediated particle network beyond what is achievable in laboratory rheometers. Their diagnostic value lies less in absolute numbers and more in sensitivity to instability triggered by raw material variability, dosing errors, or adverse admixture interactions.

Stability column testing is widely applied to quantify resistance to static segregation during rest periods typical of transport, queuing, and delayed placement. The test measures vertical heterogeneity after a defined rest time, often expressed as differences in coarse aggregate or density between top and bottom sections. Effective VMA systems produce minimal gradients, indicating sufficient yield stress at near-zero shear to immobilize aggregates without inducing excessive stiffness. In industrial practice, elevated segregation indices frequently correlate with insufficient polymer adsorption on fines or competitive displacement by high-range water reducers, particularly in limestone-rich or low-clay systems.

Sieve segregation testing complements static methods by evaluating stability under dynamic flow. Concrete is allowed to pass through a standardized sieve, and the proportion of mortar passing relative to total mass is determined. Higher passing fractions indicate inadequate viscosity or weak particle cohesion during shear. From a formulation standpoint, this test is highly sensitive to VMA molecular architecture and dosage window, as well as to aggregate shape and grading. For pumped concrete and high-slump SCC, sieve segregation often provides earlier warning of instability than visual inspection alone.

Visual stability indices and bleeding observations remain indispensable for rapid, on-site assessment. Although inherently qualitative, these indicators capture surface phenomena that are difficult to quantify instrumentally, including laitance formation, aggregate haloing, water sheen, and localized settlement. Their reliability increases substantially when operators are trained against standardized reference images and when observations are recorded digitally rather than informally. Consistent visual trends, when aligned with quantitative segregation data, often reveal subtle shifts in fines content, moisture correction accuracy, or cement sulfate balance.

Table 14.2.1. Field Stability Tests and Diagnostic Sensitivity

Test approach	Dominant failure mode detected	Typical industrial trigger
Stability column	Static segregation during rest	Under-dosage, low fines, delayed placement
Sieve segregation	Dynamic separation under flow	Pumping, high slump, aggregate grading drift
Visual indices	Bleeding, surface heterogeneity	Moisture error, SCM variability, thermal effects

Integration of these tests into plant quality systems is increasingly supported by digital tools aligned with Industry 4.0. Results are logged alongside mixer torque, batch temperature, and moisture sensor data, enabling correlation between observed instability and process signals. Over time, this structured dataset allows early identification of performance drift, reducing reliance on reactive field adjustments and strengthening confidence in VMA robustness across plants and seasons.

14.2.2 Mixing Energy, Torque Monitoring, and Process-Integrated Indicators

Process-integrated measurements provide a quantitative bridge between classical field tests and laboratory rheology by capturing the energetic response of fresh concrete during mixing, transport, and discharge. Unlike discrete sampling-based tests, mixing energy and torque monitoring deliver continuous data streams that reflect the evolving rheological state of the system under realistic shear histories. For viscosity-modifying admixtures, these indicators are particularly sensitive to over-viscosification, delayed polymer association, and competitive adsorption phenomena that develop during mixing or extended agitation.

Mixer and truck drum torque measurements correlate strongly with apparent plastic viscosity and structural build-up rate. VMAs that form extended polymer–particle networks typically increase power demand at constant rotational speed, while systems prone to instability often show erratic torque profiles or sudden inflections after rest periods. From an industrial control perspective, absolute torque values are less informative than normalized trends relative to reference mixes, cement lots, or seasonal baselines. This normalization enables meaningful comparison across equipment types and plant configurations.

Mixing energy, calculated as the integrated power consumption over defined mixing intervals, provides an aggregate descriptor of workability evolution. Elevated energy demand at constant slump frequently indicates excessive water immobilization or unintended synergistic effects with superplasticizers. Conversely, declining energy signatures coupled with stable slump may signal insufficient VMA adsorption, increasing segregation risk. These relationships become especially valuable in high-throughput plants where traditional stability testing is impractical for every batch.

Real-time process indicators gain predictive power when combined with temperature, moisture, and batch timing data. Thermal acceleration of hydration amplifies torque build-up in VMA-rich systems, while moisture correction errors often manifest as short-term torque suppression followed by rapid stiffening. Embedding these signals within digital control systems enables early detection of deviation from expected rheological trajectories, supporting corrective action before field performance is compromised.

Table 14.2.2. Process-Integrated Indicators for VMA Performance Control

Process signal	Rheological implication	Typical corrective response
Rising torque trend	Increasing plastic viscosity or structuration	Reduce VMA dosage or adjust HRWR balance
Erratic torque profile	Inhomogeneous mixing or instability	Extend mixing, verify moisture and fines
Elevated mixing energy	Over-viscosification, polymer over-association	Reformulate or optimize addition sequence

From a technology transfer standpoint, mixing energy and torque monitoring support scalable, operator-independent quality control. Data streams from multiple plants can be aggregated into centralized analytics platforms, enabling cross-site benchmarking and continuous model refinement. When aligned with laboratory rheology and segregation testing, these process-integrated metrics complete a closed-loop framework for robust, Industry 4.0-compliant deployment of viscosity-modifying admixtures.

14.3 Data Analytics and Predictive Modeling

Data analytics and predictive modeling convert dispersed laboratory, plant, and field measurements into structured knowledge that supports robust viscosity-modifying admixture deployment. In contrast to empirical adjustment practices, data-driven frameworks enable quantitative separation of formulation effects from raw material and process variability. For industrial organizations operating across multiple plants and cement sources, such approaches are essential to maintain consistent rheological performance while minimizing admixture overdesign.

Digitalization of testing outputs, process signals, and material descriptors forms the foundation of this methodology. Rheological parameters, segregation indices, torque trends, and environmental conditions are aggregated into harmonized datasets that support statistical analysis and model development. When implemented within Industry 4.0 architectures, these datasets evolve continuously, enabling adaptive performance management rather than static qualification.

14.3.1 Historical Mix Data Utilization

Historical mix data provide a powerful empirical basis for understanding long-term VMA performance trends under real production conditions. When structured with consistent metadata, including cement phase composition, SCM type and fineness, aggregate grading, admixture identity, dosage sequence, and ambient temperature, these datasets enable retrospective analysis of compatibility and robustness.

Statistical evaluation of multi-year datasets reveals recurring correlations that are often invisible in short-term trials. Progressive increases in required VMA dosage frequently align with gradual shifts in cement sulfate balance or increasing fines reactivity rather than with polymer degradation. Similarly, seasonal variability in segregation resistance can often be traced to temperature-driven adsorption kinetics rather than to water content alone. Such insights allow targeted intervention at the raw material or process level, reducing unnecessary reformulation.

Advanced analytics, including principal component analysis and clustering algorithms, support classification of mix families with similar rheological fingerprints. These groupings simplify technology transfer by allowing new formulations to be benchmarked against established performance envelopes rather than against isolated reference mixes. Digital dashboards that visualize historical trends in torque, slump retention, and segregation indices further enhance decision-making at both plant and corporate levels.

14.3.2 AI-Assisted Dosage Optimization

Artificial intelligence methodologies extend conventional data analytics by capturing nonlinear interactions between VMA chemistry, co-admixtures, cement composition, and processing conditions. Machine learning models trained on historical and real-time data generate dosage recommendations that balance stability, flowability, and cost efficiency across a defined operating window rather than at a single target point.

In practical implementation, AI-assisted systems ingest laboratory rheology, plant torque and energy signals, moisture sensor data, and environmental parameters to predict short- and medium-term performance. Dosage adjustments are suggested proactively when model outputs indicate rising risk of segregation or over-viscosification. This approach is particularly effective in self-consolidating and pumped concrete, where narrow stability margins amplify the consequences of small material deviations.

From an industrial deployment perspective, AI-based optimization enhances consistency across geographically distributed plants. Centralized models encode global formulation knowledge, while local data continuously refine predictions to reflect site-specific materials and equipment. The result is a scalable, adaptive control strategy that reduces operator dependency, minimizes trial-and-error adjustments, and accelerates reliable technology transfer of viscosity-modifying admixtures.

15. Case Studies from Formulation to Field

Case studies provide critical validation of viscosity-modifying admixture technologies by demonstrating performance continuity from laboratory formulation through industrial production and jobsite execution. The following examples illustrate how multi-scale testing, digital analytics, and adaptive control strategies converge to resolve real-world instability challenges. Emphasis is placed on reproducibility across plants, economic impact, and data-driven decision-making consistent with Industry 4.0 deployment models.

15.1 SCC Robustness Improvement Across Multiple Plants

Achieving consistent self-consolidating concrete performance across multiple production plants requires formulation strategies that tolerate raw material and process variability without compromising rheological control. Differences in cement phase composition, filler mineralogy, aggregate grading, moisture condition, and available mixing energy significantly influence yield stress and plastic viscosity in SCC. Under such conditions, the sensitivity of SCC to viscosity-modifying admixture chemistry is amplified, often leading to segregation, blocking, or unstable flow behavior when formulations are transferred between plants.

Robust SCC design therefore relies on VMA systems with broad compatibility windows and predictable activation across varying material inputs. Polymer architectures with moderated adsorption strength and controlled hydration kinetics help buffer fluctuations in fines content and surface chemistry. Laboratory screening using rheometry, bleeding resistance, and segregation indices must be complemented by plant-level trials to capture scale-dependent effects such as mixer efficiency and dosing accuracy. Harmonized test protocols and acceptance criteria are essential to translate laboratory robustness into production reliability.

Increasingly, multi-plant SCC deployment is supported by Industry 4.0-enabled performance monitoring and digital quality management. Centralized data platforms integrating fresh concrete rheology, batch records, and environmental conditions allow early detection of drift and rapid corrective action. Predictive analytics and digital twins facilitate proactive adjustment of VMA dosage and mix parameters, enabling reproducible SCC behavior across geographically dispersed plants while maintaining productivity, quality, and risk control.

15.1.1 Problem Definition and Solution Deployment

A regional SCC producer operating several batching plants reported inconsistent static stability and premature loss of flowability despite compliance with nominal mix design specifications. Plant audits revealed significant variability in limestone filler content, cement sulfate balance, and mixing intensity, leading to unpredictable VMA adsorption efficiency. Laboratory rheological profiling identified excessive yield stress build-up in certain cement-filler combinations, indicative of over-structuring driven by narrow polymer molecular weight distribution and high functional group density.

The deployed solution focused on reformulating the VMA toward a broader molecular architecture with moderated charge density to reduce sensitivity to fines mineralogy. Controlled adsorption studies confirmed more uniform polymer uptake across cements, while time-resolved rheometry demonstrated reduced structural build-up under rest conditions. In parallel, plant test protocols were harmonized, including standardized stability column procedures and consistent torque normalization. Digital capture of batch temperature, mixing energy, and discharge timing enabled real-time comparison across sites, accelerating validation and alignment.

15.1.2 Performance KPIs and Economic Impact

Post-deployment evaluation showed a consistent reduction in static segregation indices and stabilization of slump flow retention beyond 90 minutes across all plants. Mixer torque variability decreased significantly, indicating tighter control of apparent plastic viscosity under production shear conditions. From an operational standpoint, SCC-related rejected loads were reduced by approximately 35 to 45 percent, depending on site-specific baseline performance.

Economic assessment demonstrated a net reduction in admixture cost per cubic meter despite the use of a higher-performance VMA, driven by elimination of site-specific over-dosing and reduced corrective batching. Additional savings were realized through lower technical service intervention and improved production scheduling reliability. The case underscores how polymer-level formulation adjustments, when coupled with standardized testing and digital analytics, deliver measurable value in multi-plant SCC operations.

15.2 Stabilizing Recycled Aggregate Concrete

Concrete incorporating recycled aggregates demonstrates markedly higher variability in fresh-state rheology and segregation resistance compared with conventional mixtures, driven by irregular particle shape, residual adhered cement paste, elevated porosity, and unstable fines chemistry. Such multiscale heterogeneity alters water absorption kinetics and disrupts particle packing, thereby intensifying the response of the system to viscosity-modifying admixtures through nonuniform water immobilization and transient polymer–particle networks. Robust stabilization is achieved only when formulation design is preceded by structured quantification of aggregate-related variability and followed by adaptive admixture control frameworks capable of adjusting dosage and polymer architecture in response to real-time fluctuations in recycled material properties, consistent with digitalized and data-driven concrete production environments.

15.2.1 Variability Mapping

A structured variability mapping program was implemented to identify dominant drivers of instability in recycled aggregate concrete. Laboratory characterization included particle size distribution of recycled fines, methylene blue values as a proxy for clay activity, and water absorption kinetics of coarse recycled fractions. Rheological testing revealed pronounced scatter in yield stress development, with certain aggregate sources inducing rapid structuration despite identical nominal water content.

At plant level, historical batch data were analyzed to correlate segregation indices, mixer torque, and bleeding observations with recycled aggregate source and moisture correction accuracy. Clustering analysis grouped aggregate deliveries into distinct rheological risk categories, ranging from low-reactivity crushed concrete fines to clay-contaminated mixed demolition residues. These classifications enabled predictive identification of batches prone to instability before placement.

The integration of laboratory descriptors with plant sensor data provided a comprehensive variability map that distinguished chemical reactivity effects from purely physical water demand. This distinction proved essential for selecting appropriate mitigation strategies and avoiding indiscriminate VMA over-dosing.

15.2.2 Adaptive VMA Dosing Models

Based on the variability map, an adaptive VMA dosing framework was developed using machine learning algorithms trained on rheology, torque, moisture, and segregation datasets. The model generated dosage ranges responsive to real-time indicators of fines activity and mixing response rather than relying on fixed set points. For aggregates classified as high-reactivity, the algorithm favored incremental VMA adjustments combined with modified addition sequencing to improve polymer dispersion.

Field implementation demonstrated substantial stabilization of yield stress and plastic viscosity evolution, reflected in consistent visual stability and reduced sieve segregation across variable aggregate supplies. Notably, the adaptive approach limited overall VMA consumption increases to less than 10 percent while reducing corrective water additions and batch rejections by more than 40 percent.

From a technology transfer perspective, the adaptive dosing model reduced operator dependency and enabled consistent performance across recycling streams with fluctuating composition. The case confirms that data-driven VMA control, grounded in polymer–particle interaction understanding, is an effective pathway to industrially viable recycled aggregate concrete.

15.3 Digital Twin Validation of Rheology Control

Digital twin architectures provide a structured pathway for continuous, physics-informed supervision of fresh concrete rheology by coupling mechanistic models of particle–polymer interactions with live data streams from batching, mixing, and transport operations. In the context of viscosity-modifying admixtures, such virtual replicas enable forward prediction of yield stress drift, segregation risk, and robustness margins before deviations translate into field-level non-compliance or rejected deliveries. Industrial validation under real production constraints is therefore a critical step to confirm model fidelity, data integrity, scalability across variable raw material inputs, and the quantifiable economic benefits of closed-loop rheology control aligned with Industry 4.0 manufacturing principles.

15.3.1 Sensor Data Integration

In the validated deployment, a digital twin was constructed using a hybrid modeling architecture combining mechanistic rheology relationships with data-driven calibration. Real-time inputs were collected from mixers, truck drums, and batching systems, including torque, rotational speed, power consumption, batch temperature, moisture content, and timing of admixture addition. These signals were synchronized with laboratory-derived yield stress and plastic viscosity benchmarks to anchor the model in physically meaningful parameters.

Data ingestion followed standardized digital structures to ensure compatibility across equipment suppliers and plant locations. Signal normalization accounted for mixer geometry and drive efficiency, allowing cross-site comparison of rheological indicators. The digital twin continuously updated predictions of structuration rate, apparent viscosity evolution, and stability risk during mixing, transport, and discharge. Scenario simulations, such as altered VMA dosage or delayed discharge, were executed virtually to support operational decision-making without interrupting production.

15.3.2 Reduction in Rejected Loads

Performance validation focused on the ability of the digital twin to reduce rejected loads associated with workability loss, segregation, and pumping difficulty. Early-warning thresholds triggered alerts when predicted rheological trajectories deviated from acceptable envelopes, enabling corrective actions such as minor dosage adjustment or extended mixing. Over a six-month evaluation period, rejected loads decreased by approximately 45 to 55 percent compared with historical baselines.

Beyond direct quality gains, the system delivered secondary benefits, including reduced on-site adjustments, lower water addition frequency, and improved consistency of placement behavior. From an economic standpoint, savings derived from reduced waste, fewer returns, and enhanced scheduling reliability outweighed digital infrastructure costs within the first year of operation. The case demonstrates that digital twin technology, when grounded in polymer science and validated against field outcomes, represents a practical and scalable tool for rheology control in modern concrete production.

15.4 Pumpability Control in Long-Distance and High-Rise Concrete

Concrete pumping over long horizontal distances or significant vertical lifts subjects fresh mixes to sustained shear, high pressure gradients, and progressive temperature increase within the pipeline. Under these conditions, rheological stability becomes critical, as excessive viscosity leads to pressure spikes and blockage risk, while insufficient cohesion promotes segregation and collapse of the lubrication layer. Viscosity-modifying admixtures are therefore central to maintaining a stable, low-friction paste film at the pipe wall while preserving bulk homogeneity under continuous energy input.

Effective pumpability control requires VMA formulations that exhibit shear-stable viscosity and rapid structural recovery once shear is reduced. Polymer systems must withstand prolonged mechanical stress without irreversible chain degradation or excessive adsorption onto fines. In high-rise pumping, elevated temperatures generated by friction further accelerate hydration and polymer activation, narrowing the operational window. Tailored molecular weight distributions, associative thickening mechanisms, and controlled hydration rates are commonly applied to stabilize yield stress and plastic viscosity throughout the pumping cycle.

From an operational and digital control perspective, pumpability optimization increasingly integrates formulation design with real-time process monitoring. Adjusted addition sequences, pre-conditioning of the mix, and split dosing of VMAs are combined with Industry 4.0-enabled pressure sensing, flow analytics, and predictive modeling to anticipate rheological drift along the pipeline. These strategies support reliable concrete delivery over extended distances and heights while minimizing energy consumption, downtime, and quality variability at the point of placement.

15.4.1 Rheological Failure Mechanism Identification

Field projects involving long-distance pumping reported intermittent pressure spikes and unstable discharge despite acceptable slump flow at the plant. Laboratory vane rheometry and tribological testing indicated excessive shear-thinning behavior combined with insufficient low-shear viscosity, resulting in breakdown of the lubrication layer at pipe walls. Torque data from truck drums showed delayed viscosity recovery after rest, correlating with pipeline blockage risk.

The root cause was traced to a VMA formulation optimized for static stability but insufficiently resistant to prolonged high-shear exposure. Polymer chain scission sensitivity and weak associative interactions limited structural resilience during pumping.

15.4.2 Formulation Adjustment and Operational Impact

The VMA was reformulated to incorporate shear-resilient associative segments while maintaining moderate charge density. Laboratory testing confirmed improved hysteresis recovery and reduced viscosity decay under sustained shear. Field validation demonstrated smoother pressure profiles, reduced pump energy demand, and elimination of unplanned pumping interruptions.

Operationally, pump output variability decreased, and contractor feedback indicated improved placement control at discharge points. The case highlights the importance of coupling rheology characterization with real pumping energy signatures during formulation optimization.

15.5 Hot-Weather Concreting and Delayed Placement Stability

High ambient temperatures fundamentally alter early-age cementitious kinetics by accelerating clinker hydration, intensifying ionic release, and increasing the rate of polymer adsorption onto reactive mineral surfaces. Under such conditions, viscosity-modifying admixtures exhibit faster network development and more rapid water immobilization, which can translate into premature viscosity rise, loss of workability, and reduced placement tolerance during extended transport or queuing. Stabilization in hot-weather production therefore relies on coordinated control of VMA molecular architecture, dosage evolution, and compatibility with retarders or hydration-moderating admixtures, supported by real-time temperature and rheology monitoring. Integration of predictive models and sensor-driven feedback loops enables proactive adjustment of admixture strategies to preserve rheological stability up to delayed placement, aligning hot-weather concreting practice with digitally enabled, resilient production frameworks.

15.5.1 Thermal Sensitivity Mapping

A series of hot-weather placements exhibited rapid slump loss and late-stage viscosity increase despite normal early-age workability. Temperature-conditioned rheological testing revealed strong temperature dependence of polymer–cement interactions, with accelerated network formation above 30 °C. Plant data showed sharp increases in mixer torque during transport delays.

Correlation analysis demonstrated that hydration-driven structuration dominated over water evaporation effects, indicating a chemistry-driven instability mechanism rather than simple water loss.

15.5.2 Temperature-Adaptive Control Strategy

Mitigation combined two measures: reformulation of the VMA toward reduced thermal sensitivity and implementation of a temperature-indexed dosage adjustment model. Digital monitoring of batch temperature triggered small, pre-emptive dosage modifications within validated limits.

Field performance improved markedly, with stable flowability maintained beyond two hours even under peak summer conditions. Rejected loads due to delayed stiffening were reduced by more than 50 percent during the hot-weather season, confirming the value of temperature-aware rheology control.

15.6 High-Fines and Limestone-Rich Binder Systems

Binder systems enriched with limestone fillers and elevated fines content introduce a dense population of reactive surfaces that significantly modify polymer adsorption equilibria and capillary water partitioning. Calcium carbonate surfaces promote rapid VMA uptake and competitive adsorption with dispersants, while the high specific surface area of fines accelerates water immobilization and narrows the effective dosage window for rheology control. Stable performance in such formulations is achieved through tailored VMA molecular design, including controlled charge density and associative group distribution, combined with data-driven analytics that track fines variability and adjust admixture strategies in real time. The convergence of formulation science with sensor-based monitoring and predictive modeling supports robust rheology management in limestone-rich systems while maintaining production flexibility under modern low-clinker, Industry 4.0-oriented concrete technologies.

15.6.1 Adsorption Competition and Over-Structuring

Concrete mixes with elevated limestone filler content exhibited inconsistent viscosity response, ranging from under-stabilization to excessive stiffening. Adsorption studies showed preferential polymer uptake by fine carbonate particles, reducing availability for broader particle network formation. Rheometry revealed sharp yield stress increases at relatively low VMA dosages.

These effects were exacerbated by simultaneous use of highly charged polycarboxylate superplasticizers, intensifying competitive adsorption.

15.6.2 Dosage Window Expansion Through Polymer Design

The VMA was redesigned with reduced affinity for carbonate surfaces and increased steric stabilization capability. Laboratory testing confirmed flatter dosage–response curves and reduced sensitivity to filler fluctuations. Digital tracking of fines content and torque trends allowed continuous validation at plant scale.

The expanded robustness window simplified mix design management and reduced the need for frequent on-site adjustments, particularly in blended cement systems aligned with low-clinker sustainability targets.

15.7 Precast and Low-Water Binder Systems

Precast concrete manufacturing operates within narrowly defined rheological windows at low water-to-binder ratios, where minor fluctuations in viscosity or yield stress rapidly propagate into surface defects, incomplete mold filling, or compaction inefficiencies. In such dense binder matrices, viscosity-modifying admixtures govern lubrication layers and particle network stability under high shear, while simultaneously interacting with accelerators, release agents, and thermal curing regimes. Effective deployment therefore depends on precise VMA molecular tuning, stringent raw material consistency, and digitally integrated process control that couples mixer energy signatures, inline rheology proxies, and curing feedback. The combination of formulation robustness with real-time analytics enables defect minimization and repeatable quality in precast environments characterized by high throughput, automation, and limited tolerance for variability.

15.7.1 Casting Window Control

Precast elements exhibited variability in mold filling and surface finish when binder composition or moisture correction deviated slightly from target values. Rheological analysis identified narrow yield stress margins separating optimal filling from incomplete consolidation.

Torque-based monitoring during mixing provided early indication of deviation from target rheological envelopes, outperforming conventional slump-based checks.

15.7.2 Digital Setpoint Enforcement

A closed-loop control system linked mixer torque to predefined rheological setpoints derived from laboratory optimization. VMA dosage and mixing duration were adjusted automatically within validated limits. Casting consistency improved, surface defect rates declined, and cycle time variability was reduced.

This case demonstrates that VMA performance, when embedded in digitally enforced process control, supports high-throughput, low-variability precast manufacturing.

15.8 Ultra-Low Clinker and High-SCM Binder Systems

The shift toward low-carbon concrete is increasingly driven by ultra-low clinker binders incorporating high proportions of supplementary cementitious materials such as calcined clays, ground granulated blast furnace slag, and natural pozzolans. These binders present heterogeneous surface chemistries, elevated specific surface areas, and modified ionic environments, all of which substantially alter the adsorption, hydration, and activation behavior of viscosity-modifying admixtures. As a result, conventional VMA formulations optimized for Portland cement-rich systems often exhibit reduced efficiency or delayed response in high-SCM matrices.

High-SCM systems typically display slower early hydration kinetics and extended dormant periods, influencing the timing and magnitude of viscosity development. Clay-rich binders, particularly those based on metakaolin or LC^3 concepts, introduce platelet-shaped particles and high surface charge densities that intensify polymer adsorption and water demand. Slag- and pozzolan-dominated systems further modify pore solution composition, affecting polymer solubility and competitive interactions with polycarboxylate ether superplasticizers. These effects can manifest as delayed thickening, increased dosage sensitivity, or unstable rheological evolution over time.

Effective application of VMAs in ultra-low clinker binders requires tailored polymer architectures and process adaptation. Formulation strategies include controlled molecular weight distributions, reduced anionic charge density, and hybrid polymer systems designed to moderate adsorption strength and activation rate. From a production standpoint, optimized addition sequences, split dosing, and temperature-compensated mixing protocols are increasingly combined with Industry 4.0-enabled analytics, such as real-time rheology monitoring and digital hydration modeling, to stabilize fresh-state behavior. Such integrated approaches are essential to unlock the workability, robustness, and durability potential of next-generation low-carbon concrete systems.

15.8.1 SCM Reactivity and Adsorption Imbalance

Concrete mixes incorporating more than 50 percent SCM replacement showed inconsistent viscosity development and erratic segregation resistance. Laboratory adsorption analysis demonstrated preferential VMA uptake on high-surface-area SCM particles, particularly calcined clays, leading to local over-structuring and depletion of free polymer. Time-resolved rheometry revealed delayed but abrupt yield stress increase, incompatible with extended transport requirements.

Plant data confirmed elevated mixer torque after rest periods, often misinterpreted as water deficiency. Traditional dosage escalation aggravated the issue by intensifying adsorption imbalance.

15.8.2 Formulation and Digital Mitigation Strategy

The VMA was reformulated with reduced polar functionality and increased steric stabilization to moderate adsorption strength. In parallel, a digital classifier distinguished SCM blends based on rheological response rather than nominal replacement level. Dosage guidance was adjusted dynamically according to SCM reactivity category.

Field performance stabilized across diverse SCM supply streams, enabling consistent workability retention beyond 120 minutes. The case demonstrates that SCM-rich systems require both chemistry-aware formulation and digitally assisted classification to achieve industrial robustness.

15.9 Early-Age Crack Risk Reduction in Flatwork and Slabs

Viscosity-modifying admixtures play a decisive role in controlling bleeding, settlement, and plastic deformation, all of which directly influence early-age cracking risk in slabs and flatwork. By increasing the cohesiveness of the fresh concrete matrix, VMAs reduce upward water migration and particle settlement, thereby minimizing capillary pressure gradients that drive plastic shrinkage and settlement cracking. Properly balanced formulations contribute to more uniform paste distribution beneath reinforcement and across large surface areas, improving dimensional stability during the critical early hours after placement.

Optimization of VMA dosage must be integrated with mix design parameters, curing practices, and placement conditions to avoid unintended stiffening or surface crusting. Excessive viscosity can restrict internal water redistribution, potentially exacerbating autogenous shrinkage or delaying finishing operations. Conversely, insufficient viscosity control allows localized bleeding and differential settlement, particularly in high-slump flatwork and large-area slabs exposed to wind and elevated evaporation rates. Field validation through bleeding rate measurement, settlement monitoring, and early-age strain tracking is essential to define the optimal operating window.

As part of a comprehensive crack mitigation strategy, VMAs are increasingly combined with controlled addition sequences, evaporation control measures, and curing optimization. Industry 4.0-enabled monitoring platforms, incorporating ambient condition sensors, maturity modeling, and real-time surface moisture tracking, enable predictive adjustment of admixture dosage and finishing timing. Such data-driven integration enhances early-age crack resistance while preserving workability, finishability, and long-term performance in industrial floors, pavements, and architectural slabs.

15.9.1 Bleeding Control Versus Plastic Shrinkage

High-performance floor concretes exhibited reduced bleeding but increased plastic shrinkage cracking under windy and low-humidity conditions. Rheological assessment showed excessively high near-zero shear viscosity, restricting internal water redistribution during evaporation. Field observations linked crack initiation timing to rapid surface stiffening rather than total water loss.

These findings highlighted the need to balance anti-bleeding performance with controlled permeability of the fresh matrix.

15.9.2 Optimized Structuration Window

VMA chemistry was adjusted to delay network percolation while maintaining sufficient stability during placement. Laboratory evaporation simulations confirmed improved surface moisture resilience without compromising segregation resistance. Digital weather data integration enabled proactive adjustment of VMA dosage during high-risk casting windows.

Crack density in treated slabs decreased significantly, and finishing operations became more predictable. The case illustrates how rheology control contributes directly to durability outcomes beyond immediate workability.

15.10 Carbon Capture and Mineralized Binder Compatibility

Binder systems derived from carbon capture and utilization routes, including carbonated calcium silicates, mineralized industrial residues, and CO_2-cured precursors, introduce fundamentally different surface chemistries compared with Portland-based binders. Altered zeta potential, higher carbonate surface density, and accelerated early-age reactions modify polymer adsorption pathways and water structuring, frequently leading to unpredictable viscosity evolution in VMA-containing formulations. Achieving reliable rheology control under these conditions requires reassessment of VMA charge architecture, associative functionality, and competitive adsorption with dispersants, supported by real-time monitoring of reaction kinetics and surface reactivity. Integration of formulation science with data-driven process supervision enables compatibility optimization as carbon-mineralized binders transition from pilot scale to industrial concrete production.

15.10.1 Interaction with Carbonated Phases

Concrete mixes containing mineralized aggregates or carbon-cured binders showed unpredictable viscosity response, including sudden stiffening during mixing. Adsorption testing revealed strong affinity of conventional VMAs for carbonated surfaces, resulting in rapid polymer depletion. Rheological curves showed steep yield stress escalation within minutes.

Plant operators responded with water addition, which restored flow temporarily but compromised stability.

15.10.2 Compatibility-Oriented Control Framework

A compatibility-focused VMA formulation with reduced carbonate affinity was developed. Digital monitoring of torque rise rate served as an early indicator of adverse interaction, triggering corrective dosage adjustments within validated limits.

Field validation demonstrated stable mixing behavior and consistent placement, enabling use of mineralized components without sacrificing fresh-state control. This case underscores the need for admixture design aligned with emerging low-carbon binder technologies.

15.11 Lessons Learned and Key Takeaways

1. System-Level Robustness as a Design Criterion

   Field-oriented case studies confirm that effective viscosity-modifying admixture performance is governed by system-specific variability rather than nominal concrete classification alone. Recycled aggregates, SCM heterogeneity, fines mineralogy, and climatic exposure introduce distinct perturbations to water balance and particle interactions, requiring VMAs engineered with tolerance margins rather than laboratory-optimized peak efficiency. Robust formulations consistently demonstrate moderated adsorption sensitivity and controlled network formation across fluctuating raw material inputs.

2. Predictive Rheology through Digital Enablement

   Digital instrumentation and analytics elevate VMA functionality from passive rheology adjustment to predictive process control. Inline sensors, mixer energy tracking, and digital twins allow early detection of instability trajectories, enabling preemptive dosage or formulation adjustments before non-compliance occurs. Such closed-loop strategies reduce waste, improve delivery reliability, and align admixture deployment with Industry 4.0 manufacturing paradigms.

3. Alignment of Polymer Architecture with Process Dynamics

   Industrial use cases reinforce that VMA molecular design must reflect application-specific deformation histories and time scales. Rapidly associating polymers with strong rebuild characteristics support shape retention in extrusion-based processes, while shear-resilient, low-affinity architectures favor long pumping distances and delayed placement. Attempts to generalize VMA chemistry across incompatible process regimes consistently result in compromised performance.

4. Management of Thermal and Temporal Sensitivity

   Temperature elevation and extended idle periods accelerate hydration and polymer adsorption kinetics, magnifying the risk of premature stiffening or rheology drift. Successful stabilization strategies combine thermally stable VMA chemistries with coordinated use of retarders or accelerators, supported by temperature-aware control logic. Proactive management of time–temperature effects proves more reliable than corrective intervention at placement.

5. Dynamic Compatibility in Complex Binder Systems

   Low-clinker and high-SCM binders introduce evolving competitive adsorption landscapes that challenge fixed admixture strategies. VMAs must be co-optimized with dispersants, air-control agents, and activators to maintain predictable rheology without destabilizing flow or air content. Compatibility is therefore an adaptive property, requiring continuous validation as binder compositions and sourcing change.

6. Sustainability as a Formulation Constraint

   Low-carbon concrete technologies highlight the enabling role of VMAs in stabilizing systems with elevated fines content, recycled materials, and variable water demand. Sustainability-driven formulations succeed when rheology control is treated as a structural requirement rather than an auxiliary adjustment, allowing cement reduction without sacrificing processability or quality consistency.

7. Technology Transfer as Controlled Adaptation

   Transitioning from laboratory development to field implementation emerges as an iterative optimization process grounded in data transparency and cross-disciplinary collaboration. The most effective VMAs are conceived, validated, and refined within the full production ecosystem, integrating material science, digital oversight, and operational feedback into a coherent technology transfer framework.

16. Health, Safety, and Environmental Considerations

Health, safety, and environmental management of viscosity-modifying admixtures must be addressed with the same rigor applied to performance and durability. VMAs are increasingly complex polymer systems supplied in powder or liquid form, often integrated into automated dosing and storage infrastructure. Effective control therefore requires alignment between polymer chemistry, plant operations, regulatory frameworks, and digital documentation systems consistent with Industry 4.0 principles.

Beyond conventional hazard identification, modern VMA stewardship encompasses lifecycle thinking, process safety, and data-driven risk management. Exposure scenarios evolve as admixtures transition from manual handling toward automated bulk delivery, enclosed transfer, and real-time dosing. While automation reduces direct operator contact, it introduces new dependencies on sensor reliability, software integrity, and maintenance discipline, making digital oversight a critical safety component rather than a purely operational tool.

Environmental considerations are similarly multi-dimensional. The direct footprint of polymer admixtures is typically small relative to concrete production, yet their indirect influence on cement reduction, waste minimization, and production efficiency is substantial. Evaluating VMAs therefore requires a functional-unit-based perspective that captures both upstream material impacts and downstream performance benefits. This approach aligns safety and environmental objectives with formulation efficiency and production robustness.

Regulatory expectations continue to expand toward greater transparency, traceability, and documentation accuracy. Compliance is no longer limited to static safety data sheets or periodic audits, but increasingly relies on continuously updated digital records linking formulation, production, logistics, and use. Within this framework, health, safety, and environmental management becomes an integrated element of technology transfer, ensuring that advanced VMA formulations deliver not only technical performance but also safe, compliant, and sustainable deployment across diverse industrial contexts.

16.1 Polymer Handling and Storage Safety

Safe handling of polymer-based VMAs depends on physical form, concentration, and exposure pathways rather than intrinsic chemical hazard alone. Most commercial VMAs are classified as low acute toxicity materials, yet improper storage or handling can introduce operational risks that affect both personnel safety and product integrity.

From an industrial perspective, risk profiles evolve across the supply chain, from bulk delivery and on-site storage to dosing, cleaning, and maintenance operations. Each stage presents distinct exposure mechanisms, including dust generation during powder transfer, slip hazards from liquid spills, and confined-space risks during tank inspection. Effective safety management therefore requires a system-level approach that integrates engineering controls, procedural safeguards, and operator training rather than relying solely on material classification.

Storage systems must prevent moisture ingress, temperature excursions, and cross-contamination. Liquid VMAs require freeze–thaw stability management and corrosion-resistant tanks, while powdered products demand controlled humidity and dust containment. Deviations from recommended storage conditions not only compromise safety but also alter polymer rheology, leading to viscosity drift, microbial growth, or irreversible agglomeration. Such changes can propagate downstream as dosing errors or inconsistent concrete performance.

Automation increasingly reshapes polymer handling practices. Automated transfer and closed dosing systems significantly reduce exposure risk and improve consistency in high-throughput plants, yet they introduce new dependencies on sensor accuracy, valve integrity, and software interlocks. Preventive maintenance, alarm validation, and digital logging of handling events become integral safety measures, ensuring that reduced manual exposure does not translate into hidden systemic risk. Within this context, polymer handling and storage safety emerges as a convergence point between chemical stewardship, process reliability, and Industry 4.0-enabled operational control.

16.1.1 Dust, Slip, and Microbial risks

Dust Generation and Inhalation Control

Powdered viscosity-modifying admixtures present inhalation and housekeeping risks during bag opening, pneumatic transfer, silo charging, and cleaning operations. Fine polymer particles readily become airborne, increasing the likelihood of respiratory exposure and surface contamination. In systems containing organic carriers, dispersed dust clouds may also present low-probability but high-consequence explosibility scenarios under unfavorable ignition conditions. Risk mitigation relies primarily on engineering controls, including enclosed conveying systems, localized exhaust ventilation at transfer points, and antistatic grounding of equipment. Administrative measures such as controlled handling procedures and routine housekeeping support these controls, while personal protective equipment serves as a complementary safeguard rather than the principal risk barrier.

Slip Hazards Associated with Liquid Admixtures

Liquid VMAs introduce a distinct set of physical hazards related to spill-induced loss of traction. Even small leaks or splashes can create hazardous walking surfaces, particularly on smooth concrete or epoxy-coated floors commonly found in admixture handling areas. Effective risk reduction depends on rapid spill response protocols, clearly defined cleanup responsibilities, and the use of containment trays or bunding beneath storage tanks and dosing units. Floor surface treatments with enhanced slip resistance further reduce residual risk in high-traffic zones.

Microbial Growth and Product Hygiene Risks

Aqueous polymer-based VMAs can support microbial proliferation during extended storage, especially under warm conditions or when stagnant zones develop within tanks. Microbial growth manifests as odor formation, viscosity instability, gas generation, or biofilm accumulation, each of which compromises product performance and can disrupt dosing systems. Preventive strategies include incorporation of compatible preservative systems, periodic tank circulation to eliminate dead zones, and scheduled microbiological monitoring as part of routine quality control. Integration of storage temperature monitoring and digital maintenance logs further enhances early detection and mitigation of hygiene-related risks.

16.1.2 Thermal Stability, Shelf-Life, and Degradation Risks

Thermal exposure and storage duration exert a direct influence on the safety, reliability, and performance consistency of viscosity-modifying admixtures. Polymer-based VMAs are generally formulated to withstand typical ambient conditions, yet prolonged exposure to elevated or sub-zero temperatures can initiate physical or chemical degradation pathways that compromise both handling safety and functional efficacy.

Temperature Excursions and Physical Instability

Liquid VMAs are particularly sensitive to freeze–thaw cycles and sustained high-temperature storage. Freezing can induce irreversible phase separation, polymer precipitation, or changes in molecular association that alter viscosity and dosing accuracy. Conversely, elevated temperatures accelerate hydrolysis, oxidative degradation, and preservative depletion, increasing the risk of microbial growth. Temperature-controlled storage, insulation of tanks and pipelines, and automated temperature alarms are therefore essential safeguards in climates with large seasonal variation.

Shelf-Life Management and Product Aging

Shelf life is governed by polymer backbone stability, formulation pH, preservative system robustness, and storage conditions. Aging effects may manifest as gradual viscosity drift, sediment formation, or reduced adsorption efficiency in cementitious systems. These changes are rarely hazardous in a toxicological sense but can create operational risks through inconsistent dosing and unpredictable concrete performance. Clearly defined shelf-life specifications, first-in–first-out inventory management, and periodic quality verification testing are critical to maintaining product integrity.

Degradation Byproducts and Operational Implications

Polymer degradation may generate low-molecular-weight byproducts or volatile compounds that contribute to odor formation or corrosion of storage and dosing equipment. While typically present at low concentrations, such byproducts can trigger operator complaints or maintenance issues if left unmanaged. Selection of compatible construction materials for tanks and seals, combined with routine inspection and digital logging of storage conditions, reduces the likelihood of degradation-related incidents.

By integrating thermal monitoring, shelf-life tracking, and preventive maintenance into digital plant management systems, producers align polymer handling safety with Industry 4.0 principles. This approach ensures that VMAs remain stable, predictable, and safe throughout their intended service life, supporting both operational reliability and regulatory compliance.

16.1.3 Spill Response, Waste Handling, and Emergency Preparedness

Effective spill response and emergency preparedness are essential components of polymer handling safety, particularly as viscosity-modifying admixtures are increasingly stored and dosed in bulk quantities. Although most VMAs do not present acute toxicity hazards, uncontrolled releases can create slip risks, environmental contamination, and operational disruption if not managed systematically.

Spill Containment and Immediate Response Measures

Spill response strategies must reflect the physical form of the VMA. Liquid admixture spills require rapid containment using bunding, absorbent barriers, or dedicated spill kits to prevent spreading across walkways and into drainage systems. Powdered spills demand dust suppression techniques, such as controlled wetting or vacuum recovery with explosion-rated equipment, rather than dry sweeping, which can re-aerosolize fine particles. Clearly defined response procedures, supported by signage and readily accessible equipment, ensure timely and consistent action by plant personnel.

Waste Segregation and Disposal Practices

Recovered VMA material, contaminated absorbents, and cleaning residues must be segregated and disposed of in accordance with local waste regulations and SDS guidance. Although polymer VMAs are typically classified as non-hazardous waste, improper disposal can lead to regulatory non-compliance or environmental impact, particularly if aqueous residues enter surface water systems.

Documentation of waste streams, quantities, and disposal routes supports compliance and facilitates continuous improvement in waste minimization.

Emergency Preparedness and Training

Emergency preparedness extends beyond spill cleanup to include fire response, equipment failure, and loss of containment scenarios. While VMAs generally exhibit low flammability, packaging materials and organic carriers may contribute to fire load. Emergency response plans should therefore integrate admixture storage areas into broader site safety strategies, including fire suppression systems and evacuation procedures. Regular training, drills, and digital incident reporting reinforce preparedness and ensure that response protocols remain effective as formulations, storage volumes, or plant layouts evolve.

By embedding spill response and emergency management into digital safety systems, including incident logging and corrective action tracking, producers strengthen alignment with Industry 4.0 safety frameworks. This proactive approach minimizes risk escalation, supports regulatory compliance, and safeguards both personnel and operational continuity.

16.2 Environmental Footprint and Life Cycle Assessment

Environmental evaluation of viscosity-modifying admixtures encompasses far more than the renewable or fossil origin of the polymer backbone. Meaningful assessment must account for functional efficiency at use level, since VMA dosage, compatibility, and robustness directly influence cement demand, water consumption, energy intensity, and waste generation throughout the concrete value chain. From an industrial perspective, the environmental relevance of a VMA is therefore inseparable from its ability to deliver stable rheology under variable raw material and production conditions.

Life cycle assessment provides a quantitative framework to translate formulation and performance attributes into measurable environmental indicators. Upstream stages include polymer synthesis routes, energy sources, solvent or water usage, and packaging formats, while downstream stages capture impacts associated with transport, storage losses, and operational efficiency in batching plants. In the use phase, VMAs exert disproportionate leverage by enabling reduced paste content, improved segregation resistance with lower cementitious fines, and higher utilization of supplementary cementitious materials or recycled aggregates. These indirect effects often dominate the overall carbon and resource footprint relative to the admixture mass fraction.

For technology transfer and Industry 4.0 implementation, LCA outcomes must be linked to real production data rather than generic datasets alone. Integration of digital batch records, dosage histories, and performance KPIs enables dynamic environmental modeling that reflects actual operating conditions. Such data-driven LCAs support informed selection between alternative VMA chemistries, guide formulation optimization toward low-impact solutions, and provide credible documentation for environmental product declarations and low-carbon concrete strategies.

16.2.1 Bio-Based Versus Synthetic VMAs

Bio-based viscosity-modifying admixtures, commonly derived from polysaccharides, cellulose ethers, starch derivatives, or other modified natural polymers, are often positioned as lower-impact alternatives due to reduced fossil carbon input and partial renewability. From a life cycle perspective, however, their environmental profile is governed by the full value chain rather than feedstock origin alone. Agricultural cultivation introduces land-use change, fertilizer demand, water consumption, and seasonal variability, while downstream chemical modification, purification, and drying steps can be energy-intensive and materially significant in LCA inventories.

At the use phase, bio-based VMAs frequently exhibit higher sensitivity to cement chemistry, fines mineralogy, and temperature, which can translate into elevated dosage requirements or narrower operating windows. When assessed per functional unit of concrete, such as achieving a defined level of static stability or rheology retention, increased dosage and corrective batching may neutralize or exceed upstream carbon advantages. Variability-driven over-dosing also elevates transport emissions and storage losses, particularly in decentralized production networks.

Synthetic VMAs, typically based on tailored polyethers, polyacrylamides, or associative polymers, present higher fossil-derived input at the polymer synthesis stage but deliver consistent performance at significantly lower dosages. Their controlled molecular architecture enables predictable adsorption behavior, reduced sensitivity to raw material fluctuations, and improved robustness across temperature and shear regimes. As a result, life cycle impacts normalized to concrete performance often favor synthetic systems, especially in high-throughput industrial environments where stability and repeatability are critical.

Recent comparative LCA studies emphasize that performance efficiency, dosage stability, and the ability to enable clinker reduction or higher SCM incorporation outweigh polymer origin as primary environmental drivers. Hybrid VMA systems that combine bio-based backbones with selectively engineered synthetic functional groups increasingly represent an optimal compromise. These systems leverage partial renewability while maintaining industrially reliable rheology control, aligning sustainability targets with practical manufacturing and placement requirements under Industry 4.0-driven optimization frameworks.

16.2.2 Contribution to Low-Carbon Concrete

Viscosity-modifying admixtures contribute to carbon reduction primarily through their enabling role rather than through their intrinsic environmental footprint. By stabilizing fresh concrete rheology under conditions of reduced clinker content, VMAs allow the effective deployment of binders with high proportions of supplementary cementitious materials, calcined clays, limestone fillers, and other low-carbon constituents. These systems often exhibit elevated sensitivity to segregation, bleeding, and early-age instability due to altered particle size distributions and surface chemistries. Appropriately engineered VMAs mitigate these effects by controlling water mobility and particle network formation, thereby preserving workability and placement reliability at significantly lower cement intensities.

At the concrete system level, the embodied CO_2 savings achieved through clinker reduction typically exceed the carbon footprint of the VMA by one to two orders of magnitude when normalized per cubic meter. Life cycle assessments consistently show that even modest reductions in clinker factor, enabled by improved rheological stability, dominate environmental outcomes relative to admixture-related impacts. In practical terms, VMAs expand the feasible formulation space for low-carbon concrete, allowing producers to push SCM substitution rates or filler contents beyond thresholds that would otherwise compromise fresh-state performance.

A second, often underestimated, contribution lies in waste avoidance and process efficiency. Low-carbon mixes with narrow robustness margins are more susceptible to variability in raw materials, temperature, and mixing energy, increasing the risk of rejected loads, corrective water addition, or on-site rework. VMAs that deliver consistent stability across these fluctuations reduce such losses, translating directly into avoided embodied emissions associated with disposal, reprocessing, and replacement of non-conforming concrete. From an LCA standpoint, these avoided burdens can rival or exceed the direct benefits associated with clinker reduction alone.

Digital integration further amplifies the environmental value of VMAs in low-carbon concrete systems. Industry 4.0-enabled dosage optimization, informed by real-time sensor data and predictive models, minimizes systematic over-dosing while maintaining performance envelopes. This precision not only reduces admixture consumption but also stabilizes low-carbon formulations at scale, supporting reliable production across multiple plants and climatic conditions. When evaluated holistically, VMAs emerge as critical enablers of decarbonized concrete technologies, with their environmental contribution driven by functional performance, robustness, and data-driven deployment rather than material mass or origin.

16.3 Regulatory Compliance and Documentation

Regulatory compliance for viscosity-modifying admixtures spans chemical safety classification, occupational exposure control, transport regulations, environmental disclosure, and full product traceability across the supply chain. For polymer-based VMAs, compliance obligations are directly influenced by formulation chemistry, physical form, preservative systems, and regional market requirements. The progressive tightening of chemical legislation and the globalization of construction supply chains have transformed compliance from a documentation exercise into a core operational discipline.

Meeting these expectations increasingly requires structured, auditable documentation systems embedded within digital quality management, manufacturing execution, and supply chain platforms. Integration of regulatory data with formulation control, production records, and distribution information enables consistent compliance across sites while supporting rapid response to audits, regulatory inquiries, and customer assessments. Such digitally enabled compliance architectures align with Industry 4.0 principles by combining transparency, traceability, and data-driven governance into routine industrial practice.

16.3.1 SDS Management

Role of Safety Data Sheets in VMA Regulatory Control

Safety Data Sheets represent the foundational mechanism for regulatory communication of chemical hazards, safe handling instructions, storage requirements, and emergency response protocols across the lifecycle of viscosity-modifying admixtures. Within polymer-based VMA portfolios, SDS accuracy is directly linked to formulation complexity, physical state, and intended use conditions. Effective SDS content ensures that downstream users, transport operators, and regulatory authorities receive consistent and technically sound information, reducing liability and operational risk.

Hazard Classification Challenges in Polymer-Based VMAs

Polymeric VMAs present specific classification challenges not always addressed by generic polymer exemptions. Powdered products require careful evaluation of dust explosibility, inhalable particulate fractions, and chronic exposure considerations. Liquid or high-solids VMAs often contain preservatives or stabilizing additives that trigger additional hazard statements under CLP and GHS. Residual monomer levels, although typically low, must be assessed in the context of cumulative exposure and regional regulatory thresholds.

Regulatory Alignment and Continuous Update Obligations

Maintaining SDS validity requires systematic monitoring of regulatory developments under REACH, CLP, GHS, and national occupational safety frameworks. Updates to harmonized substance classifications, biocide approvals, or workplace exposure limits can necessitate immediate SDS revision. For VMA producers operating across multiple jurisdictions, delayed or inconsistent updates represent a significant compliance vulnerability, particularly during customer audits or regulatory inspections.

Digital SDS Management and Industry 4.0 Integration

Digital SDS management systems provide structured control over document lifecycle, including revision history, approval workflows, and language localization. Automated regulatory alerts and centralized data repositories reduce dependence on manual updates and fragmented document storage. Integration of SDS platforms with batch records, formulation databases, and delivery documentation enables traceable linkage between safety information and actual product supply, supporting rapid response during audits or incident investigations. Such digitally enabled SDS governance aligns with Industry 4.0 principles by embedding compliance into routine manufacturing and supply chain operations.

16.3.2 Digital Traceability Systems

Expansion of Compliance from Documentation to Accountability

Digital traceability systems transform regulatory compliance from a static documentation requirement into an active, real-time accountability framework. Rather than relying solely on archived records, traceability platforms capture and correlate operational data across the full lifecycle of viscosity-modifying admixtures. This approach enables manufacturers to demonstrate not only regulatory intent but also verifiable execution under defined manufacturing and quality conditions.

Batch-Level Data Integration and Root-Cause Resolution

At the core of digital traceability is batch-level tracking that links raw material lots, formulation versions, process parameters, and finished product identifiers. For VMA production, this integration is essential for verifying preservative dosing accuracy, confirming declared shelf-life windows, and maintaining formulation consistency across multiple plants. In the event of a quality deviation or safety concern, such structured datasets enable rapid root-cause analysis and targeted corrective actions, minimizing operational disruption and regulatory exposure.

Multi-Site Consistency and Regulatory Confidence

Traceability platforms play a critical role in harmonizing compliance across geographically distributed manufacturing sites. Standardized data models ensure that formulation changes, process adjustments, and quality limits are uniformly applied and documented. This consistency supports regulatory inspections, customer audits, and certification reviews by providing comparable, auditable records regardless of production location.

Industry 4.0 Integration and Closed-Loop Compliance

Integration of traceability systems with plant automation, laboratory information management systems, and enterprise resource planning platforms establishes closed-loop compliance architectures consistent with Industry 4.0 principles. Real-time data capture, automated deviation alerts, and synchronized reporting enable continuous regulatory oversight rather than retrospective verification. Beyond meeting formal regulatory requirements, such digitally enabled transparency strengthens confidence among customers and authorities by evidencing systematic, data-driven control over product safety, quality stability, and environmental performance.

16.3.3 Batch Traceability Framework

End-to-End Data Continuity Across the VMA Lifecycle

Batch traceability frameworks are designed to establish an uninterrupted data chain linking raw material intake, formulation execution, production operations, and customer delivery. For viscosity-modifying admixtures, this continuity is essential due to the sensitivity of performance and compliance to raw material variability, preservative systems, and processing conditions. Each batch is uniquely identified and associated with supplier lot numbers, formulation codes, production timestamps, and packaging records, creating a complete digital fingerprint for every delivered product.

Supplier Integration and Formulation Control

Effective batch traceability begins at the supplier interface. Integration of incoming raw material data, including certificates of analysis and regulatory status, ensures that only compliant inputs enter VMA production. Formulation revisions, even when performance-neutral, are recorded at batch level to maintain regulatory transparency. This level of control is particularly relevant when preservatives, stabilizers, or auxiliary additives are adjusted to meet regional requirements or raw material availability constraints.

Incident Isolation and Regulatory Defense

In the event of a quality deviation, safety concern, or customer complaint, structured batch traceability enables rapid identification and isolation of affected material. Rather than broad, precautionary recalls, manufacturers can implement targeted corrective actions based on verifiable data. Such precision significantly reduces response time, financial impact, and reputational risk while providing defensible evidence during regulatory inspections, customer audits, or contractual dispute resolution.

Alignment with Industry 4.0 Manufacturing Systems

Modern batch traceability frameworks are tightly integrated with manufacturing execution systems, laboratory information management systems, and enterprise resource planning platforms. Automated data capture minimizes manual intervention and ensures real-time accuracy. This integration supports Industry 4.0 strategies by embedding traceability into routine operations, enabling predictive quality analytics and reinforcing data-driven governance across multi-site VMA production networks.

16.3.4 ERP-Centered Documentation Integration

ERP Systems as Compliance Control Hubs

Enterprise resource planning systems function as the central coordination layer for regulatory documentation within VMA manufacturing organizations. By consolidating production planning, quality management, inventory control, and logistics data, ERP platforms provide a single source of truth for compliance-relevant information. This centralized role is critical for ensuring that regulatory documentation accurately reflects real manufacturing and distribution activities rather than parallel, manually maintained records.

Automated Linkage of Regulatory Artifacts

Modern ERP configurations enable automated linkage between production orders, batch records, quality control results, approved SDS versions, and transport documentation. For VMA producers, this linkage ensures that each shipment is accompanied by the correct and current regulatory information, aligned with the specific formulation and destination market. Automation eliminates manual reconciliation steps that commonly introduce transcription errors, version mismatches, or incomplete documentation.

Reduction of Administrative Burden and Compliance Risk

By embedding documentation workflows directly within ERP processes, administrative workload associated with compliance is significantly reduced. Routine activities such as SDS assignment, certificate generation, and regulatory reporting become rule-based and auditable. This systematic integration lowers the risk of non-compliance during inspections or customer audits while freeing technical and quality resources for higher-value activities such as formulation optimization and process improvement.

Industry 4.0 Alignment and Data Consistency

ERP-centered documentation integration supports Industry 4.0 strategies by synchronizing regulatory artifacts with real-time manufacturing and logistics data. Interfaces with manufacturing execution systems, laboratory information management systems, and digital SDS platforms enable closed-loop data consistency across the organization. Such architectures enhance transparency, facilitate rapid regulatory response, and demonstrate robust governance over product safety and compliance across multi-site VMA operations.

16.3.5 Automation and Process Data Capture

Role of Automation in Compliance-Critical Manufacturing

Plant automation systems play a central role in ensuring that VMA production operates within validated and documented process windows. Continuous monitoring of critical parameters such as mixing energy, shear intensity, temperature profiles, residence time, and additive dosing accuracy provides a real-time representation of manufacturing conditions. For polymer-based VMAs, where rheological performance and stability are highly sensitive to processing history, such automation is essential for maintaining both product quality and regulatory conformity.

Data Historians and Regulatory-Grade Record Retention

Automation data achieves compliance value only when archived in structured, tamper-resistant data historians designed for long-term retention. These systems store time-stamped process records linked to specific batch identifiers, enabling reconstruction of production conditions months or years after manufacture. For regulatory audits or customer investigations, historian data provides objective, verifiable evidence that declared manufacturing parameters were consistently achieved.

Evidence-Based Demonstration of Process Control

Automation-derived datasets enable manufacturers to demonstrate process control using quantitative evidence rather than procedural descriptions alone. Statistical analysis of parameter variability supports validation of formulation robustness, scale-up consistency, and multi-site equivalence. Such evidence is increasingly expected by regulators and large construction clients seeking assurance that VMA performance claims are supported by controlled manufacturing practice.

Industry 4.0 Integration and Predictive Compliance

Integration of automation systems with manufacturing execution systems, ERP platforms, and analytics layers enables closed-loop compliance aligned with Industry 4.0 principles. Real-time deviation detection, automated alarm management, and trend-based risk indicators allow proactive intervention before regulatory thresholds are exceeded. This shift from retrospective verification to predictive compliance strengthens governance over product safety, quality stability, and environmental performance across industrial VMA operations.

16.3.6 Compliance Analytics and Risk Monitoring

Consolidation of Compliance-Relevant Data Streams

Compliance analytics platforms function as integrative layers that consolidate structured and unstructured data from SDS management systems, batch traceability frameworks, ERP databases, and automation data historians. For VMA producers, this aggregation enables a unified view of regulatory status across formulations, production sites, and customer markets. By correlating safety documentation, operational records, and quality outcomes, analytics systems convert dispersed compliance data into actionable intelligence.

Trend Analysis and Early Deviation Detection

Longitudinal analysis of compliance indicators allows early identification of emerging risks that may not be apparent at batch level. Recurrent deviations in process parameters, increasing frequency of SDS revisions, or repeated documentation gaps can be detected before they escalate into regulatory non-conformities. For polymer-based VMAs, such trend visibility supports preventive adjustments in formulation control, preservative strategies, or manufacturing discipline.

Predictive Risk Assessment and Proactive Mitigation

Advanced analytics tools apply statistical models and rule-based logic to predict compliance vulnerabilities based on historical patterns. Scenario analysis and risk scoring enable prioritization of corrective actions, focusing resources on high-impact areas. This predictive approach reduces dependence on post-incident investigations and supports more efficient regulatory preparedness across multi-site operations.

Industry 4.0 Alignment and Continuous Improvement

Integration of compliance analytics within Industry 4.0 architectures shifts regulatory management from reactive correction to continuous improvement. Automated dashboards, real-time alerts, and management review interfaces enable data-driven governance at both operational and strategic levels. By embedding compliance intelligence into routine decision-making, VMA manufacturers demonstrate sustained control over product safety, quality stability, and environmental performance, reinforcing confidence among regulators, customers, and certification bodies.

17. Technical Challenges in VM Admixture Selection

The selection of viscosity-modifying admixtures has evolved into a high-risk formulation decision as concrete systems become increasingly multicomponent, low-clinker, and performance-driven. VMAs are no longer simple stabilizers added at the end of mix design; they actively interact with cement phases, supplementary cementitious materials, water chemistry, and other chemical admixtures throughout the fresh-state lifecycle. Small deviations in raw material characteristics or processing conditions can therefore translate into disproportionate rheological and durability consequences at plant and jobsite scale.

Industrial experience shows that many VMA-related failures arise not from insufficient thickening efficiency, but from limited robustness under real production variability. Cement mineralogy drift, recycled water usage, temperature fluctuations, and dosing inaccuracies expose narrow operational windows that are rarely captured during laboratory qualification. As concretes increasingly incorporate calcined clays, fillers, and recycled constituents, traditional selection criteria based on segregation resistance alone become inadequate.

A further challenge lies in aligning rheological control with workability retention, pumpability, and finishing requirements across extended transport and placement times. VMAs influence both static and dynamic yield stress development, often in non-linear ways when combined with PCE superplasticizers. Without mechanistic understanding and digital support tools, these interactions remain largely empirical, leading to conservative overdosing or frequent mix redesign.

The following discussion addresses the dominant technical barriers encountered during VMA selection, spanning multicomponent compatibility, material variability sensitivity, dosage control, environmental conditions, and digital integration gaps. Emphasis is placed on challenges that directly affect industrial scalability, quality consistency, and technology transfer from laboratory development to ready-mix production environments aligned with Industry 4.0 practices.

17.1 Multicomponent Compatibility in Modern Concrete Formulations

Complexity of Admixture Interactions in High-Performance Mix Designs

Modern concrete formulations rarely rely on a single chemical admixture. VMAs are introduced into systems already containing PCE superplasticizers, set modifiers, air-entraining agents, shrinkage reducers, and sometimes corrosion inhibitors. Each component competes for solvation water, ionic species, and cement surface adsorption sites. VMAs with high anionic charge density or strong hydrogen-bonding capacity may partially displace PCEs from C_3S and C_3A surfaces, reducing dispersion efficiency and altering yield stress development. From an industrial perspective, incompatibility often manifests as unexpected slump loss, excessive viscosity build-up, or delayed strength development rather than obvious phase separation.

Competitive Adsorption and Polymer–Polymer Interference

At the molecular level, multicomponent incompatibility is driven by competitive adsorption and solution-phase complexation. VMAs based on polyacrylamide or associative polymers can interact with PCE side chains through calcium bridging or hydrophobic association, modifying effective molecular conformation. Such interactions can suppress the steric repulsion mechanism essential for PCE performance. In low water-to-binder ratios, even weak polymer–polymer associations become rheologically significant. Selection protocols must therefore evaluate combined adsorption isotherms and rheological fingerprints rather than relying on isolated admixture testing.

Influence of Supplementary Cementitious Materials and Fillers

The inclusion of SCMs such as slag, fly ash, calcined clays, and limestone fillers adds another layer of compatibility risk. High surface area SCMs can preferentially adsorb VMAs, reducing their availability in the pore solution and shifting the balance between static stability and flow. Calcined clays with heterogeneous surface charges are particularly challenging, as they can selectively bind cationic or weakly anionic polymer segments. Industrially robust VMAs must maintain functionality across SCM loadings without requiring formulation-specific redesign.

Water Chemistry and Recycled Water Effects

Process water quality is frequently overlooked during VMA selection. Recycled wash water introduces dissolved solids, residual admixtures, and fine particles that can pre-condition polymer behavior before cement contact. Elevated sulfate, carbonate, or organic residues may alter VMA hydration kinetics and associative strength. In multicomponent systems, these effects are amplified, leading to batch-to-batch variability. Industrial qualification should therefore include sensitivity testing against realistic water chemistries representative of ready-mix operations.

Compatibility Testing and Industry 4.0-Oriented Screening

Traditional pass–fail compatibility tests are insufficient for complex admixture systems. Advanced rheometry, calorimetry, and adsorption analysis provide deeper insight but remain underutilized at plant scale. Emerging Industry 4.0 approaches combine inline viscosity sensors, digital material passports, and machine-learning models to predict incompatibility risks based on raw material fingerprints. Integrating these tools into VMA selection workflows enables proactive mitigation rather than reactive troubleshooting during production.

Industrial Implications for VMA Selection

From a formulation and manufacturing standpoint, multicomponent compatibility must be treated as a core performance criterion equal to segregation resistance or pumpability. VMAs optimized for single-admixture systems often fail when exposed to the full chemical complexity of modern concrete. Successful selection strategies prioritize tolerance, predictability, and digital traceability over maximum laboratory performance.

17.2 Sensitivity to Cement Mineralogy and SCM Variability

Influence of Cement Phase Composition on VMA Performance

Viscosity-modifying admixtures exhibit pronounced sensitivity to cement mineralogy due to their reliance on adsorption, ionic bridging, and solution-phase structuring mechanisms. Variations in C_3A content, sulfate form, and alkali availability directly affect polymer anchoring kinetics and conformation at solid–liquid interfaces. High C_3A cements tend to accelerate VMA adsorption, often leading to premature viscosity build-up and reduced workability windows, while low-aluminate systems may delay polymer activation and compromise early segregation resistance. Such mineralogical dependencies complicate cross-plant formulation transfer and require cement-specific validation during industrial deployment.

Role of Sulfate Balance and Early Hydration Chemistry

The sulfate balance governs early aluminate reactions and indirectly controls the availability of calcium ions critical for VMA association. Changes in gypsum type, fineness, or sulfate dosage can shift the timing of ettringite formation, altering pore solution chemistry during the first minutes after mixing. VMAs based on polyacrylamide or associative backbones are particularly sensitive to these dynamics, as their structuring efficiency depends on calcium-mediated interactions. Inconsistent sulfate control therefore manifests as batch-to-batch variability in viscosity development, even at constant admixture dosage.

Impact of Supplementary Cementitious Materials

SCM incorporation amplifies mineralogical sensitivity through increased surface area, heterogeneous charge distributions, and delayed dissolution behavior. Ground granulated blast-furnace slag introduces latent hydraulic phases that modify calcium availability over time, while fly ash variability affects early adsorption equilibria. Calcined clays and metakaolin present a more severe challenge due to their high reactivity and mixed surface chemistry, which can selectively bind VMA segments and reduce effective concentration in the pore solution. Limestone fillers further influence VMA response by diluting clinker phases and altering particle packing density, thereby shifting rheological balance.

Variability in SCM Quality and Supply Chains

Industrial SCMs often display wider compositional and physical variability than Portland cement. Fluctuations in fineness, amorphous content, loss on ignition, and impurity levels translate into inconsistent VMA demand and response. From a formulation standpoint, VMAs optimized for a nominal SCM specification may fail when exposed to supply chain drift or seasonal changes. Robust VMA selection must therefore account for worst-case SCM variability rather than average properties alone.

Mitigation Strategies and Industry 4.0 Approaches

Addressing mineralogical sensitivity requires a combination of polymer design, testing strategy, and digital support. VMAs with broader molecular weight distributions and moderated charge density generally exhibit higher tolerance to mineralogical shifts. Advanced qualification protocols increasingly integrate adsorption measurements, time-resolved rheology, and calorimetry under variable cement and SCM scenarios. Industry 4.0 tools, including raw material fingerprinting, predictive modeling, and adaptive dosing algorithms, offer practical pathways to stabilize performance across mineralogical variability without continuous manual recalibration.

17.3 Rheology Control Versus Workability Retention Trade-Off

Static Stability Requirements Versus Dynamic Flow Demands

Viscosity-modifying admixtures are primarily introduced to increase static yield stress and suppress segregation, bleeding, and particle settlement. However, the same structuring mechanisms that enhance static stability often penalize dynamic flow under shear. VMAs based on high molecular weight or strongly associative polymers can induce excessive plastic viscosity, leading to reduced slump flow, impaired self-leveling, and increased pumping pressures. In practical formulations, achieving sufficient static stability without sacrificing dynamic flow remains one of the most delicate balance points in VMA selection.

Time-Dependent Rheological Evolution

The trade-off between rheology control and workability retention is strongly time-dependent. Many VMAs continue to hydrate, swell, or associate after initial mixing, progressively increasing viscosity during transport and placement. When combined with PCE superplasticizers, this evolution can become non-linear due to competitive adsorption and delayed steric stabilization. Concrete that appears workable at discharge may therefore stiffen prematurely, creating finishing difficulties and increasing the risk of cold joints. Industrial qualification must extend beyond initial rheology and capture viscosity evolution over realistic transport durations.

Interaction with Superplasticizers and Shear History

Workability retention is governed not only by admixture chemistry but also by shear history. High-shear mixing can temporarily disrupt VMA networks, masking their true structuring potential. As shear is reduced during rest periods or low-energy pumping segments, viscosity may rapidly recover or overshoot. The presence of PCE superplasticizers further complicates this behavior, as VMA-induced viscosity can limit the effective dispersion of cement particles. Selection strategies must therefore evaluate rheological response under cyclic shear conditions representative of actual processing and placement.

Influence of Fines Content and Particle Packing

Modern concretes increasingly rely on high fines content and optimized particle packing to achieve performance and sustainability targets. While such systems benefit from improved cohesiveness, they also amplify the impact of VMAs on rheology. Fine fillers and SCMs increase surface area and water demand, narrowing the margin between adequate stability and excessive viscosity. VMAs that perform well in conventional mixes may thus become overly restrictive in high-fines formulations, necessitating chemistry-specific tuning.

Industrial Strategies for Managing the Trade-Off

Mitigating the rheology–workability conflict requires a combination of polymer design and process control. Weakly associative VMAs or systems with delayed activation profiles can provide stability without immediate viscosity penalties. From an operational standpoint, optimized dosing sequences, split additions, and temperature-adjusted formulations improve workability retention. Industry 4.0-enabled rheology monitoring, coupled with predictive models, allows dynamic adjustment of admixture dosage to maintain target flow profiles throughout production and placement.

17.4 Dosage Window Narrowness and Overdose Risk

Steep Dose–Response Behavior of VMAs

Many viscosity-modifying admixtures exhibit highly non-linear dose–response characteristics, where small changes in addition level produce disproportionate rheological effects. This behavior is particularly pronounced for high molecular weight synthetic polymers and strongly associative systems. Once a critical concentration is exceeded, polymer chain overlap and network formation accelerate rapidly, resulting in sharp increases in yield stress and plastic viscosity. In industrial practice, such steep response curves translate into narrow operational dosage windows that are difficult to control under normal batching variability.

Consequences of Overdosing in Industrial Concrete

Overdosing VMAs rarely manifests as an immediately obvious defect but instead produces a cascade of secondary issues. Excessive viscosity can trap air, reduce consolidation efficiency, and impair surface finishing. In pump applications, overdosing increases pressure demand and energy consumption while elevating the risk of line blockage. In extreme cases, strong polymer structuring can retard hydration kinetics, delay setting, and compromise early strength development. These effects are often misattributed to cement quality or superplasticizer inefficiency, obscuring the root cause.

Underdosing and False-Negative Performance Assessment

While overdose risks are more visible, underdosing presents its own technical challenge. In highly optimized or high-fines systems, small VMA deficits may not be apparent in laboratory tests but can lead to segregation, bleeding, or washout resistance failure under field conditions. The proximity of under- and over-effective dosage levels compresses the usable range, increasing sensitivity to plant-scale dosing accuracy and water content variability.

Interaction with Dosing Accuracy and Process Variability

Ready-mix plants operate under inherent dosing tolerances influenced by equipment calibration, admixture viscosity, temperature, and delivery logistics. Narrow VMA dosage windows amplify the impact of these tolerances, converting minor metering deviations into measurable performance shifts. Recycled water, residual admixture carryover, and sequence of addition further contribute to effective dosage uncertainty. Robust VMA selection must therefore consider plant realities rather than idealized laboratory precision.

Mitigation Strategies and Digital Control Approaches

Reducing overdose risk requires both formulation and process-level interventions. VMAs with broader molecular weight distributions and moderated association strength offer smoother dose–response behavior. Split dosing, delayed addition, and temperature-compensated formulations can further stabilize performance. Industry 4.0-enabled dosing systems, inline rheology sensors, and model-based control algorithms provide practical tools to detect early signs of over-structuring and adjust dosage in real time, improving consistency and reducing waste.

17.5 Temperature and Mixing Energy Dependence

Temperature-Driven Polymer Activation Kinetics

Viscosity-modifying admixtures rely on polymer hydration, swelling, and associative interactions that are inherently temperature dependent. At low temperatures, polymer chain mobility and water uptake are reduced, delaying viscosity development and weakening early static stability. Conversely, elevated temperatures accelerate polymer hydration and association, often leading to premature viscosity build-up and shortened workability windows. In hot climates, this effect is amplified by increased cement dissolution rates and higher calcium ion availability, intensifying VMA structuring beyond laboratory expectations.

Interaction Between Temperature and Cement Hydration

Temperature not only affects the polymer itself but also modifies early cement hydration chemistry. Faster aluminate and silicate reactions at higher temperatures alter pore solution composition during the critical first minutes after mixing.

VMAs that depend on calcium-mediated interactions or hydrogen bonding respond sensitively to these changes, resulting in temperature-specific rheological profiles. Without temperature-adapted selection or dosing strategies, identical formulations can exhibit divergent performance across seasonal conditions.

Influence of Mixing Energy and Shear History

Mixing energy governs the dispersion and activation state of VMAs. High shear promotes rapid polymer distribution and temporarily disrupts associative networks, reducing apparent viscosity during mixing. As shear decreases during transport or rest, polymer networks may reform, leading to viscosity recovery or overshoot. Low-energy mixing, common in some ready-mix operations, may fail to fully disperse the VMA, causing localized over-concentration and inconsistent rheology. These effects highlight the importance of evaluating VMAs under realistic mixing regimes rather than standardized laboratory protocols alone.

Coupled Effects in Ready-Mix Operations

In industrial environments, temperature and mixing energy rarely vary independently. Hot weather often coincides with shorter mixing times to control slump loss, while cold conditions may involve extended mixing to compensate for slow hydration. These coupled variables can shift VMA performance outside the intended operating window. Plant-to-plant differences in mixer type, rotational speed, and truck agitation further complicate reproducibility, increasing the risk of inconsistent field behavior.

Mitigation Strategies and Industry 4.0 Integration

Managing temperature and mixing sensitivity requires both chemical and digital solutions. VMAs with controlled hydration rates and reduced temperature sensitivity provide a wider performance envelope. Operational measures such as temperature-adjusted dosing, optimized addition sequences, and minimum mixing energy thresholds improve consistency. Industry 4.0 tools, including real-time temperature monitoring, mixer energy tracking, and predictive rheology models, enable adaptive control strategies that stabilize VMA performance across varying environmental and process conditions.

17.6 Digitalization and Industry 4.0 Integration Gaps

Empirical Nature of Current VMA Selection Practices

Despite the increasing chemical and mineralogical complexity of modern concrete, VMA selection remains largely empirical. Laboratory trials are typically based on static segregation tests and single-point rheology measurements that fail to capture time-dependent behavior and plant-scale variability. The absence of structured data integration across raw materials, admixture chemistry, and process parameters limits the ability to predict VMA performance beyond controlled test conditions. As a result, formulation decisions rely heavily on experience and conservative safety margins rather than quantitative, transferable models.

Limited Deployment of Real-Time Rheology Monitoring

While rheometers and inline sensors capable of tracking viscosity, yield stress, and flow evolution exist, their penetration into ready-mix and precast operations remains limited. High cost, maintenance requirements, and lack of standardized interpretation frameworks restrict widespread adoption. Without real-time feedback, VMA overdosing or underperformance is often detected only after quality deviations occur, increasing rework, waste, and customer dissatisfaction. The gap between available sensor technology and its practical deployment represents a critical bottleneck in digital transformation.

Data Fragmentation Across the Concrete Value Chain

Concrete production data are typically fragmented across batching systems, laboratory databases, and quality reports. Raw material variability, admixture lot changes, and environmental conditions are rarely linked in a unified digital framework. This fragmentation prevents the development of reliable cause–effect correlations between VMA dosage and fresh concrete behavior. For VMAs, which exhibit narrow operational windows and strong interaction effects, the lack of integrated data severely constrains predictive capability and process optimization.

Absence of Predictive and Adaptive Control Models

True Industry 4.0 implementation requires predictive models capable of anticipating rheological drift and adapting admixture dosage in real time. For

VMAs, such models remain at an early stage of development. Machine-learning approaches show promise but are often trained on limited datasets that do not reflect long-term raw material variability. Without validated digital twins of fresh concrete rheology, plants continue to operate in reactive mode, adjusting VMA dosage only after deviations are observed.

Pathways Toward Digital Integration

Bridging the digitalization gap demands coordinated advances in sensing, data management, and modeling. Digital material passports capturing cement mineralogy, SCM properties, and admixture chemistry provide a foundation for predictive control. Coupling inline rheology and temperature sensors with adaptive dosing algorithms enables dynamic VMA adjustment within safe operating windows. From a technology transfer perspective, success depends on translating advanced tools into robust, user-friendly systems compatible with existing plant infrastructure.

17.7 Sustainability and Regulatory Constraints

Environmental Footprint of VMA Chemistries

Sustainability considerations increasingly influence the selection of viscosity-modifying admixtures, extending beyond performance into environmental impact and regulatory compliance. Many conventional VMAs are based on synthetic polymers derived from petrochemical feedstocks with non-negligible carbon footprints. Life cycle assessments indicate that, while VMA dosages are low, their upstream emissions and contribution to overall admixture complexity can become relevant in low-clinker and low-carbon concrete systems. Industrial users are therefore under growing pressure to justify VMA selection not only on technical merit but also on quantified environmental performance.

Shift Toward Bio-Based and Low-Impact VMAs

Bio-based VMAs derived from cellulose ethers, starch derivatives, and modified polysaccharides are gaining attention due to their renewable origin and favorable regulatory perception. However, these materials often exhibit broader molecular weight distributions, higher sensitivity to water chemistry, and reduced robustness under temperature and shear variability. From a formulation standpoint, the challenge lies in achieving performance consistency comparable to synthetic VMAs while maintaining acceptable shelf life, microbial stability, and compatibility with modern admixture packages.

Regulatory Constraints and Chemical Compliance

Regulatory frameworks increasingly restrict residual monomer content, impurity profiles, and hazard classifications of chemical admixtures. Requirements related to worker safety, transport, and environmental release place constraints on polymer synthesis routes and additive formulations. In some jurisdictions, biodegradability claims and eco-labeling criteria further limit the use of certain synthetic backbones or functional groups. These constraints reduce formulation flexibility and may necessitate trade-offs between rheological performance and regulatory acceptance.

Impact on Recycling and Circular Concrete Strategies

VMAs influence not only fresh concrete behavior but also downstream recycling and circularity. Polymer residues can affect the quality of recycled aggregates and process water, potentially interfering with subsequent admixture performance. As concrete recycling intensifies, regulatory scrutiny of cumulative polymer loading increases. Selecting VMAs compatible with closed-loop water systems and mechanical recycling processes becomes a strategic requirement rather than an optional sustainability feature.

Balancing Compliance, Performance, and Industrial Robustness

Achieving alignment between sustainability goals, regulatory compliance, and industrial robustness remains a central technical challenge. Hybrid VMA systems combining bio-based components with synthetic stabilizers offer one pathway to balance performance and environmental impact. Industry 4.0 tools, including life cycle data integration and digital compliance tracking, support informed decision-making by linking VMA selection to quantified sustainability metrics and regulatory status across markets.

17.8 Summary of Key Technical Challenges and Mitigation Strategies

Overview of Dominant Technical Barriers

Viscosity-modifying admixture selection faces multifaceted challenges driven by increasing concrete formulation complexity, production variability, and stringent performance demands. Key issues include multicomponent compatibility, sensitivity to cement and SCM variability, narrow and steep dosage windows, temperature and mixing energy dependence, limited digital integration, and evolving sustainability and regulatory pressures. These factors interact synergistically, complicating the reliable transfer of laboratory-optimized VMAs to industrial-scale ready-mix and precast operations.

Mitigation Strategies and Formulation Best Practices

Table 17.8 summarizes the principal technical challenges encountered in viscosity-modifying admixture selection, their impacts on fresh concrete performance, and corresponding mitigation strategies. Multicomponent compatibility issues, which can cause loss of dispersion and unpredictable rheology shifts, are addressed through matrix-based screening and tailored polymer chemistry. Variability in cement mineralogy and SCMs leads to batch inconsistencies, mitigated by employing broad molecular weight distributions and adaptive dosing guided by raw material fingerprinting. The narrow dosage windows characteristic of many VMAs increase quality risks, which can be controlled by split dosing, moderating polymer association strength, and implementing inline rheology monitoring. Temperature fluctuations and mixing energy variations introduce seasonal and process-driven performance variability; these are managed via temperature-adjusted formulations, minimum mixing energy protocols, and digital process monitoring. Limited digital integration results in reactive control and inefficiencies, which can be overcome by adopting inline sensors, predictive AI models, and digital twins to enable proactive rheology management. Finally, sustainability and regulatory constraints narrow formulation options and elevate environmental impact concerns, encouraging the development of hybrid bio-synthetic VMAs combined with life cycle assessments and regulatory compliance tracking to balance performance with environmental stewardship.

Table 17.8: Technical Challenges. Impact on Concrete Performance, and Mitigation Strategies

Technical Challenge	Impact on Concrete Performance	Mitigation Strategies
Multicomponent Compatibility	Loss of dispersion, unexpected rheology shifts	Matrix-based screening, adsorption analysis, polymer chemistry tuning
Cement Mineralogy and SCM Variability	Batch-to-batch inconsistency, performance drift	Broad molecular weight distributions, raw material fingerprinting, adaptive dosing
Dosage Window Narrowness and Overdose Risk	Quality risk, mixing difficulties	Split dosing, moderated association strength, inline rheology monitoring
Temperature and Mixing Energy Dependence	Seasonal and process variability	Temperature-adjusted formulations, minimum mixing energy protocols, digital monitoring
Limited Digitalization and Industry 4.0 Gaps	Reactive control, inefficiencies	Integration of inline sensors, predictive AI models, digital twins of fresh concrete rheology
Sustainability and Regulatory Constraints	Restricted formulation space, environmental impact	Hybrid bio-synthetic systems, life cycle assessment integration, regulatory compliance tracking

Integration of Digital Tools for Enhanced Control

Industry 4.0 technologies provide critical leverage to overcome these challenges through enhanced data collection, real-time monitoring, and model-based control. Digital twins, machine learning algorithms, and adaptive dosing systems enable dynamic optimization of VMA dosage, ensuring consistent fresh concrete properties despite raw material and environmental variability. Implementation of such tools is pivotal for scalable, robust, and sustainable concrete production.

Future Outlook

Continued advances in polymer chemistry, coupled with expanding digitalization and increasing environmental regulation, will progressively reshape VMA selection paradigms. Emphasis will shift toward designing admixtures with intrinsic robustness, sustainability credentials, and compatibility with intelligent process control frameworks. These developments will support the broader goals of low-carbon, high-performance concrete manufacturing aligned with Industry 4.0 principles.

18. Technical Challenges in VMA Application

Applying viscosity-modifying admixtures (VMAs) at industrial scale extends beyond chemical formulation, encompassing practical issues related to plant operations, environmental conditions, equipment capabilities, and workforce expertise. The transition from controlled laboratory environments to dynamic field conditions reveals multiple application-specific challenges that directly affect VMA efficacy and concrete performance. These challenges include variability in site and process parameters, complex dosing and addition protocols, water quality impacts, equipment limitations, and gaps in operator training. Addressing these issues is critical to achieving consistent rheology control, minimizing waste, and ensuring quality assurance across production and placement stages.

18.1 Variability in Field Conditions and Operational Practices

Influence of Mixer Type and Mixing Regime

The design and operational parameters of concrete mixers significantly affect the dispersion and activation of viscosity-modifying admixtures. Drum mixers, pan mixers, and planetary mixers each impose different shear profiles and mixing intensities. High-shear mixers typically enhance polymer dispersion and accelerate network formation, resulting in faster viscosity development. Conversely, low-shear or incomplete mixing can lead to localized polymer agglomeration or under-dispersion, causing heterogeneity in rheological properties. Mixing time also plays a critical role; insufficient mixing may prevent full polymer activation, while excessive mixing can degrade polymer chains or accelerate hydration, complicating workability control.

Water Quality Variations and Chemical Environment

Water source variability, including the use of potable, recycled, or soft water, introduces fluctuations in ionic strength, pH, and dissolved solids content. These parameters influence the hydration state of VMAs, affecting polymer chain conformation, associative interactions, and adsorption on cementitious particles. High sulfate or carbonate levels, for instance, can compete with polymer functional groups for calcium ions, weakening the VMA's thickening effect. Variations in water temperature also affect polymer swelling kinetics and viscosity evolution. Inconsistent water quality across batches or sites often manifests as unpredictable rheological behavior and unstable workability retention.

Ambient Temperature Effects on Rheology

Environmental temperature directly impacts both polymer chemistry and cement hydration kinetics. At low temperatures, polymer chain mobility and swelling are reduced, delaying viscosity buildup and potentially causing insufficient static stability. Elevated temperatures accelerate polymer association and cement hydration, often leading to rapid stiffening and shorter workability windows. Seasonal temperature swings, common in many geographic regions, require adaptive VMA dosing or formulation strategies to maintain consistent fresh concrete behavior.

Transport Distance and Time Impact on VMA Performance

The time elapsed between mixing and placement influences the time-dependent rheological evolution governed by VMA hydration and network formation. Long haul distances or extended holding times in transit can lead to excessive viscosity increase, reducing pumpability and finishing ease. Conversely, short transport durations may not allow sufficient viscosity development for adequate segregation resistance. Variable transport conditions, such as intermittent agitation or vibration, further complicate the kinetics of viscosity evolution, necessitating precise control of admixture dosing and timing.

Operational Practice Variability and Batch-to-Batch Consistency

Differences in batching sequence, operator skill, and equipment maintenance contribute to inconsistent admixture dosing and mixing effectiveness. In some plants, changes in shift or operator result in different mixing times, addition sequences, or dosing accuracy. These operational inconsistencies produce variability in fresh concrete rheology and workability, complicating quality control and customer satisfaction. Standardized operating procedures and rigorous process control are often lacking, especially in smaller or decentralized production sites.

Table 18.1: Impact of Field Condition Variability on VMA Performance and Concrete Quality

Variable Factor	Effect on VMA Performance	Impact on Concrete Fresh Properties	Industrial Mitigation Strategies
Mixer Type and Mixing Time	Incomplete polymer dispersion; network variability	Inconsistent viscosity; segregation risk	Standardized mixing protocols; mixer capability assessment
Water Quality and Chemistry	Altered polymer hydration and adsorption	Variable viscosity; workability fluctuations	Water quality monitoring; pre-treatment or blending
Ambient Temperature	Altered polymer kinetics and cement hydration	Viscosity delay or acceleration; slump loss	Temperature-adjusted dosing; seasonal formulations
Transport Distance and Time	Time-dependent viscosity evolution	Over- or under-structured concrete; pumpability issues	Optimized transport planning; adjusted admixture timing
Operational Practices	Variability in dosing accuracy and mixing	Batch-to-batch inconsistency; quality deviations	Staff training; process standardization; automation

18.2 Dosing Complexity and Addition Sequence Optimization

Criticality of Precise VMA Dosing

Viscosity-modifying admixtures require dosing within narrow concentration ranges due to their steep dose–response curves. Minor deviations from the optimal dosage can lead to either insufficient viscosity modification, risking segregation and bleeding, or excessive thickening, causing workability loss and pumping difficulties. Achieving precise dosing in industrial settings is challenged by equipment calibration variability, admixture viscosity changes due to temperature, and operator inconsistencies. These factors necessitate stringent quality assurance measures and routine calibration protocols to maintain dosing accuracy.

Importance of Addition Sequence

The sequence in which VMAs are introduced relative to other admixtures, especially polycarboxylate ether (PCE) superplasticizers, significantly influences polymer adsorption and rheological behavior. Adding VMAs prior to superplasticizers may inhibit PCE adsorption on cement particles due to competitive surface coverage or altered ionic environments, diminishing dispersion effectiveness. Conversely, delayed VMA addition may result in incomplete polymer dispersion or rapid viscosity increase, impairing homogeneity. Optimizing addition sequences requires understanding polymer–cement interactions and mixing dynamics to balance dispersion and viscosity development.

Impact of Mixing Stage on Polymer Activation

The mixing stage at which VMAs are added determines their hydration, swelling, and associative behavior. Early addition during high-shear mixing promotes uniform polymer distribution but may accelerate viscosity increase before placement. Late addition may preserve workability but risks uneven polymer activation and localized over-thickening. Split dosing strategies, dividing total VMA dosage across different mixing stages, are increasingly used to extend workability windows and enhance stability while minimizing adverse effects.

Lack of Standardized Protocols and Operational Risks

Variation in dosing and addition practices across plants and operators introduces significant operational risks. Absence of standardized protocols leads to inconsistent polymer activation, uneven fresh concrete properties, and challenges in process reproducibility. This risk is amplified when coupled with narrow VMA dosage windows, where small procedural deviations cause outsized performance impacts. Establishing clear, plant-specific dosing guidelines and operator training programs is essential to mitigate these risks.

Table 18.2: VMA Dosing and Addition Sequence Factors

Factor	Impact on VMA Performance	Recommended Mitigation
Dosing Precision	Over- or under-dosing; workability loss or segregation	Routine pump calibration; automated dosing controls
Sequence Relative to Superplasticizers	Competitive adsorption; impaired dispersion	Add superplasticizers first, followed by VMAs; test plant-specific sequences
Mixing Stage Timing	Polymer dispersion vs. premature viscosity build-up	Use split dosing; adjust addition timing per mix design
Operator Variability	Process inconsistency; batch variability	Standardized protocols; operator training

18.3 Influence of Water Quality and Residual Admixtures

Impact of Water Quality Variability on Polymer Performance

Water is a critical component in concrete mixing, directly influencing the behavior of viscosity-modifying admixtures (VMAs) through its chemical composition and physical properties. Variations in water quality, such as differences in hardness, pH, ionic strength, and dissolved solids, alter polymer hydration, swelling, and associative interactions. For example, high sulfate or carbonate content competes with polymer functional groups for calcium ions, disrupting polymer–cement particle interactions and reducing thickening efficiency. Similarly, soft or deionized water can lead to enhanced polymer swelling but may destabilize cement particle dispersion. Temperature and microbial content in water also influence VMA behavior by affecting polymer conformation and stability over time. These fluctuating parameters cause variability in fresh concrete rheology, complicating consistent performance.

Challenges Posed by Recycled and Process Water

The increasing use of recycled process water from concrete washing, batching, or return flows introduces additional chemical complexity. Such waters typically contain elevated levels of dissolved solids, fines, residual admixtures, and organic matter, which can modify ionic strength and chemical equilibria in the mix. The presence of residual polymers or surfactants may lead to competitive adsorption phenomena, affecting VMA action and interaction with superplasticizers. Variability in recycled water composition over time further amplifies unpredictability, demanding robust monitoring and adaptive process control.

Effects of Residual Admixtures from Previous Batches

Residual admixtures remaining in mixing equipment, trucks, or pipelines from prior batches introduce another layer of complexity. These carryovers change the effective chemical environment by altering ionic balance, adsorption sites, or polymer concentration. Residual superplasticizers can inhibit VMA adsorption or modify dispersion behavior, while leftover VMAs may cause unexpected viscosity spikes. The cumulative effect often results in non-linear rheology changes that are difficult to anticipate without detailed tracking of admixture usage and cleaning cycles.

Monitoring and Control Strategies for Water and Residual Effects

To mitigate the influence of water quality and residual admixtures on VMA performance, comprehensive monitoring of water chemistry and admixture carryover is essential. Implementing routine water testing for key parameters such as sulfate, chloride, hardness, and organic content allows timely adjustments to VMA dosing or formulation. Process water treatment or blending can stabilize chemical conditions. Rigorous cleaning and flushing protocols reduce residual admixture accumulation, while digital tracking systems facilitate real-time assessment of chemical environment changes. Combining these practices with inline rheology sensors enables dynamic control of VMA dosage to maintain target concrete consistency.

Table 18.3: Water Quality and Residual Admixture Influences on VMA Performance

Factor	Impact on VMA Behavior	Concrete Rheology Consequences	Mitigation Strategies

High Sulfate/Carbonate Content	Competitive ion binding; reduced polymer adsorption	Lower viscosity; increased segregation risk	Water quality monitoring; use of sulfate-tolerant VMAs
Soft or Deionized Water	Enhanced polymer swelling; destabilized dispersion	Excessive viscosity or unstable workability	Water blending; formulation adjustments
Recycled Process Water	Variable dissolved solids and organics; admixture carryover	Rheology variability; inconsistent viscosity	Process water treatment; monitoring and adaptive dosing
Residual Admixtures	Altered ionic balance; competitive adsorption	Non-linear viscosity changes; premature stiffening	Rigorous cleaning; admixture tracking systems

18.4 Equipment Limitations and Process Control Constraints

Limitations in Metering Accuracy and Admixture Delivery

One of the primary technical challenges in the industrial application of viscosity-modifying admixtures (VMAs) stems from the inherent limitations of dosing equipment commonly used in ready-mix plants. Traditional dosing pumps often lack the precision necessary to maintain tight admixture concentration windows, especially for VMAs exhibiting steep dose–response curves. Variations in pump calibration, mechanical wear, temperature-dependent changes in admixture viscosity, and inconsistent pump priming contribute to dosing inaccuracies. Such fluctuations can lead to under- or overdosing, compromising concrete rheology and workability. Moreover, manual interventions and infrequent maintenance exacerbate these issues, making it difficult to achieve consistent dosing across shifts or multiple production sites.

Absence of Real-Time Rheological Feedback and Adaptive Control

Most ready-mix operations currently operate without inline rheological monitoring tools capable of measuring viscosity, yield stress, or segregation resistance in real time. Without continuous feedback, admixture dosage adjustments are reactive, often based on delayed laboratory testing or visual slump assessments. This reactive approach reduces the ability to promptly respond to raw material variability, temperature fluctuations, or process disturbances. Consequently, plants frequently adopt conservative dosing strategies, adding excess admixture to safeguard against segregation or bleeding, which elevates costs and environmental impact. Lack of adaptive control systems also limits optimization of VMA dosage for specific mix designs or changing operational conditions, reducing process efficiency.

Integration Challenges with Digital Technologies

Although Industry 4.0 advancements provide opportunities for improved process control, many ready-mix plants face barriers to digital integration. These include lack of compatible sensor infrastructure, fragmented data management systems, and insufficient expertise to implement predictive models or digital twins. Equipment retrofitting costs and concerns about operational downtime also deter adoption. Without integrated control platforms linking admixture dosing, mixer operation, environmental sensors, and quality analytics, the potential for automated, model-driven VMA optimization remains unrealized.

Consequences for Quality Consistency and Operational Efficiency

Equipment and control limitations translate directly into variability in fresh concrete properties, undermining quality assurance and customer satisfaction. Inconsistent VMA dosing leads to fluctuations in viscosity, segregation resistance, and workability retention, complicating pumping, placing, and finishing operations. Additionally, overdosing increases admixture consumption and costs, while underdosing risks structural defects or durability issues. The cumulative effect reduces plant productivity, increases waste, and heightens risk of non-compliance with project specifications.

Strategies for Overcoming Equipment and Process Control Constraints

Addressing these challenges requires a multi-faceted approach combining equipment upgrades, process standardization, and digital transformation. Investing in precision dosing pumps with temperature-compensated flow control improves admixture delivery accuracy. Installation of inline rheometers and sensors capable of measuring key rheological parameters enables real-time monitoring and early detection of deviations. Integration of these data streams into centralized control systems facilitates adaptive dosing adjustments using machine learning algorithms or model predictive control. Regular calibration, preventive maintenance, and operator training further enhance reliability. Collaboration with technology providers to develop plant-specific solutions ensures compatibility with existing infrastructure and workflows.

Table 18.4: Equipment and Process Control Limitations Impacting VMA Application

Limitation	Impact on VMA Application and Concrete Quality	Mitigation Strategies
Dosing Pump Inaccuracy	Over/under-dosing; inconsistent rheology	High-precision dosing pumps; temperature-compensated flow control
Lack of Real-Time Rheological Feedback	Delayed response to process changes; reactive dosing	Inline rheometers; real-time viscosity and yield stress sensors
Limited Digital Integration	Fragmented data; inability to implement adaptive control	Centralized data platforms; digital twins; AI-based process optimization
Equipment Maintenance Deficiencies	Mechanical wear; pump drift; dosing variability	Preventive maintenance; routine calibration programs
Operator Dependency and Manual Controls	Inconsistent procedures; human error	Standard operating procedures; comprehensive training; automation

18.5 Training and Knowledge Transfer Challenges

Importance of Skilled Personnel in VMA Application

The effective use of viscosity-modifying admixtures (VMAs) in concrete production relies heavily on the expertise of plant operators, quality control personnel, and mix designers. Understanding the complex interactions of polymer chemistry, cement hydration, and concrete rheology is essential for interpreting fresh concrete behavior and adjusting admixture dosing accordingly. Skilled operators can identify subtle changes in workability, recognize signs of improper polymer activation, and implement corrective measures to ensure consistent performance. However, such expertise requires both foundational scientific knowledge and practical experience, which are often lacking in many production environments.

Knowledge Gaps and High Staff Turnover

Industrial settings frequently face challenges related to workforce turnover, variable educational backgrounds, and uneven technical training. New or temporary staff may lack familiarity with VMA mechanisms, sensitive dosage windows, and the impact of operational variability. This gap leads to inconsistent admixture handling, dosing errors, and misinterpretation of rheological signs, directly impacting concrete quality. High turnover exacerbates the problem by continuously resetting accumulated experience, impeding the development of a stable knowledge base within plants.

Limited Access to Up-to-Date Technical Resources

Access to current, scientifically validated information on VMA chemistry and application best practices is often restricted by organizational barriers, time constraints, or lack of structured training programs. Many operators rely on informal knowledge transfer or legacy procedures that may be outdated given recent advances in admixture formulations and Industry 4.0 tools. Without continuous education and easy access to technical support, plants risk persisting with suboptimal dosing protocols or misaligned mixing sequences, contributing to performance variability and inefficiency.

Challenges in Technology Transfer and Standardization

Technology transfer from admixture suppliers and research institutions to plant personnel is frequently hindered by communication gaps and the complexity of scientific content. Transferring laboratory-scale findings to large-scale production involves adaptation to site-specific conditions, requiring tailored training that bridges theory and practice. The absence of standardized operating procedures or digital training platforms limits scalability and repeatability of knowledge transfer. Additionally, cross-plant variability and decentralized operations complicate consistent implementation of best practices.

Strategies for Enhancing Training and Knowledge Transfer

To address these challenges, a multi-pronged approach is necessary. Structured training programs combining theoretical instruction with hands-on practice improve operator understanding of VMA behavior and dosing precision. Digital learning tools, including interactive modules, virtual simulations, and mobile-accessible technical databases, enhance knowledge retention and accessibility. Regular refresher courses and certification schemes reinforce competence and motivate continuous improvement. Supplier-led technical support and collaborative problem-solving workshops foster ongoing technology transfer. Finally, embedding standardized operating procedures and decision-support systems within plant control frameworks reduces operator dependency and minimizes human error.

Table 18.5: Training and Knowledge Transfer Challenges and Solutions

Challenge	Impact on VMA Application and Concrete Quality	Mitigation Strategies
Workforce Skill Deficiency	Inconsistent dosing; poor rheology control	Comprehensive training programs; competency certification
High Staff Turnover	Loss of accumulated expertise; variability	Cross-training; retention strategies; digital training platforms
Limited Access to Technical Resources	Reliance on outdated methods; performance drift	Digital knowledge bases; real-time technical support
Communication and Technology Transfer Gaps	Misalignment of lab findings and field practices	Tailored on-site training; supplier-plant collaboration

Lack of Standardized Protocols	Variable procedures; operational inconsistency	Standard operating procedures; embedded decision support systems

18.6 Key Takeaways

Applying viscosity-modifying admixtures (VMAs) industrially requires managing complex interactions beyond formulation, including operational variability, equipment constraints, and workforce expertise. Addressing these factors ensures consistent concrete performance and process efficiency.

Table 18.6: Summary of Technical Challenges in VMA Application and Mitigation Strategies

Challenge Area	Impact on VMA Performance & Concrete Quality	Recommended Mitigation Strategies
Variability in Field Conditions and Operational Practices	Inconsistent polymer dispersion and network formation due to mixer type, mixing time, water quality, temperature, transport time; results in variable viscosity, workability, and segregation risk.	Standardize mixing protocols, monitor and treat water quality, adjust dosing seasonally, optimize transport logistics, train staff, and implement process controls.
Dosing Complexity and Addition Sequence Optimization	Narrow dosage windows sensitive to under- or overdosing; improper addition sequence leads to competitive adsorption and uneven polymer activation, causing workability loss or instability.	Routine pump calibration, automated dosing, add superplasticizers before VMAs, use split dosing, establish standardized protocols, and train operators.
Influence of Water Quality	Variability in ionic strength, pH, and residual admixtures	Monitor water chemistry, use sulfate-tolerant VMAs, blend

and Residual Admixtures	disrupt polymer hydration and adsorption, causing rheology fluctuations and premature stiffening.	water sources, treat recycled water, enforce cleaning protocols, and track admixture residues digitally.
Equipment Limitations and Process Control Constraints	Inaccurate dosing pumps, absence of inline rheological feedback, limited digital integration cause reactive dosing, conservative admixture use, quality variability, and increased costs.	Upgrade to precision dosing pumps, install inline rheometers, integrate data platforms with AI-driven control, maintain equipment rigorously, automate processes, and train staff.
Training and Knowledge Transfer Challenges	Knowledge gaps, high turnover, outdated procedures lead to dosing errors, inconsistent rheology control, and performance variability.	Implement comprehensive, ongoing training programs; leverage digital learning tools; develop standardized operating procedures; foster supplier collaboration; embed decision support systems.

This summary underscores the multidimensional nature of VMA application challenges, emphasizing the need for integrated approaches combining technical, operational, and educational solutions to optimize fresh concrete rheology and ensure industrial consistency.

19. Troubleshooting VMA Manufacturing Issues

Manufacturing viscosity-modifying admixtures demands a level of process discipline that extends beyond formulation chemistry, as polymer architecture, processing history, and plant-specific operating conditions interact to determine final performance. Deviations observed in finished VMA products are rarely attributable to a single root cause; instead, they emerge from coupled effects involving shear exposure, thermal gradients, raw material variability, and time-dependent structural evolution of the polymer system. Effective troubleshooting therefore requires a structured, mechanistic understanding of how molecular-level changes introduced during manufacturing translate into macroscopic rheological instability, storage failure, or inconsistent concrete performance at point of use.

From an industrial and technology transfer perspective, troubleshooting frameworks must be scalable, data-driven, and reproducible across sites with differing equipment and utilities. Traditional corrective actions based on operator experience are increasingly insufficient in modern plants operating at high throughput and narrow specification windows. Integration of inline sensing, statistical process control, and Industry 4.0-enabled analytics allows early detection of failure signatures, differentiation between formulation and process-induced defects, and rapid stabilization during scale-up or change management. By aligning polymer science with digital manufacturing intelligence, robust VMA production can be achieved with reduced waste, improved consistency, and accelerated deployment across diverse concrete applications.

19.1 Process-Specific Failure Modes in VMA Production

Process-specific failure modes in viscosity-modifying admixture manufacturing arise from the complex interaction between polymer chemistry and mechanical processing conditions. During dissolution, neutralization, and dilution, VMAs are exposed to shear fields, temperature gradients, and concentration transitions that can irreversibly alter polymer conformation or associative behavior. Inadequate wetting, non-uniform hydration, or poorly controlled order of addition frequently introduce microstructural heterogeneity that remains undetected by routine quality checks yet manifests later as inconsistent rheological performance in concrete. Understanding these failure modes requires correlating molecular integrity with processing history rather than relying solely on end-product viscosity measurements.

From an industrial standpoint, distinguishing intrinsic formulation weaknesses from process-induced defects is essential for effective troubleshooting and technology transfer. Failure modes often intensify during scale-up, where laboratory mixing regimes cannot replicate industrial residence time distributions or energy dissipation profiles. Structured analysis of unit operations, supported by mechanistic polymer knowledge and process analytics, enables targeted corrective actions that stabilize production without unnecessary reformulation.

19.1.1 Polymer Hydration and Wetting Inefficiencies

Mechanistic Origin of Hydration Failures

Polymer hydration and wetting represent the first critical transformation step in VMA manufacturing, where dry or concentrated polymer phases must transition into a uniformly solvated, functional state. Inefficiencies at this stage originate from unfavorable surface energy balance between polymer particles and the aqueous phase, rapid surface gelation, or insufficient dispersion energy. Cellulose ethers and high-molecular-weight associative polymers are particularly prone to forming impermeable gel shells upon premature contact with water, which effectively isolates the particle core and arrests complete hydration. Once formed, these partially hydrated domains persist throughout downstream processing and cannot be fully corrected by extended mixing alone.

From a molecular perspective, incomplete hydration results in heterogeneous chain mobility and uneven availability of hydrophilic substituents. This translates directly into erratic rheological contribution, delayed viscosity development, and poor robustness against shear in concrete applications. Such failures often remain invisible in standard Brookfield viscosity checks yet emerge as instability during pumping, recirculation, or field dilution.

Processing Conditions Driving Wetting Defects

Industrial wetting inefficiencies are strongly influenced by order of addition, solids loading, and local shear intensity. High polymer concentration zones formed during powder charging or rapid addition promote agglomeration and localized gel formation. Low-shear mixing systems exacerbate this effect, while excessive shear applied too early can mechanically damage fragile polymer structures before full hydration is achieved. Temperature also plays a dual role; insufficient temperature slows hydration kinetics, whereas elevated temperature accelerates surface swelling and gel locking.

Scale-up amplifies these risks due to larger vessel diameters, non-ideal flow patterns, and broader residence time distributions. Laboratory hydration protocols that appear robust often fail under industrial conditions unless explicitly re-engineered for energy dissipation and mass transfer equivalence.

Industrial Manifestations and Diagnostic Signatures

Hydration and wetting failures manifest downstream as inconsistent batch viscosity, delayed thickening upon dilution, and sensitivity to minor formulation or process variations. In concrete, these defects translate into poor segregation resistance, unstable yield stress, and erratic finishing behavior. Diagnostic signatures include time-dependent viscosity rise during QC testing, visible microgels under optical microscopy, and high variance in rheological measurements despite nominally identical formulations.

Advanced plants increasingly rely on inline torque monitoring, acoustic emission sensors, or focused beam reflectance measurement to detect agglomerate persistence during hydration. Such tools provide early warning signals that conventional off-line testing cannot capture.

Table 19.1.1: Polymer Hydration Failure Modes, Causes, and Industrial Impact

Failure Mode	Primary Cause	Manufacturing Indicator	Concrete Performance Impact
Surface gelation	Rapid water contact	Torque spike during charging	Delayed viscosity build-up
Incomplete core hydration	Low dispersion energy	Stable low viscosity plateau	Poor segregation control

Agglomerate formation	High local solids	Visible microgels	Rheology variability
Shear damage pre-hydration	Excess early shear	Viscosity loss after mixing	Reduced yield stress
Temperature mismatch	Non-optimal hydration temperature	Extended hydration time	Inconsistent workability

Corrective and Preventive Strategies

Effective mitigation of hydration inefficiencies requires a combination of formulation-aware process design and digitally assisted control. Pre-slurrying, staged addition, or use of dispersing aids can dramatically improve wetting behavior. From an Industry 4.0 perspective, recipe-driven automated dosing combined with real-time torque and temperature feedback enables reproducible hydration across batches and sites. Digital twins of the hydration step are increasingly applied to simulate gel formation risk and optimize mixing energy profiles prior to scale-up.

For technology transfer, explicit documentation of hydration protocols, including energy input, temperature ramps, and addition sequences, is essential to avoid misattributing process failures to polymer chemistry.

19.1.2 Molecular Weight Degradation During Processing

Degradation Pathways Under Industrial Conditions

Molecular weight integrity is a primary determinant of VMA efficiency, as chain length directly governs hydrodynamic volume, entanglement density, and associative network formation. During manufacturing, polymer chains are exposed to mechanical, thermal, and chemical stressors that can induce irreversible molecular weight degradation. High shear rates in rotor–stator mixers, prolonged recirculation, and cavitation effects in pumps promote mechanical scission, particularly in high-molecular-weight cellulose ethers and acrylic-based VMAs.

Thermal exposure, especially under alkaline or oxidative conditions, further accelerates backbone cleavage and side-chain loss.

Chemical degradation pathways are frequently overlooked during processing. Elevated pH during neutralization promotes alkaline hydrolysis of ester or ether linkages, while trace oxidants introduced through process water or residual initiators catalyze radical-driven chain scission. These mechanisms act cumulatively, meaning that even subcritical stress levels can significantly reduce effective molecular weight over extended processing time.

Process Variables Governing Chain Scission Risk

The likelihood of molecular weight degradation scales with energy density rather than absolute mixing time. Excessive localized shear, often introduced during rapid neutralization or aggressive homogenization, disproportionately damages polymer chains before uniform distribution is achieved. Temperature excursions above recommended processing windows exacerbate degradation kinetics, particularly when combined with high pH or metal ion contamination. Equipment selection therefore plays a critical role; pumps and mixers designed for low-viscosity fluids can impose destructive shear profiles when handling structured polymer solutions.

Scale-up intensifies these risks as larger volumes demand higher power input to maintain homogeneity. Without compensatory design adjustments, industrial reactors unintentionally subject polymers to stress regimes far exceeding laboratory conditions, leading to unexplained viscosity loss despite nominal formulation equivalence.

Industrial Symptoms and Diagnostic Indicators

Molecular weight degradation manifests as permanent viscosity loss that cannot be recovered by aging or formulation adjustment. Unlike hydration-related defects, degraded VMAs show stable but systematically lower viscosity across all shear rates. In concrete applications, this translates into reduced yield stress contribution, diminished segregation resistance, and higher dosage demand. Diagnostic confirmation typically involves gel permeation chromatography or intrinsic viscosity measurements, supported by comparative rheology before and after controlled shear exposure.

In Industry 4.0-enabled plants, indirect indicators such as declining torque at constant solids or reduced pressure drop across recirculation loops provide early warning of chain scission. Trend-based analytics are particularly effective in distinguishing gradual degradation from batch-to-batch variability.

Table 19.1.2: Molecular Weight Degradation: Causes, Indicators, and Impact

Degradation Driver	Processing Origin	Manufacturing Signal	Functional Consequence
Mechanical shear	High-speed mixers	Irreversible viscosity loss	Reduced thickening efficiency
Thermal exposure	Local overheating	Color change, odor	Lower yield stress
Alkaline hydrolysis	Over-neutralization	pH drift, viscosity drop	Increased dosage demand
Oxidative scission	Impure water, metals	Gradual viscosity decay	Long-term instability
Cavitation	Inappropriate pumps	Noise, pressure fluctuation	Polymer fragmentation

Mitigation and Control Strategies

Preventing molecular weight degradation requires deliberate limitation of energy input and strict control of chemical environment. Low-shear mixing designs, staged neutralization, and controlled temperature profiles significantly reduce scission risk. Antioxidant stabilization and metal ion control further protect sensitive polymers during processing. From an Industry 4.0 perspective, coupling power draw, temperature, and pH data into predictive degradation models enables proactive intervention before irreversible damage occurs.

For technology transfer, specifying maximum allowable shear rates, energy densities, and residence times is as critical as defining formulation composition. Failure to communicate these constraints often results in silent performance loss when production is relocated or scaled.

19.1.3 Incomplete Neutralization and pH Control Errors

Chemical Basis of Neutralization-Driven Failure Modes

Neutralization is a defining step in the activation of many VMA chemistries, particularly acrylic-based, carboxylated, and partially ionizable polymer systems. The degree and uniformity of neutralization govern polymer solubility, chain expansion, and electrostatic repulsion, all of which directly influence rheological efficiency. Incomplete or non-uniform neutralization leaves segments of the polymer in a collapsed or partially associated state, suppressing hydrodynamic volume and reducing thickening performance. Conversely, localized over-neutralization creates highly ionized zones that promote excessive swelling, gel formation, or sensitivity to multivalent ions.

These chemical imbalances are rarely apparent from bulk pH measurements alone. Macroscopic pH may fall within specification while microenvironments within the polymer matrix remain under- or over-neutralized. Such hidden heterogeneity translates into unstable rheological behavior during dilution, pumping, or exposure to cementitious systems.

Process-Related Origins of pH Control Errors

pH control errors typically arise from rapid base addition, inadequate mixing energy during neutralization, or poor buffering capacity of the system. High-viscosity zones formed early in neutralization hinder mass transfer, preventing uniform distribution of neutralizing agents. Temperature fluctuations further complicate control by altering dissociation equilibria and reaction kinetics. In continuous processes, short residence times and plug-flow deviations can generate partially neutralized fractions that escape detection until blended into the final product.

Scale-up exacerbates these challenges due to increased diffusion distances and non-ideal flow patterns. Laboratory titration protocols often fail to translate directly to industrial reactors unless mixing intensity and addition rates are explicitly engineered.

Manifestation in Manufacturing and Application Performance

Incomplete neutralization manifests as delayed viscosity development, non-linear dose response, and sensitivity to minor formulation or water quality changes. In concrete, these VMAs exhibit erratic yield stress contribution and reduced robustness against shear and temperature variation. Quality control data often show acceptable initial viscosity but poor reproducibility across batches or after storage.

Advanced diagnostics include potentiometric titration, zeta potential analysis, and rheology profiling across controlled pH ranges. Industry 4.0 systems increasingly integrate inline pH probes with high spatial resolution and model-based estimation of neutralization degree rather than reliance on single-point measurements.

Table 19.1.3: Neutralization Errors, Root Causes, and Industrial Impact

Neutralization Issue	Process Origin	Manufacturing Indicator	Functional Impact
Under-neutralization	Insufficient base addition	Low viscosity plateau	Poor thickening efficiency
Over-neutralization	Excess base, hot spots	Gel formation	Pumping instability
Non-uniform neutralization	Inadequate mixing	Batch variability	Inconsistent concrete rheology
Temperature-driven pH drift	Poor thermal control	pH instability	Dosage sensitivity
Buffer depletion	Weak formulation buffering	Storage pH change	Shelf-life reduction

Control, Prevention, and Digital Optimization

Robust neutralization control requires synchronized management of addition rate, mixing energy, and temperature. Staged or distributed base addition minimizes local pH extremes, while enhanced mixing designs improve homogeneity without excessive shear. Buffer systems tailored to the target pH window provide additional stability against process and storage-induced drift.

Industry 4.0 approaches integrate inline pH measurement, conductivity, and viscosity data into adaptive control loops that adjust neutralization in real time. Digital twins of neutralization kinetics support scale-up and technology transfer by predicting pH gradients and polymer response under site-specific conditions.

19.1.4 Shear-Induced Structural Damage

Polymer Structural Sensitivity to Shear Fields

Viscosity-modifying admixtures derive their functionality from intact polymer chains and, in many cases, from reversible associative networks formed in solution. Exposure to excessive shear during manufacturing disrupts these structures through chain scission, disentanglement, or irreversible rearrangement of hydrophobic associations. High-molecular-weight cellulose ethers, starch derivatives, and associative acrylic polymers are particularly vulnerable, as their rheological efficiency depends on extended chain conformation and transient junction zones that collapse under extreme deformation rates.

Unlike reversible shear thinning observed during application, shear-induced structural damage represents a permanent loss of functional architecture. Once critical deformation thresholds are exceeded, the polymer system cannot fully recover its original rheological contribution, even after extended resting or post-processing aging.

Processing Operations Driving Structural Damage

Structural damage is most frequently introduced during high-energy mixing, aggressive homogenization, and recirculation through pumps not designed for viscous polymer solutions. Rotor–stator mixers, narrow clearance gear pumps, and cavitation-prone centrifugal pumps generate localized shear rates several orders of magnitude higher than bulk averages. These zones of extreme stress disproportionately affect polymer chains, creating fragmented populations that dilute the effective molecular weight distribution.

Scale-up intensifies this risk, as increased power input is often used to compensate for poor mixing efficiency. Without careful energy management, industrial processes unintentionally exceed the shear tolerance of the polymer system, leading to unexplained viscosity loss despite correct formulation and neutralization.

Manufacturing and Application-Level Consequences

Shear-induced damage manifests as irreversible viscosity reduction, diminished yield stress contribution, and reduced thixotropic recovery. In manufacturing, affected batches show stable but depressed rheological profiles that cannot be corrected by dilution or post-addition of fresh polymer. In concrete applications, these VMAs require higher dosages to achieve target performance and exhibit poor robustness under pumping and vibration.

Diagnostic differentiation from hydration or neutralization defects relies on controlled shear testing and comparative rheology before and after simulated processing. Inline indicators include declining torque at constant solids content and reduced pressure drop across recirculation loops.

Table 19.1.4: Shear Damage Mechanisms and Industrial Implications

Shear Source	Equipment Origin	Manufacturing Signal	Performance Impact
Rotor–stator mixing	High-speed dispersers	Sudden viscosity drop	Reduced thickening efficiency
Pump cavitation	Centrifugal pumps	Noise, vibration	Polymer fragmentation
Excessive recirculation	Loop mixing	Gradual viscosity decay	Poor thixotropic recovery
Narrow clearances	Gear pumps	Elevated temperature	Yield stress loss
Scale-up overcompensation	High power density	Batch inconsistency	Increased dosage demand

Mitigation Strategies and Industry 4.0 Controls

Preventing shear-induced structural damage requires deliberate limitation of maximum shear exposure rather than simple reduction of mixing time. Selection of low-shear mixers, progressive cavity pumps, and optimized impeller designs significantly reduces damage risk. Process envelopes defining maximum allowable shear rates and energy densities should be formally specified during technology transfer.

Industry 4.0 implementations enable continuous monitoring of power draw, torque, vibration, and pressure differentials to detect early signs of structural damage. Coupled with predictive models, these data streams support adaptive control strategies that preserve polymer integrity while maintaining manufacturing efficiency.

19.1.5 Thermal Hot Spots and Localized Overheating

Thermo-Physical Origins of Localized Overheating in VMA Manufacturing

Thermal hot spots in viscosity-modifying admixture production arise from non-uniform energy dissipation during mixing, neutralization, and dilution of polymer systems. Exothermic neutralization reactions, viscous dissipation under high shear, and poor heat transfer at vessel walls or internal baffles generate localized temperature excursions that remain masked by acceptable bulk temperature readings. These microthermal gradients are particularly damaging for high-molecular-weight and partially substituted polymers, whose conformational stability and associative behavior are highly temperature sensitive. Once exceeded, critical thermal thresholds induce irreversible changes in chain flexibility, substitution distribution, and intermolecular association density.

From a polymer science perspective, localized overheating accelerates degradation kinetics disproportionately compared with uniform heating. Reaction rates governing hydrolysis, oxidation, and backbone scission increase exponentially with temperature, meaning that short-lived hot spots can cause permanent loss of molecular integrity even when average process temperatures appear benign. Such damage is rarely recoverable and often misdiagnosed as formulation inconsistency rather than a process-induced defect.

Processing Conditions Promoting Thermal Hot Spot Formation

Industrial hot spots most commonly develop in high-viscosity regions where convective heat transfer is limited. Rapid base addition during neutralization creates locally concentrated reaction zones, while rotor–stator mixers and narrow-clearance pumps convert mechanical energy into heat faster than it can be dissipated. Poorly designed jacket systems, fouled heat transfer surfaces, and inadequate temperature sensor placement further exacerbate these conditions by providing false assurance of thermal control.

Scale-up significantly increases thermal risk due to larger characteristic heat transfer distances and higher power input requirements. Laboratory reactors with favorable surface-area-to-volume ratios rarely reproduce the thermal inertia of industrial vessels. Without deliberate thermal design, industrial processes unintentionally operate under transient temperature spikes that remain invisible to standard control schemes.

Manifestation in Product Quality and Concrete Performance

Thermal hot spot damage manifests as subtle but systematic loss of rheological efficiency. Affected VMAs exhibit lower-than-expected viscosity at equivalent solids content, reduced yield stress contribution, and diminished thixotropic recovery. Unlike hydration or neutralization defects, these changes are typically permanent and stable over time, misleading quality control efforts that rely on short-term viscosity checks.

In concrete applications, overheated VMAs demonstrate poor robustness under temperature variation and shear, increased dosage demand, and inconsistent segregation resistance. Field performance often deteriorates under summer conditions or during extended pumping, revealing latent thermal damage introduced during manufacturing rather than during application.

Diagnostic Signatures and Advanced Detection Methods

Conventional bulk temperature monitoring is insufficient to detect localized overheating. Diagnostic signatures include unexplained viscosity loss correlated with high power draw, color darkening, or subtle odor formation indicative of thermal degradation. Rheological profiling across temperature ramps often reveals abnormal sensitivity, confirming compromised polymer architecture.

Industry 4.0-enabled facilities increasingly deploy distributed temperature sensing, infrared thermal mapping of vessels and recirculation lines, and real-time correlation of torque, power consumption, and temperature data. Machine learning models trained on historical batch data can identify overheating signatures long before final product specifications are breached, enabling preventive intervention.

Table 19.1.5: Thermal Hot Spots, Root Causes, and Industrial Consequences

Thermal Issue	Process Origin	Manufacturing Indicator	Functional Impact
Neutralization hot spots	Rapid base addition	Local temperature spike	Polymer degradation

Viscous dissipation heating	High shear mixing	Rising power draw	Viscosity loss
Poor heat transfer	Fouled jackets	Slow cooling response	Structural instability
Pump-induced heating	Narrow clearances	Elevated outlet temperature	Yield stress reduction
Scale-up thermal inertia	Large vessel volume	Batch-to-batch drift	Increased dosage demand

Mitigation Strategies and Digital Thermal Management

Effective control of thermal hot spots requires integration of polymer-aware thermal design with real-time process analytics. Staged neutralization, controlled shear application, and enhanced heat exchange capacity reduce peak temperatures without compromising throughput. Equipment selection should prioritize low energy dissipation per unit volume and uniform flow patterns to minimize localized heating.

From an Industry 4.0 standpoint, digital twins incorporating heat generation, transfer, and polymer degradation kinetics enable predictive thermal management during scale-up and technology transfer. Coupling distributed temperature data with adaptive control algorithms allows proactive adjustment of mixing intensity and addition rates, preserving polymer integrity while maintaining manufacturing efficiency.

19.1.6 Mixing Sequence and Order-of-Addition Errors

Molecular and Colloidal Basis of Sequence-Dependent Failures

Mixing sequence and order of addition govern the pathway by which polymers transition from discrete phases into functional, solvated architectures. In viscosity-modifying admixtures, this pathway determines polymer conformation, associative junction formation, and the spatial distribution of ionized groups. Deviations in sequence alter local concentration fields and transient interactions, producing kinetic traps such as premature gelation, polymer collapse, or irreversible aggregation. These effects are pathway-dependent and cannot be inferred from final composition alone.

At the molecular level, early exposure of partially solvated polymers to high ionic strength, extreme pH, or intense shear restricts chain expansion and suppresses hydrodynamic volume development. Associative polymers are especially sensitive, as hydrophobic domains can lock into non-productive clusters if introduced before adequate hydration. Once formed, these structures persist through downstream processing and degrade rheological efficiency in application.

Process Scenarios That Induce Order-of-Addition Errors

Common industrial errors include adding neutralizing agents before complete polymer wetting, introducing salts or preservatives ahead of chain expansion, and charging high-molecular-weight polymers into already viscous media. Such sequences create localized environments that favor gel shell formation, uneven ionization, or mechanical damage during attempted dispersion. In batch systems, manual interventions and recipe drift further increase variability, while continuous systems suffer when residence time distributions are not aligned with reaction and hydration kinetics.

Scale-up amplifies sequence sensitivity because mixing energy and mass transfer are non-uniform across larger vessels. Laboratory protocols that tolerate minor sequence deviations often fail industrially when local concentration extremes persist longer. Without explicit sequencing control, plants experience sporadic failures that evade root-cause analysis based on formulation audits alone.

Manufacturing and Field-Level Consequences

Order-of-addition errors manifest as inconsistent batch viscosity, delayed viscosity build-up after dilution, and heightened sensitivity to water quality or temperature. Quality control results may meet initial specifications yet display poor reproducibility, particularly after storage or transport. In concrete, these VMAs show erratic yield stress contribution, unstable thixotropy, and reduced segregation resistance under shear.

Unlike molecular weight degradation, sequence-induced defects often present as nonlinear dose response and time-dependent rheology. Field complaints frequently include finishing variability and loss of robustness during pumping, which are traced back to irreversible microstructural heterogeneity introduced during manufacturing.

Diagnostic Approaches and Failure Fingerprints

Identifying sequence errors requires correlating rheological evolution with process history rather than relying on single-point viscosity checks. Time-resolved rheology during dilution, optical microscopy for microgel detection, and conductivity or turbidity profiling reveal hidden heterogeneity. Comparative trials that alter only the addition order provide definitive confirmation of pathway dependence.

Industry 4.0 implementations increasingly employ recipe enforcement via MES, inline viscosity and torque monitoring, and event-based analytics to flag deviations in real time. Pattern recognition across batches can distinguish sequence-related anomalies from raw material variability or equipment-induced shear damage.

Table 19.1.6: Order-of-Addition Errors, Root Causes, and Industrial Impact

Sequence Error	Process Origin	Manufacturing Indicator	Functional Impact
Neutralization before hydration	Incorrect addition order	Rapid viscosity spike	Suppressed thickening
Salt addition too early	Preservative dosing	Turbidity increase	Rheology instability
Polymer into viscous medium	High solids charging	Poor dispersion	Batch variability
Shear before solvation	Early high-speed mixing	Irreversible viscosity loss	Yield stress reduction
Manual recipe deviation	Operator intervention	Inconsistent torque profile	Field performance drift

Control Strategies, Automation, and Technology Transfer

Preventing sequence-related failures requires codifying the kinetic pathway as part of the product definition. Staged addition, controlled pre-dilution, and delayed activation steps ensure that polymers encounter chemical and mechanical environments only after achieving adequate solvation. Equipment should support gentle initial mixing followed by progressive energy input aligned with structural development.

From an Industry 4.0 perspective, recipe-driven automation, interlocked dosing permissions, and digital twins that simulate pathway sensitivity provide robust safeguards during scale-up and cross-plant deployment. For technology transfer, documenting not only quantities but also timing, energy input, and conditional triggers is essential. Omission of these parameters is a frequent cause of silent performance loss when production is replicated at new sites.

19.1.7 Scale-Up Effects from Lab to Industrial Reactors

Fundamental Scale-Up Mismatch Between Polymer Science and Process Reality

Scale-up in viscosity-modifying admixture manufacturing represents a transition from kinetically well-controlled laboratory systems to industrial reactors dominated by non-ideal flow, heterogeneous energy dissipation, and thermal inertia. Polymer hydration, neutralization, and associative network formation are pathway-dependent phenomena that respond to local shear, temperature, and concentration fields rather than to average process conditions. As reactor size increases, these local environments diverge sharply from laboratory benchmarks, producing functional outcomes that cannot be predicted by formulation equivalence alone.

From a polymer chemistry standpoint, laboratory-scale success often reflects favorable surface-area-to-volume ratios, short diffusion paths, and uniform energy input. Industrial reactors, by contrast, impose broad residence time distributions and spatial gradients that selectively damage or underactivate polymer populations. These mismatches explain why VMAs that perform reproducibly in the laboratory frequently exhibit inconsistent rheology, storage instability, or elevated dosage demand after scale-up.

Hydrodynamic and Mixing Regime Transitions During Scale-Up

Hydrodynamic similarity is rarely preserved when transitioning from beakers or pilot vessels to full-scale reactors. Reynolds number matching alone is insufficient for polymer systems, as viscoelastic effects, shear sensitivity, and hydration kinetics dominate behavior. Industrial impellers generate zones of extreme shear near blades and quiescent regions elsewhere, leading to simultaneous over-processing and under-processing within the same batch.

Residence time distribution widens significantly with scale, particularly in semi-batch and continuous systems. Portions of the polymer experience prolonged exposure to high shear or elevated temperature, while others bypass critical activation steps. Such heterogeneity creates a broadened effective molecular weight and activation distribution that manifests as unpredictable rheological performance.

Thermal and Chemical Non-Equivalence at Larger Scale

Thermal control degrades with increasing scale due to reduced heat transfer efficiency and higher power input. Exothermic neutralization and viscous dissipation generate localized hot spots that are absent or negligible at laboratory scale. These thermal excursions accelerate degradation pathways and alter ionization equilibria, compounding the effects of hydrodynamic non-uniformity.

Chemical non-equivalence further arises from delayed reagent dispersion and pH gradients. Neutralizing agents, preservatives, or stabilizers require significantly longer to achieve homogeneity in large vessels, during which polymers are exposed to transient chemical extremes. Laboratory titration protocols therefore systematically underpredict degradation and activation risks encountered industrially.

Observable Industrial Failure Patterns

Scale-up-induced failures typically present as batch-to-batch variability rather than outright specification failure. Manufacturing data reveal increased variance in viscosity, torque, and power consumption, even when raw materials and recipes remain unchanged. In concrete applications, these VMAs exhibit inconsistent yield stress contribution, reduced robustness to shear and temperature, and heightened sensitivity to minor mix design changes.

Diagnostic differentiation from formulation defects requires comparative trials across scales, supported by rheological fingerprinting and controlled shear simulations. Without such analysis, scale-up failures are often misattributed to raw material quality or operator error, delaying effective corrective action.

Table 19.1.7: Scale-Up Drivers, Process Signals, and Functional Consequences

Scale-Up Driver	Industrial Origin	Manufacturing Signal	Performance Consequence

Non-uniform shear fields	Large impellers	Torque variability	Rheology inconsistency
Broad residence time	Semi-batch reactors	Viscosity scatter	Unstable yield stress
Thermal inertia	Poor heat removal	Temperature lag	Polymer degradation
Delayed reagent mixing	Large diffusion length	pH gradients	Activation inefficiency
Energy overcompensation	High power input	Elevated power draw	Molecular weight loss

Mitigation Strategies and Industry 4.0–Enabled Scale-Up

Effective mitigation of scale-up effects requires redefining scale-up criteria around polymer integrity rather than geometric similarity alone. Energy density limits, maximum shear exposure, and allowable thermal gradients should be specified as critical quality parameters. Pilot-scale trials must be designed to replicate industrial stress regimes, not merely to confirm formulation viability.

Industry 4.0 approaches integrate digital twins that couple hydrodynamics, heat transfer, and polymer degradation kinetics to predict failure modes before physical scale-up. Real-time analytics linking torque, temperature, and viscosity trends enable adaptive control during commissioning and technology transfer. Explicit documentation of process envelopes and stress limits is essential to ensure reproducible VMA performance across sites and scales.

19.1.8 Entrapped Air and Foam Formation During Processing

Physicochemical Basis of Air Entrainment in VMA Systems

Entrapped air and foam formation during viscosity-modifying admixture manufacturing originate from the interfacial activity of polymer chains and formulation auxiliaries under dynamic mixing conditions. Many VMAs contain hydrophilic backbones with associative or amphiphilic segments that reduce surface tension and stabilize gas–liquid interfaces. When exposed to agitation, recirculation, or high shear, these systems readily trap air bubbles that become kinetically stabilized by polymer adsorption and weak network formation at the interface. Once stabilized, bubble coalescence is suppressed, leading to persistent foam structures rather than transient air release.

At the molecular level, air stabilization is amplified when polymers are partially hydrated or unevenly neutralized. In such states, chain segments preferentially orient at interfaces, forming elastic films that resist rupture. These effects are highly pathway-dependent and are intensified by impurities such as surfactants, residual monomers, or preservatives with surface-active character.

Processing Conditions Promoting Air Entrapment and Foam Stability

Industrial foam formation is commonly induced by top-entry mixing, vortex formation during liquid charging, and high-speed recirculation loops that entrain air at pump inlets. Low-viscosity early-stage mixtures are particularly vulnerable, as air is easily incorporated before viscosity development suppresses bubble ingress. Conversely, once viscosity increases, trapped air becomes increasingly difficult to remove due to reduced buoyancy-driven rise and enhanced interfacial stabilization.

Scale-up exacerbates air entrapment risks through larger free surfaces, higher volumetric flow rates, and longer residence times. Poorly designed baffles, excessive headspace, and inadequate deaeration capacity further promote foam persistence. Laboratory processes often underestimate these effects because of lower gas exposure and shorter processing times.

Impact on Manufacturing Consistency and Concrete Performance

Entrapped air manifests in manufacturing as apparent viscosity variability, inaccurate density-based dosing, and unstable inline sensor readings. Foamed batches may falsely meet viscosity specifications while containing reduced active solids per unit volume. Storage stability is often compromised, with foam collapse leading to phase separation or apparent viscosity drift over time.

In concrete applications, air-contaminated VMAs exhibit inconsistent rheological response, erratic yield stress contribution, and sensitivity to mixing energy at the job site. Interaction with air-entraining agents becomes unpredictable, increasing the risk of excessive total air content, strength loss, and surface defects. These issues are frequently misattributed to incompatibility with cement or aggregates rather than to upstream processing defects.

Diagnostic Indicators and Advanced Detection Techniques

Traditional quality control methods rarely detect entrapped air directly. Diagnostic indicators include unexplained density reduction, oscillatory viscosity readings, and visual foam persistence after mixing cessation. Rheological testing under controlled pressure often reveals apparent viscosity collapse as bubbles compress or dissolve.

Industry 4.0-enabled plants increasingly deploy inline density meters, acoustic sensors, and optical bubble detection to quantify air content in real time. Correlation of these signals with torque and power draw provides a robust fingerprint of foam-related instability. Data-driven models can distinguish air-induced variability from true polymer degradation or hydration failure.

Table 19.1.8: Entrapped Air and Foam Formation: Causes, Indicators, and Effects

Foam Driver	Process Origin	Manufacturing Indicator	Functional Impact
Vortex-induced air ingress	Top-entry mixing	Density reduction	Dosage inconsistency
Surface-active polymers	Associative segments	Persistent foam	Apparent viscosity drift
Pump suction aeration	High flow velocity	Noise, cavitation	Rheology variability
Partial hydration	Early processing stage	Stable microbubbles	Poor robustness
Excess headspace	Reactor design	Slow foam collapse	Storage instability

Mitigation Strategies and Digital Process Control

Mitigating air entrapment requires coordinated control of formulation, equipment design, and processing sequence. Bottom-entry or low-vortex impeller designs, controlled fill levels, and reduced recirculation velocities significantly limit air ingress. Strategic timing of viscosity development, ensuring adequate hydration before high-energy mixing, reduces interfacial stabilization of bubbles. Chemical defoamers should be selected with caution, as incompatibility with polymer networks can induce secondary rheological defects.

From an Industry 4.0 perspective, closed-loop control integrating density, acoustic, and torque data enables real-time foam detection and corrective action. Digital twins that incorporate gas–liquid mass transfer and bubble stability models support scale-up and technology transfer by predicting air entrainment risk under site-specific conditions. Explicit specification of maximum allowable air content and deaeration protocols is essential for reproducible VMA performance across plants.

19.2 Rheology Drift During Batch and Continuous Processing

Rheology drift represents one of the most operationally disruptive issues in VMA manufacturing, as it directly undermines batch consistency and dosing reliability at the concrete plant. Associative polymers and modified polysaccharides are inherently sensitive to shear history and thermal exposure, making viscosity a dynamic rather than static property during processing. Variations in agitation intensity, recirculation time, or heat removal efficiency can progressively weaken transient polymer networks or, conversely, promote excessive structuring that leads to gelation or pumping difficulties.

In continuous production environments, rheology drift is further complicated by residence time distribution effects and transient conditions during startup or shutdown. Small fractions of under- or over-processed material can disproportionately influence bulk rheology, especially in high-solids formulations. Industrial troubleshooting increasingly relies on inline rheological monitoring and digital trend analysis to identify early drift signatures and implement corrective feedback before product release.

19.2.1 Shear History Sensitivity of Associative Polymers

Molecular Origin of Shear-Dependent Rheological Evolution

Associative polymers derive their thickening efficiency from transient physical junctions formed between hydrophobic groups distributed along a hydrophilic backbone. These junctions are continuously formed and broken under flow, rendering the rheological response highly sensitive to cumulative shear exposure. Moderate shear promotes optimal network rearrangement and uniform chain extension, whereas excessive or prolonged shear progressively disrupts junction density, reduces effective molecular connectivity, and lowers apparent viscosity. Crucially, this degradation of associative structure can occur without detectable molecular weight loss.

From a polymer physics standpoint, shear history alters the population balance between productive and non-productive associations. Once hydrophobic clusters are repeatedly disrupted beyond their reformation kinetics, the system relaxes into a lower-viscosity steady state. This process is path-dependent and cannot be reversed simply by resting or reducing shear intensity, distinguishing shear history effects from reversible shear thinning during application.

Processing Variables Governing Shear History Accumulation

In manufacturing environments, shear history accumulates through multiple unit operations including mixing, recirculation, filtration, and pumping. Even low-shear operations contribute cumulatively when residence times are long. High-solids formulations are particularly vulnerable, as elevated viscosity amplifies local shear rates at impeller tips, valves, and pump clearances. Continuous processes further intensify shear sensitivity by subjecting material to repeated stress cycles rather than a single defined mixing event.

Scale-up magnifies these effects due to increased recirculation demand and higher energy input required to maintain homogeneity. Laboratory trials rarely reproduce the cumulative shear exposure encountered industrially, leading to systematic underestimation of shear-induced rheology drift during process design and technology transfer.

Industrial Manifestation and Operational Consequences

Shear history–driven drift typically presents as gradual, irreversible viscosity loss during processing or holding, despite stable temperature and composition. Quality control measurements taken at different sampling points show consistent downward trends rather than random variability. In concrete applications, such VMAs deliver reduced yield stress contribution, diminished segregation resistance, and increased dosage sensitivity, particularly under extended pumping or high-energy mixing conditions.

In continuous production, shear-sensitive drift often appears as a slow baseline shift in inline viscosity or pressure drop, which may remain unnoticed until product performance complaints arise. Differentiating this failure mode from molecular weight degradation requires controlled shear simulations and comparison of rheological recovery behavior.

Diagnostic Tools and Industry 4.0 Analytics

Effective diagnosis of shear history sensitivity relies on time-resolved rheology rather than single-point viscosity checks. Step-shear and loop-shear tests reveal irreversible viscosity decay characteristic of associative network fatigue. Inline torque, pressure drop, and power draw trends provide indirect but highly sensitive indicators of accumulating shear damage.

Industry 4.0 frameworks integrate these signals into multivariate control charts and predictive models that correlate shear exposure with rheological drift. Machine learning algorithms trained on historical production data can forecast drift onset and recommend corrective actions such as reducing recirculation intensity or adjusting throughput before specification limits are breached.

Table 19.2.1: Shear History Effects on Associative Polymer Rheology

Shear Factor	Process Origin	Manufacturing Signal	Functional Impact
Prolonged recirculation	Holding loops	Gradual viscosity decay	Reduced yield stress
High local shear	Pumps, valves	Stable low viscosity	Dosage increase
Repeated stress cycles	Continuous processing	Baseline drift	Poor robustness
High solids loading	Concentrated formulations	Elevated power draw	Network fatigue
Scale-up energy density	Large reactors	Batch-to-batch trend	Performance inconsistency

Control Strategies and Technology Transfer Considerations

Mitigating shear history–induced rheology drift requires defining allowable cumulative shear exposure as a critical process parameter. Limiting recirculation time, selecting low-shear pumping technologies, and redesigning flow paths to eliminate unnecessary stress points are effective preventive measures. Associative polymer selection should consider not only peak rheological efficiency but also resistance to shear fatigue under realistic industrial conditions.

From an Industry 4.0 perspective, cumulative shear indices derived from power input and residence time data provide a transferable metric for process control across sites. For technology transfer, explicitly specifying maximum allowable shear energy and stress cycles is essential, as omission of these constraints is a common cause of silent rheology drift when production is scaled or relocated.

19.2.2 Residence Time Distribution Effects

Mechanistic Significance of Residence Time Distribution in VMA Rheology

Residence time distribution governs the duration and variability of exposure of viscosity-modifying admixtures to shear, thermal, and chemical environments during processing. Unlike ideal plug-flow assumptions often implicit in laboratory design, industrial batch and continuous reactors exhibit broad residence time spectra where different fractions of the polymer population experience markedly different stress histories. For associative polymers and modified polysaccharides, rheological response is a nonlinear function of exposure time, making even small fractions of over- or under-processed material disproportionately influential on bulk viscosity.

From a polymer science perspective, RTD-driven heterogeneity alters the balance between chain activation, associative junction formation, and irreversible structural fatigue. Long-residence fractions accumulate shear and thermal stress, leading to weakened networks, while short-residence fractions remain under-hydrated or partially neutralized. When recombined, these populations produce unstable rheology that drifts over time rather than converging to a steady state.

RTD Broadening in Industrial Batch and Continuous Operations

In batch reactors, RTD effects arise from non-uniform mixing, dead zones, and stratification, particularly in high-viscosity systems. Material near impeller discharge zones experiences repeated high-shear cycles, while regions near walls or baffles remain relatively quiescent. Holding periods and recirculation loops further widen effective residence time by repeatedly reprocessing portions of the batch.

Continuous systems inherently exhibit RTD governed by reactor design and flow regime. Deviations from ideal plug flow, such as axial mixing, back-mixing, or short-circuiting, generate long tails in the RTD curve. During startup and shutdown, transient conditions introduce extreme RTD distortion, allowing poorly conditioned material to enter the product stream and seed long-term rheological drift.

Manifestation of RTD-Induced Rheology Drift

RTD effects typically manifest as slow viscosity drift during steady-state operation rather than abrupt specification failure. Inline rheological measurements show gradual baseline shifts or oscillations correlated with throughput changes, valve adjustments, or temperature fluctuations. Batch-to-batch variability increases despite consistent raw materials and recipes, confounding traditional quality control logic.

In concrete applications, RTD-driven drift appears as inconsistent yield stress contribution and sensitivity to minor changes in mixing energy or water content. These VMAs often perform acceptably in short laboratory trials but fail to maintain robustness during prolonged pumping or recirculation, revealing the latent heterogeneity introduced during manufacturing.

Diagnostic Characterization of RTD Effects

Diagnosing RTD-induced issues requires explicit measurement or inference of residence time spectra. Tracer studies using inert salts or dyes provide direct RTD characterization, while indirect methods correlate inline viscosity, pressure drop, and temperature data with throughput and residence time estimates. Rheological fingerprinting of samples collected at different points in the process often reveals bimodal or broadened response indicative of mixed polymer populations.

Industry 4.0 platforms increasingly integrate RTD models with real-time process data, enabling continuous estimation of effective residence time distribution. These tools support early identification of drift risk, particularly during transient operations that traditional steady-state controls fail to capture.

Table 19.2.2: Residence Time Distribution Effects on VMA Rheology

RTD Feature	Process Origin	Manufacturing Signal	Rheological Consequence
Long RTD tail	Back-mixing zones	Gradual viscosity decay	Network fatigue
Short-circuiting	Poor flow design	Viscosity scatter	Under-activation
Dead zones	Vessel geometry	Batch inconsistency	Drift during storage
Startup transients	Continuous lines	Inline viscosity spikes	Field variability
Excess recirculation	Loop mixing	Baseline shift	Reduced robustness

Control and Optimization Strategies Using Industry 4.0

Mitigation of RTD-driven rheology drift requires narrowing residence time spectra through reactor and process design. Improved impeller selection, elimination of dead zones, and reduction of unnecessary recirculation minimize over-processing of sensitive polymer fractions. In continuous systems, staged reactors or static mixers can approximate plug-flow behavior and reduce RTD tails.

Industry 4.0 implementations couple RTD models with digital twins that simulate polymer response under realistic residence time distributions. Real-time optimization algorithms adjust throughput, mixing intensity, and purge strategies during transients to prevent off-spec material from propagating downstream. For technology transfer, explicit specification of acceptable RTD characteristics, rather than nominal residence time alone, is essential to preserve rheological stability across sites.

19.2.3 Temperature-Driven Viscosity Instability

Thermodynamic and Kinetic Basis of Temperature Sensitivity

Temperature exerts a first-order influence on the rheological behavior of viscosity-modifying admixtures by governing polymer solvation, chain mobility, and associative junction lifetime. For modified polysaccharides and associative acrylic polymers, viscosity reflects a balance between entropic chain expansion and enthalpically driven intermolecular interactions, both of which are temperature dependent. Increasing temperature reduces solvent viscosity and accelerates junction dissociation kinetics, weakening transient networks and lowering apparent viscosity. Conversely, cooling can promote over-structuring, phase association, or incipient gelation, particularly in systems with lower cloud points or strong hydrophobic interactions.

From a kinetic standpoint, temperature also controls the rate of irreversible degradation pathways. Hydrolysis, oxidation, and backbone scission accelerate exponentially with temperature, meaning that even modest thermal excursions during processing or storage can permanently alter rheological performance. Distinguishing reversible thermal thinning from irreversible temperature-induced damage is therefore critical in industrial troubleshooting.

Processing Scenarios Leading to Thermal Rheology Drift

In batch manufacturing, viscosity instability frequently arises from inadequate heat removal during high-shear mixing or exothermic neutralization. Bulk temperature measurements may remain within specification while localized hot spots introduce disproportionate structural damage. During holding, slow cooling or ambient temperature fluctuations induce gradual viscosity drift that is often misinterpreted as storage instability rather than a thermal control deficiency.

Continuous processes amplify temperature-driven effects through cumulative exposure and thermal lag. Heat generated by pumps and static mixers, combined with insufficient cooling capacity, causes progressive baseline shifts in inline viscosity. Startup and shutdown transients introduce abrupt temperature changes that expose fractions of the product to extreme conditions, seeding long-term instability in downstream storage or application.

Manifestation in Manufacturing Control and Concrete Performance

Temperature-driven viscosity instability manifests as time-dependent drift rather than random variability. Inline viscosity and pressure drop measurements show systematic correlation with temperature trends, while off-line QC tests reveal inconsistent results depending on sampling time and conditioning. Products may pass initial release criteria yet fail to maintain specification after transport or seasonal temperature changes.

In concrete applications, thermally unstable VMAs exhibit inconsistent yield stress contribution, reduced robustness under hot weather placement, and sensitivity to mixing energy. Cold conditions may induce excessive structuring or delayed dissolution, compromising workability and finishing. These performance issues are frequently misattributed to cement variability or job-site conditions, obscuring the manufacturing origin.

Diagnostic Methodology and Industry 4.0 Monitoring

Effective diagnosis requires rheological characterization across controlled temperature ramps to separate reversible thermal effects from irreversible degradation. Arrhenius-type analysis of viscosity decay provides insight into activation energies associated with degradation processes. Complementary techniques such as dynamic mechanical analysis and differential scanning calorimetry support identification of thermal transitions relevant to processing stability.

Industry 4.0-enabled facilities integrate distributed temperature sensing with inline rheology and multivariate analytics to identify early instability signatures. Predictive models correlate temperature exposure history with viscosity drift, enabling proactive intervention through adjustment of mixing intensity, cooling rates, or throughput before product quality is compromised.

Table 19.2.3: Temperature Effects on VMA Rheology and Stability

Thermal Factor	Process Origin	Manufacturing Indicator	Rheological Impact
Elevated bulk temperature	Insufficient cooling	Gradual viscosity loss	Network weakening
Local hot spots	High shear zones	Power–temperature mismatch	Irreversible damage
Rapid cooling	Shutdown transients	Viscosity spike	Over-structuring

Ambient fluctuation	Storage conditions	Seasonal drift	Inconsistent dosing
Thermal aging	Extended holding	Time-dependent decay	Reduced robustness

Control Strategies and Technology Transfer Implications

Stabilizing temperature-driven rheology requires defining acceptable thermal exposure envelopes as critical quality attributes. Adequate heat exchange capacity, controlled neutralization rates, and minimized pump-induced heating reduce thermal stress. Conditioning protocols specifying sampling temperature and equilibration time are essential for reliable quality control interpretation.

From an Industry 4.0 perspective, digital twins incorporating heat generation, transfer, and polymer response kinetics enable simulation of thermal scenarios across batch and continuous operations. For technology transfer, explicit communication of temperature limits, allowable gradients, and transient handling procedures is as critical as formulation disclosure. Failure to transfer these constraints is a common root cause of rheology instability when production is relocated or scaled.

19.2.4 Irreversible Network Breakdown Under High Shear

Mechanistic Basis of High-Shear-Induced Network Failure

Associative polymers and modified polysaccharides rely on transient physical junctions to establish three-dimensional networks responsible for thickening and yield stress in high-solids formulations. Under moderate shear, these networks dynamically reform, preserving bulk viscosity. However, exposure to extreme or prolonged shear overwhelms reformation kinetics, causing irreversible dissociation of hydrophobic clusters or interchain entanglements. Unlike reversible shear thinning, this breakdown results in permanent loss of network connectivity without detectable chain scission in most cases. The effect is accentuated in polymers with low hydrophobic group density or short chain backbones, which offer limited redundancy in junction formation.

Processing Conditions Promoting Network Degradation

High-shear events arise from multiple sources in industrial operations: high-speed impellers, narrow-gap pumping, high-pressure filtration, or static mixers in continuous lines. Recirculation loops and prolonged exposure exacerbate cumulative damage, particularly in concentrated formulations where local shear rates at impeller tips or valve restrictions can exceed nominal estimates by an order of magnitude. Scale-up further amplifies shear exposure because maintaining homogeneity in larger reactors requires higher energy input, translating directly to increased mechanical stress on associative networks. Laboratory-scale trials often fail to reproduce these extreme conditions, making irreversible network breakdown a hidden risk during technology transfer.

Industrial Manifestation and Operational Implications

High-shear network breakdown manifests as permanent viscosity loss, increased flowability, and reduced yield stress. Inline rheology may show a sudden downward shift in baseline viscosity during high-energy operations, while batch samples collected before and after pumping demonstrate irrecoverable differences. In concrete applications, VMAs affected by irreversible breakdown exhibit decreased slump retention, lower segregation resistance, and heightened sensitivity to dosing errors, especially under extended mixing or pumping cycles. The phenomenon can also propagate variability downstream in continuous processes, where small fractions of degraded polymer disproportionately influence bulk rheology.

Diagnostic and Monitoring Strategies

Controlled loop-shear or step-shear tests allow quantification of irreversible viscosity loss, distinguishing it from reversible shear thinning. Inline power, pressure drop, and torque measurements act as indirect yet sensitive indicators of accumulating network fatigue. Advanced Industry 4.0 solutions integrate these signals into predictive models, correlating applied shear energy with rheological degradation. Machine learning approaches trained on historical high-shear events can forecast risk windows and recommend operational adjustments, such as reducing recirculation intensity or implementing low-shear transfer devices before product quality is compromised.

Table 19.2.4: High-Shear Network Breakdown Effects on Polymer Rheology

Shear Factor	Process Origin	Manufacturing Signal	Functional Impact
High-velocity impellers	Batch mixing	Sudden viscosity drop	Reduced yield stress
Narrow-gap pumping	Recirculation loops	Irrecoverable low viscosity	Dosage increase
Prolonged stress cycles	Continuous processing	Baseline drift	Poor robustness
High solids concentration	Concentrated formulations	Elevated torque/power	Network fatigue
Scale-up energy input	Large reactors	Batch-to-batch trend	Performance inconsistency

Control Measures and Technology Transfer Considerations

Mitigation requires defining maximum allowable shear energy and stress exposure as critical process parameters. Implementing low-shear pumps, optimized impeller design, and avoiding unnecessary recirculation minimize irreversible network damage. For technology transfer, explicit specification of shear limits, energy density thresholds, and validation of scalable mixing strategies is essential to maintain rheological stability across production sites. Digital twins can simulate high-shear scenarios and evaluate network resilience under industrial conditions, providing predictive control and reducing off-spec production risk.

19.2.5 Thixotropy Loss During Recirculation

Polymer Mechanisms Governing Thixotropic Behavior

Thixotropy in associative polymers and modified polysaccharides arises from a dynamic equilibrium between network formation and breakdown under shear. When flow is applied, physical junctions transiently dissociate, allowing deformation, and gradually rebuild once shear is reduced. Recirculation subjects the polymer to repeated low-to-high shear cycles, which can progressively fatigue the network. Over multiple cycles, hydrophobic associations fail to reform fully, leading to a permanent reduction in time-dependent structural recovery and diminished thixotropic behavior. This effect is most pronounced in high-solids or highly hydrophobic systems, where network entanglement is a key contributor to yield stress and viscosity recovery.

Industrial Conditions Contributing to Thixotropy Loss

Recirculation loops, high-speed transfer lines, and continuous pumping are primary contributors to thixotropic degradation. Even moderate shear in holding loops can cumulatively reduce network integrity if residence time is extended. Variability in impeller speed, pump pressure, or valve restriction can cause localized over-shearing, accelerating irreversible loss of structural recovery. Scale-up intensifies these effects, as energy input per unit volume increases, and longer recirculation paths amplify cumulative shear exposure. Laboratory-scale trials often underestimate these losses due to shorter residence times and lower throughput, presenting a challenge during process scale-up and technology transfer.

Operational Manifestation and Performance Consequences

Thixotropy loss manifests as slower viscosity recovery following shear, decreased yield stress, and reduced slump retention in concrete applications. Inline rheology may show a lower plateau viscosity during rest periods, and repeated measurements reveal progressive degradation rather than reversible shear thinning. VMAs experiencing thixotropy loss exhibit higher sensitivity to pumping energy and dosing errors, with diminished ability to maintain segregation resistance and workability under prolonged handling or transport conditions.

Diagnostic Tools and Industry 4.0 Monitoring

Step-shear and loop-shear rheometry provide quantitative measures of structural recovery and cumulative thixotropic loss. Power consumption, torque fluctuations, and inline viscosity trends serve as real-time indicators of network fatigue during recirculation. Industry 4.0 frameworks enable integration of these metrics into digital twins, correlating shear exposure history with predictive thixotropy loss. Machine learning models trained on historical production data can forecast degradation trends and recommend preventive adjustments, such as reducing loop recirculation, optimizing flow paths, or switching to low-shear pumping technology before performance compromise occurs.

Table 19.2.5: Thixotropy Loss During Recirculation

Recirculation Factor	Process Origin	Manufacturing Signal	Functional Impact
Prolonged holding loops	Batch recirculation	Gradual recovery reduction	Lower yield stress
Repeated shear cycles	Continuous pumping	Decreasing plateau viscosity	Reduced workability
High solids concentration	Concentrated formulations	Torque increase	Network fatigue
Localized high shear	Valves, impellers	Non-uniform recovery	Dosing variability
Scale-up energy density	Large reactors	Progressive viscosity decline	Performance inconsistency

Control and Technology Transfer Strategies

Mitigation of thixotropy loss involves limiting recirculation duration, optimizing flow geometry, and selecting low-shear pumps or impellers to reduce network fatigue. Defining cumulative shear as a critical process parameter ensures reproducible recovery behavior during scale-up and technology transfer. Digital twins and inline monitoring allow predictive adjustment of recirculation intensity, enabling consistent product performance across sites. Explicitly specifying allowable shear cycles and residence times during technology transfer prevents silent degradation that can compromise VMA function in concrete applications.

19.2.6 Startup and Shutdown Transient Effects

Polymer Sensitivity to Transient Processing Conditions

Startup and shutdown phases impose non-steady-state conditions that are fundamentally different from nominal production and disproportionately stressful for associative polymers and modified polysaccharides. During these transitions, polymer systems experience abrupt changes in shear rate, residence time, temperature, and concentration. Associative networks require controlled hydration, neutralization, and junction formation kinetics to reach a stable rheological state. Transient exposure interrupts these processes, creating polymer populations that are partially activated, over-sheared, or thermally stressed, which cannot equilibrate once blended into the bulk stream.

From a polymer physics standpoint, transient conditions disturb the balance between network formation and irreversible fatigue. Short-residence fractions exiting during startup lack sufficient structural development, while long-residence fractions retained during shutdown accumulate excessive shear and thermal exposure. When recombined, these heterogeneous populations introduce persistent rheology drift rather than converging toward a stable viscosity plateau.

Startup-Induced Rheological Instability

During startup, flow rates, mixing intensity, and temperature control systems ramp asynchronously. Initial material often encounters high shear at low fill levels, elevated local energy density, and poor heat dissipation. Associative polymers introduced under these conditions may suffer early network disruption before optimal junction formation occurs. In continuous lines, short-circuiting during startup allows under-processed material to bypass conditioning zones and enter finished product streams.

Industrial manifestations include early-batch low viscosity, delayed thixotropic recovery, and increased variability in inline rheology during the first production window. These effects are frequently misinterpreted as raw material inconsistency rather than transient process-induced structural damage, complicating root-cause analysis.

Shutdown and Hold-Over Effects

Shutdown introduces a different but equally critical stress profile. Flow reduction increases effective residence time, while continued agitation or recirculation subjects remaining material to repeated shear cycles. Thermal gradients develop as heat removal declines, accelerating irreversible degradation pathways. Concentration gradients may also form due to evaporation or phase separation in partially filled equipment.

Material retained during shutdown often re-enters the process during restart, seeding the system with rheologically compromised polymer. This carryover effect explains why post-restart viscosity drift can persist long after steady-state conditions are re-established, particularly in high-solids VMA formulations.

Manufacturing Signals and Field Performance Consequences

Transient-induced instability typically appears as step changes or oscillations in inline viscosity, pressure drop, or power draw correlated with startup or shutdown events. Batch-to-batch variability increases despite stable formulations and raw materials. Off-line QC data show time-dependent drift linked to sampling order rather than composition.

In concrete applications, VMAs affected by transient damage exhibit inconsistent yield stress contribution, reduced slump retention after transport, and heightened sensitivity to mixing energy. These products often perform acceptably in short trials but fail under extended pumping or variable job-site conditions, reflecting the latent heterogeneity introduced during manufacturing transitions.

Table 19.2.6: Startup and Shutdown Transient Effects on VMA Rheology

Transient Factor	Process Origin	Manufacturing Indicator	Rheological Impact
Low fill-level shear	Startup mixing	Early viscosity drop	Under-developed network
Short-circuiting	Continuous startup	Inline variability	Incomplete activation
Extended residence	Shutdown hold	Gradual viscosity decay	Network fatigue
Thermal gradients	Cooling loss	Temperature–viscosity mismatch	Irreversible degradation
Material carryover	Restart blending	Persistent drift	Long-term instability

Industry 4.0 Control and Technology Transfer Strategies

Mitigation requires treating startup and shutdown as critical processing phases rather than operational exceptions. Controlled ramp profiles for shear, flow, and temperature minimize exposure to damaging conditions. Dedicated diversion or purge strategies prevent transient material from contaminating on-spec product.

Industry 4.0 platforms enable transient-aware digital twins that simulate polymer response during non-steady-state operations. Real-time analytics correlate viscosity, temperature, and power signals with transient events, allowing predictive isolation of at-risk material. For technology transfer, explicit definition of startup and shutdown protocols, including purge volumes, ramp rates, and maximum allowable transient exposure, is essential. Failure to transfer these constraints is a frequent root cause of unexplained rheology drift when production is scaled or relocated.

19.2.7 Inline Rheology Measurement Deviations

Fundamental Limitations of Inline Rheological Measurements

Inline rheology sensors provide continuous insight into viscosity evolution during VMA manufacturing, yet their outputs are inherently influenced by local flow geometry, shear rate distribution, and thermal conditions at the measurement point. Unlike laboratory rheometers operating under well-defined shear protocols, inline devices measure apparent rheological responses that integrate complex velocity gradients and pressure fields. Associative polymers and modified polysaccharides exhibit strong shear and time dependence, meaning that inline readings reflect both material state and measurement history. Deviations therefore arise when sensor conditions drift from the assumptions embedded in calibration models.

From a polymer science perspective, inline deviations are often linked to mismatch between the sensor-imposed shear field and the polymer's characteristic relaxation times. When relaxation kinetics are slower than the effective shear exposure within the sensor, measured viscosity underrepresents bulk structural recovery. Conversely, excessive residence time within the sensor can induce local network fatigue, biasing readings downward relative to true process rheology.

Process-Induced Sources of Measurement Drift

Measurement deviations frequently originate from changes in flow regime, temperature gradients, or solids content near the sensor. Fouling or partial coating of sensor surfaces by associative polymers alters effective geometry and increases apparent shear stress without reflecting bulk viscosity. Gas entrainment, common during startup or recirculation, further distorts pressure-based rheological signals.

In continuous processes, throughput adjustments shift Reynolds number and shear distribution within the sensor, invalidating fixed calibration factors. Batch processes are equally vulnerable, as declining fill levels and changing circulation paths modify local flow conditions. These effects lead to apparent viscosity drift that mirrors process evolution but does not necessarily represent true material degradation.

Operational Consequences and Misinterpretation Risks

Inline rheology deviations can trigger inappropriate corrective actions if misinterpreted as formulation instability. Operators may increase polymer dosage, reduce throughput, or modify mixing intensity based on biased signals, inadvertently exacerbating shear or thermal stress. Over time, this feedback loop introduces real rheological drift driven by control response rather than material behavior.

In concrete applications, reliance on uncorrected inline data can lead to inconsistent VMA dosing recommendations, reduced robustness across plants, and apparent seasonal variability that originates from measurement artifacts rather than polymer chemistry. Distinguishing sensor deviation from true rheology change is therefore critical for both manufacturing control and downstream performance consistency.

Diagnostic Validation and Cross-Correlation Strategies

Effective validation combines inline measurements with periodic off-line rheological testing under controlled shear and temperature conditions. Time-aligned comparison of inline viscosity, pressure drop, torque, and temperature trends helps isolate sensor-specific artifacts. Step-change experiments in flow rate or temperature reveal whether measured viscosity responds according to polymer physics or sensor bias.

Industry 4.0 platforms enable sensor fusion, integrating multiple inline signals into multivariate models that detect deviations inconsistent with known material behavior. Digital twins incorporating sensor geometry and flow conditions can predict expected measurement response and flag divergence in real time, preventing erroneous process intervention.

Table 19.2.7: Sources and Impacts of Inline Rheology Measurement Deviations

Deviation Source	Process Origin	Measurement Signal	Manufacturing Risk

Shear field mismatch	Sensor geometry	Apparent low viscosity	False degradation
Thermal gradients	Inadequate insulation	Temperature-correlated drift	Misadjusted control
Sensor fouling	Polymer adsorption	Gradual signal bias	Overdosing
Flow regime change	Throughput variation	Sudden viscosity shift	Unnecessary intervention
Gas entrainment	Startup, recirculation	Signal noise	Control instability

Industry 4.0 Control and Technology Transfer Considerations

Robust use of inline rheology requires dynamic calibration frameworks that account for flow, temperature, and shear history. Sensor placement should minimize exposure to extreme shear and thermal gradients while remaining representative of bulk flow. Periodic validation against reference measurements is essential, particularly after maintenance or process changes.

From a technology transfer standpoint, sensor configuration, calibration logic, and data interpretation rules must be transferred alongside formulation and process parameters. Industry 4.0 architectures support standardized digital models that travel with the process, ensuring consistent rheology interpretation across sites and preventing silent divergence driven by measurement artifacts rather than polymer performance.

19.2.8 Feedback Control Loop Instability

Dynamic Interaction Between Polymer Rheology and Control Systems

Feedback control loops regulating viscosity-modifying admixture production are inherently challenged by the time-dependent and path-dependent rheology of associative polymers and modified polysaccharides. These materials exhibit delayed structural response to changes in shear, temperature, or concentration due to finite hydration, neutralization, and network reformation kinetics. Conventional PID control strategies assume near-instantaneous and reversible process response, an assumption that is fundamentally incompatible with thixotropic and shear-sensitive polymer systems. As a result, corrective actions based on real-time viscosity signals often lag true material state, creating oscillatory or divergent control behavior.

From a control theory perspective, the combination of sensor delay, polymer relaxation time, and residence time distribution introduces phase lag and gain amplification. When loop tuning does not explicitly account for these dynamics, small disturbances are reinforced rather than damped, leading to instability that is mistakenly attributed to raw material variability or sensor noise.

Process Conditions That Promote Control Instability

Instability is most pronounced in high-solids formulations and continuous processes with long effective residence times. Adjustments to pump speed, recirculation rate, or polymer dosing alter shear exposure and concentration simultaneously, producing non-linear rheological responses. Startup, shutdown, and throughput transitions further destabilize control loops by introducing transient material populations that violate steady-state assumptions embedded in the control logic.

Scale-up exacerbates these effects because increased volume and longer transport paths extend response times. Control strategies developed at laboratory or pilot scale often overcorrect at industrial scale, inducing cyclic viscosity fluctuations that persist even under constant operating conditions.

Manufacturing Signals and Product Performance Impact

Feedback loop instability manifests as oscillations in inline viscosity, pressure drop, or dosing rate that correlate with control actions rather than external disturbances. Power draw and torque signals show periodic modulation, reflecting repeated over- and under-structuring of the polymer network. Batch quality data reveal increased variability despite consistent formulations and raw materials.

In concrete applications, such instability translates into inconsistent VMA performance, variable yield stress contribution, and sensitivity to small dosing errors. Products may meet average specification but lack robustness, leading to unpredictable field behavior during pumping, placement, or finishing.

Diagnostic and Analytical Approaches

Diagnosing control-induced instability requires separating true process disturbances from control loop artifacts. Time-domain analysis of control signals, cross-correlation between viscosity and dosing adjustments, and frequency-domain analysis reveal characteristic oscillation patterns indicative of loop interaction. Controlled step tests under steady feed conditions help quantify system dead time and polymer response kinetics.

Industry 4.0 analytics platforms integrate these diagnostics with historical production data, enabling automated detection of unstable control behavior. Predictive models identify operating regions where loop gain exceeds system stability margins and recommend parameter adjustments or alternative control architectures.

Table 19.2.8: Feedback Control Loop Instability in VMA Manufacturing

Instability Driver	Process Origin	Control Signal	Rheological Consequence
Response time mismatch	Polymer relaxation	Oscillating viscosity	Structural cycling
Excessive loop gain	Aggressive tuning	Dosing fluctuation	Network fatigue
Residence time delay	Large-scale systems	Phase-lag oscillation	Drift and variability
Transient disturbances	Startup, shutdown	Control overshoot	Heterogeneous product
Sensor bias	Inline deviation	False correction	Performance inconsistency

Stabilization Strategies and Technology Transfer Implications

Stabilizing feedback control requires incorporating polymer kinetics into control design. Feed-forward control elements based on shear energy, temperature, and throughput reduce reliance on delayed viscosity feedback. Model predictive control strategies outperform traditional PID loops by anticipating rheological response and respecting system constraints.

Industry 4.0 frameworks enable digital twins that couple polymer rheology models with control logic, allowing virtual tuning and stress testing before implementation. For technology transfer, control philosophy, tuning parameters, and stability criteria must be transferred alongside formulation and equipment specifications. Failure to replicate control logic is a frequent root cause of unexplained rheology variability when production is scaled or relocated.

19.3 Raw Material Variability and Its Amplification

Raw material variability is a primary source of hidden instability in VMA manufacturing, as polymer-based systems amplify small compositional deviations into large rheological effects. Variations in degree of substitution, molecular weight distribution, neutralization level, or counterion composition can shift polymer solubility and association behavior well beyond acceptable functional limits. These effects are often masked by broad supplier specifications that do not reflect the narrow operating window required for consistent VMA performance in modern concrete systems.

Water quality, frequently treated as a utility rather than a raw material, plays an equally critical role. Dissolved calcium, magnesium, or sulfate ions can trigger premature polymer association or partial precipitation during processing. Effective troubleshooting therefore requires systematic qualification of all inputs, combined with digital tracking of supplier variability and correlation to manufacturing performance metrics.

19.3.1 Polymer Degree of Substitution Variability

Chemical Significance of Degree of Substitution in VMA Polymers

Degree of substitution, whether referring to etherification in polysaccharides or functional group density in associative acrylic polymers, defines the balance between hydrophilicity and hydrophobic association that governs rheological efficiency. Small shifts in substitution level alter polymer solubility, hydration kinetics, and junction density, producing nonlinear effects on viscosity and yield stress. In VMA systems operating near critical association thresholds, even minor

DS variability propagates into disproportionate changes in network strength and thixotropic behavior.

From a molecular standpoint, DS controls the spatial distribution and frequency of associative sites. Below an optimal range, insufficient junction formation limits network development, while excessive substitution promotes premature association, poor dissolution, or microgel formation. These effects occur without changes in nominal molecular weight, making DS variability a particularly insidious source of hidden instability.

Sources of Degree of Substitution Variability

DS variability originates from both polymer synthesis and downstream handling. In polysaccharide derivatives, heterogeneity arises from incomplete reaction uniformity along the backbone and between production batches. For synthetic associative polymers, fluctuations in comonomer feed ratio, conversion efficiency, or post-neutralization lead to variable functional group density. Supplier specifications typically report average DS values, obscuring distribution breadth that is highly relevant to rheological response.

Storage and logistics can further modify effective DS through partial hydrolysis, oxidation, or counterion exchange, particularly in hygroscopic materials. These changes often remain within certificate-of-analysis limits yet materially alter polymer performance during processing and application.

Manufacturing Manifestation and Rheological Amplification

DS variability manifests during hydration and neutralization as inconsistent viscosity build, variable induction time, or unexpected gelation. Inline rheology frequently shows batch-dependent slopes rather than absolute shifts, reflecting altered kinetics rather than steady-state viscosity alone. In continuous processes, these differences translate into unstable control behavior and increased sensitivity to shear and temperature fluctuations.

In concrete applications, VMAs derived from polymers with variable DS exhibit inconsistent yield stress contribution and robustness. Performance may appear acceptable at one dosage yet fail catastrophically with small adjustments in water content or mixing energy, revealing the narrow functional window imposed by DS heterogeneity.

Diagnostic Characterization and Process Analytics

Routine QC methods such as Brookfield viscosity or solids content are insufficient to detect DS-driven variability. Spectroscopic techniques including NMR, FTIR, or titration-based charge density measurement provide direct insight into substitution level but are rarely implemented at sufficient frequency. Rheological fingerprinting under controlled hydration and shear protocols offers a practical surrogate, revealing characteristic deviations in build rate and recovery behavior.

Industry 4.0 platforms enable correlation of DS-related indicators with inline process data, identifying supplier or lot-dependent trends that precede specification failure. Multivariate models linking rheology evolution to raw material attributes support early intervention before downstream instability propagates.

Table 19.3.1: Degree of Substitution Variability and Its Impact on VMA Performance

DS Characteristic	Source of Variability	Manufacturing Signal	Functional Impact
Low average DS	Under-substitution	Slow viscosity build	Weak network
High average DS	Over-substitution	Rapid association	Gelation risk
Broad DS distribution	Non-uniform synthesis	Rheology scatter	Poor robustness
DS drift in storage	Hydrolysis, oxidation	Batch-to-batch trend	Performance decay
Counterion-dependent DS	Neutralization variability	Inline instability	Dosage sensitivity

Control Strategies and Technology Transfer Considerations

Managing DS variability requires narrowing acceptable specification windows beyond supplier norms and explicitly qualifying substitution distributions rather than averages. Incoming material should be characterized using rapid proxy methods tied to rheological performance, not solely chemical metrics. Dual sourcing without DS harmonization is a common root cause of silent instability.

Industry 4.0-enabled raw material tracking systems link supplier data, lot history, and inline manufacturing signals to predict DS-related risk. For technology transfer, explicit disclosure of critical DS ranges, analytical methods, and acceptance criteria is essential. Failure to transfer this knowledge frequently results in unexplained rheology drift when production is scaled or relocated, despite apparent compliance with raw material specifications.

19.3.2 Molecular Weight Distribution Drift

Role of Molecular Weight Distribution in VMA Rheology

Molecular weight distribution governs chain entanglement density, relaxation dynamics, and the efficiency of associative network formation in viscosity-modifying admixtures. While average molecular weight is commonly specified, the breadth and shape of the distribution exert a disproportionate influence on rheological behavior. High-molecular-weight fractions dominate elasticity and yield stress, whereas low-molecular-weight fractions control dissolution kinetics and early viscosity build. Small shifts in distribution balance therefore amplify into large changes in apparent viscosity, thixotropy, and shear sensitivity.

From a polymer physics perspective, associative polymers operate near a percolation threshold where connectivity depends on the longest chains. Loss or dilution of this fraction reduces effective network span even when number-average molecular weight remains unchanged. Conversely, enrichment in high-molecular-weight tails increases sensitivity to shear fatigue and irreversible breakdown.

Sources and Mechanisms of Distribution Drift

MWD drift arises from both upstream polymer synthesis and downstream processing. Variations in initiator efficiency, chain transfer reactions, or residence time during polymerization alter distribution breadth between lots. Post-production handling introduces additional risk, as thermal exposure, oxidative environments, or high shear during dissolution can selectively degrade high-molecular-weight chains.

In aqueous VMA manufacturing, cumulative shear during mixing, recirculation, and pumping does not necessarily cleave covalent backbones but can induce scission in polysaccharides or weaken associative clustering, effectively narrowing functional MWD. These changes often escape detection by standard QC methods focused on solids content or single-point viscosity.

Manufacturing Manifestation and Rheological Consequences

MWD drift typically presents as gradual viscosity loss under constant formulation and processing conditions. Inline rheology shows declining yield stress contribution and reduced structural recovery after shear. Batch-to-batch consistency deteriorates despite compliance with supplier specifications, confounding root-cause analysis.

In concrete applications, VMAs affected by MWD drift display reduced segregation resistance, shorter slump retention, and heightened sensitivity to pumping energy. Performance decay is often time-dependent, appearing after extended storage or repeated handling, reflecting progressive loss of the high-molecular-weight fraction critical for network integrity.

Analytical Detection and Process Correlation

Gel permeation chromatography provides direct measurement of MWD but is rarely implemented as a routine incoming control due to time and cost constraints. Rheological fingerprinting under defined shear histories serves as a practical surrogate, revealing characteristic shifts in recovery and shear sensitivity associated with distribution changes.

Industry 4.0 platforms correlate these rheological signatures with process energy input, temperature history, and storage conditions to identify MWD drift drivers. Predictive models can flag abnormal degradation trends before specification limits are exceeded, enabling proactive corrective action.

Table 19.3.2: Molecular Weight Distribution Drift and Its Industrial Impact

MWD Change	**Source**	**Manufacturing Signal**	**Functional Impact**
Loss of high-MW tail	Shear, oxidation	Gradual viscosity decay	Reduced yield stress
Narrowed distribution	Processing fatigue	Faster relaxation	Poor thixotropy
Broadened distribution	Polymer synthesis	Rheology scatter	Control instability

Low-MW enrichment	Hydrolysis	Rapid build, low plateau	Weak long-term performance
Time-dependent drift	Storage aging	Delayed QC failure	Field inconsistency

Control Measures and Technology Transfer Implications

Managing MWD drift requires controlling both polymer supply quality and in-plant shear and thermal exposure. Incoming material qualification should include at least periodic MWD characterization or validated rheological proxies sensitive to distribution changes. Processing limits on cumulative shear energy and temperature must be defined as critical quality attributes to protect high-molecular-weight fractions.

Industry 4.0-based material genealogy systems link supplier lots, processing history, and performance outcomes, enabling rapid identification of MWD-related risk. For technology transfer, explicit communication of acceptable MWD profiles, degradation sensitivity, and protective processing constraints is essential. Absence of this information is a common cause of silent performance loss when VMA production is scaled or relocated.

19.3.3 Counterion and Salt Content Fluctuations

Electrochemical Role of Counterions in VMA Polymers

Counterions associated with ionizable groups on viscosity-modifying polymers govern charge screening, chain expansion, and associative junction stability. For anionic polysaccharides and acrylic-based VMAs, sodium, potassium, calcium, or ammonium counterions directly influence electrostatic repulsion along the backbone and, consequently, polymer conformation in solution. Small changes in counterion identity or concentration alter osmotic pressure and hydration shell structure, shifting the balance between extended and collapsed chain states.

From a polymer thermodynamics perspective, multivalent counterions reduce effective charge density through ion bridging, promoting premature association or partial precipitation. These effects are highly nonlinear and become amplified in high-solids formulations where ionic strength is already elevated.

Sources of Counterion and Salt Variability

Counterion fluctuations arise from inconsistent neutralization during polymer synthesis, variation in neutralizing agents, or incomplete washing of residual salts. Downstream, water quality introduces additional ionic variability through dissolved calcium, magnesium, sulfate, or bicarbonate species. Storage and transport can further modify ionic composition via CO_2 absorption, evaporation, or container leaching, subtly shifting salt balance while remaining within conventional specifications.

Dual sourcing of polymers or neutralizing agents without harmonized counterion profiles is a frequent but under-recognized cause of rheological inconsistency. Certificates of analysis typically report total solids and pH but rarely disclose ionic composition with sufficient resolution to predict performance.

Manufacturing Manifestation and Rheological Amplification

Salt and counterion variability manifests as inconsistent hydration rate, altered viscosity build kinetics, and unexpected gelation or phase separation during processing. Inline rheology often shows erratic behavior rather than monotonic drift, reflecting sensitivity to local ionic conditions. In continuous processes, minor fluctuations in make-up water quality can trigger sudden viscosity shifts that destabilize control loops.

In concrete applications, VMAs exposed to unfavorable ionic environments exhibit reduced robustness, poor tolerance to cement alkalis, and sensitivity to minor water content changes. Multivalent ions in particular diminish yield stress contribution and accelerate thixotropy loss under shear.

Analytical Characterization and Data Integration

Ion chromatography and ICP-OES provide direct quantification of counterions and residual salts but are rarely implemented as routine controls. Conductivity and alkalinity measurements serve as practical proxies when calibrated against performance data. Rheological screening in standardized high-ionic-strength media reveals susceptibility to salt-induced association and offers a functional acceptance criterion.

Industry 4.0 systems integrate water quality sensors, inline conductivity, and rheology data to establish correlations between ionic fluctuations and manufacturing performance. Predictive analytics identify ionic thresholds beyond which rheological instability becomes likely, enabling proactive adjustment of formulation or water treatment.

Table 19.3.3: Counterion and Salt Content Effects on VMA Rheology

Ionic Factor	Source	Manufacturing Signal	Functional Impact
High divalent ions	Water hardness	Rapid viscosity change	Premature association
Counterion mismatch	Neutralization variability	Batch inconsistency	Dosage sensitivity
Elevated ionic strength	Residual salts	Reduced recovery	Thixotropy loss
CO_2 absorption	Storage aging	Gradual pH shift	Viscosity drift
Seasonal water change	Utility supply	Inline variability	Control instability

Control Strategies and Technology Transfer Considerations

Controlling ionic variability requires treating water and counterions as critical raw materials rather than utilities. Specification of acceptable ionic profiles, including maximum divalent ion content, is essential. On-site water conditioning, controlled neutralization protocols, and harmonized sourcing reduce exposure to salt-induced instability.

Industry 4.0-enabled raw material governance links ionic data with rheological outcomes, allowing predictive control of salt effects across sites. For technology transfer, explicit communication of counterion requirements, water quality limits, and acceptable ionic strength ranges is mandatory. Omission of these parameters is a frequent root cause of unexplained rheology instability when VMA production is scaled or relocated.

19.3.4 Water Quality and Hardness Effects

Physicochemical Role of Water in VMA Manufacturing

Water functions as both solvent and reactive medium in viscosity-modifying admixture production, directly controlling polymer hydration, chain expansion, and associative junction stability. Variations in water hardness and dissolved ionic species modify dielectric constant, osmotic pressure, and electrostatic screening, thereby influencing polymer conformation well beyond dilution effects. For associative polymers and modified polysaccharides, water quality defines the boundary between controlled network formation and premature association or collapse.

From a polymer science perspective, hard water containing calcium and magnesium ions promotes charge neutralization and ion bridging, reducing chain repulsion and accelerating intermolecular association. These effects are amplified at high polymer concentration, where small ionic changes disproportionately affect rheological behavior.

Sources and Nature of Water Quality Variability

Industrial water supplies exhibit temporal variability driven by seasonal source changes, treatment adjustments, and infrastructure aging. Hardness, alkalinity, sulfate, and chloride levels can fluctuate within utility specifications while remaining incompatible with sensitive polymer systems. Recycled process water introduces additional complexity through accumulation of salts, organics, and trace metals that interact synergistically with polymer additives.

Because water is often classified as a utility rather than a raw material, its variability is rarely tracked with the rigor applied to polymers or additives. This oversight is a frequent root cause of unexplained batch inconsistency and long-term rheology drift in VMA manufacturing.

Manufacturing Manifestation and Rheological Consequences

Water hardness effects manifest during polymer dissolution and neutralization as variable viscosity build rates, incomplete hydration, or localized gel formation. Inline rheology may display abrupt shifts correlated with water source changes rather than formulation adjustments. Batch-to-batch variability increases despite consistent raw materials and process parameters.

In concrete applications, VMAs prepared with hard or ionically unstable water exhibit reduced yield stress contribution, poor tolerance to cement alkalis, and sensitivity to minor water dosage changes. Performance issues often intensify under hot-weather placement, where ionic effects and thermal thinning act synergistically.

Analytical Monitoring and Industry 4.0 Integration

Routine monitoring of hardness, conductivity, and alkalinity provides actionable insight into water suitability for polymer processing. Advanced facilities deploy inline conductivity and ion-selective sensors integrated with rheology measurements to detect early deviation from acceptable ionic windows. Rheological fingerprinting using standardized hard-water media supports rapid assessment of polymer robustness to water variability.

Industry 4.0 platforms correlate water quality metrics with viscosity trends, control actions, and product performance, enabling predictive identification of instability risk. Digital twins incorporating water chemistry and polymer response kinetics support proactive adjustment of formulation or conditioning strategies before quality is compromised.

Table 19.3.4: Water Quality and Hardness Effects on VMA Rheology

Water Parameter	Variability Source	Manufacturing Signal	Functional Impact
High Ca^{2+}/Mg^{2+}	Water hardness	Rapid viscosity build	Premature association
Elevated sulfate	Utility treatment	Viscosity instability	Network disruption
High alkalinity	Source fluctuation	Delayed hydration	Incomplete activation
Recycled water salts	Process reuse	Gradual drift	Reduced robustness
Seasonal variation	Supply changes	Batch inconsistency	Control instability

Control Strategies and Technology Transfer Implications

Effective control requires specification of water quality limits as critical raw material attributes, including maximum hardness, conductivity, and sulfate levels.

On-site softening, deionization, or blending strategies reduce exposure to harmful ionic species. Segregation of process water streams prevents uncontrolled salt accumulation.

Industry 4.0-enabled water management systems link real-time water chemistry with rheological performance, allowing dynamic adjustment of processing conditions. For technology transfer, explicit disclosure of acceptable water quality envelopes and conditioning requirements is essential. Failure to transfer water specifications is a common cause of VMA performance degradation when production is relocated or scaled, despite unchanged formulations.

19.3.5 Supplier-to-Supplier Consistency Gaps

Structural Origins of Inter-Supplier Variability

Supplier-to-supplier variability represents one of the most underestimated amplification mechanisms in VMA manufacturing. Polymers sourced from different producers may nominally meet identical specifications for solids content, viscosity, and pH, yet differ substantially in molecular architecture, substitution distribution, residual salt profile, and impurity content. These structural differences translate into distinct hydration kinetics, associative behavior, and shear sensitivity once formulated into high-solids VMA systems.

From a polymer chemistry standpoint, specifications based on average properties mask distribution-level differences that dominate rheological response. Variations in polymerization route, post-functionalization chemistry, and purification intensity introduce subtle but functionally critical divergences that are invisible to routine incoming quality control.

Manufacturing and Rheological Amplification Mechanisms

Inter-supplier differences become amplified during processing due to nonlinear polymer response to shear, temperature, and ionic environment. A polymer with slightly broader molecular weight distribution or higher residual salt content may behave acceptably under low-stress laboratory conditions yet exhibit accelerated network fatigue or thixotropy loss during industrial recirculation. Conversely, a more narrowly distributed material may be less tolerant to transient shear spikes, destabilizing feedback control loops.

In continuous production, switching suppliers without re-qualification often introduces step changes in inline rheology, control behavior, and product robustness. These effects propagate downstream, confounding process analytics and leading to corrective actions that further destabilize the system.

Field Performance Consequences in Concrete Applications

Supplier-induced variability frequently manifests as inconsistent yield stress contribution, segregation resistance, and slump retention at the concrete plant. VMAs from different suppliers may require different dosage windows to achieve comparable performance, reducing operational flexibility. Performance gaps become particularly evident under extreme conditions such as high pumping energy, variable cement alkali content, or seasonal temperature shifts.

These inconsistencies erode customer confidence and are often misattributed to cement or aggregate variability, delaying corrective action at the source.

Qualification, Analytics, and Industry 4.0 Enablement

Effective management of supplier consistency requires moving beyond specification compliance toward performance-based qualification. Comparative rheological fingerprinting under standardized shear, temperature, and ionic stress conditions reveals functional equivalence or divergence. Chemical characterization of substitution distribution, ionic profile, and impurity content provides complementary insight.

Industry 4.0 platforms enable supplier digital profiles that link material genealogy with manufacturing and field performance data. Multivariate analytics identify systematic supplier-dependent trends, supporting data-driven sourcing decisions and early risk detection before customer impact occurs.

Table 19.3.5: Supplier-to-Supplier Variability and Its Impact on VMA Performance

Variability Dimension	Supplier Difference	Manufacturing Signal	Functional Impact
Molecular architecture	Polymerization route	Rheology shift	Dosage mismatch

Substitution distribution	Functionalization control	Build-rate variability	Robustness loss
Residual salts	Purification efficiency	Inline instability	Thixotropy loss
Impurity profile	Raw material quality	Aging drift	Performance decay
Batch homogeneity	Process control	QC scatter	Inconsistent field behavior

Control Strategies and Technology Transfer Considerations

Mitigation requires formal equivalency protocols defining acceptable ranges for functional performance, not just chemical metrics. Dual sourcing should be implemented only after side-by-side manufacturing and application trials under representative industrial stress conditions. Change management procedures must treat supplier switches as process changes requiring validation.

Industry 4.0-enabled supplier governance systems ensure traceability of raw material origin, processing history, and performance outcomes. For technology transfer, supplier-specific constraints, qualification data, and performance envelopes must be transferred alongside formulation and process parameters. Failure to transfer this knowledge is a frequent root cause of unexplained rheology variability when production is relocated or sourcing strategies change.

19.3.6 Impact of Residual Monomers and Impurities

Chemical Nature and Origin of Residual Species

Residual monomers and low-level impurities originate from incomplete polymerization, side reactions, or insufficient purification during polymer manufacture. In associative acrylic polymers, traces of unreacted acrylic acid, methacrylic acid, or hydrophobic comonomers remain chemically active in aqueous environments. Modified polysaccharides may contain residual reagents, salts, or degradation byproducts introduced during etherification or oxidation steps. Although present at low concentrations, these species exert outsized influence on rheology due to their high mobility and chemical reactivity.

From a polymer chemistry perspective, residual monomers act as plasticizers, ionic strength modifiers, or secondary reactants that perturb chain conformation and associative junction stability. Their effects are magnified in high-solids VMA formulations where local concentration and residence time enable cumulative interaction with the polymer network.

Mechanisms of Rheological Amplification

Residual acidic or basic monomers alter local pH and degree of neutralization, shifting polymer charge density and hydration behavior. Hydrophobic impurities preferentially associate with existing junctions, disrupting network uniformity and accelerating shear fatigue. Low-molecular-weight species also migrate rapidly under shear, biasing inline rheological measurements and creating apparent instability that does not correlate with bulk composition.

In polysaccharide systems, oxidative byproducts catalyze chain scission or promote uncontrolled association, reducing molecular weight distribution integrity. These mechanisms operate silently within supplier specification limits, making impurity-driven instability difficult to detect through routine quality control.

Manufacturing Manifestation and Downstream Effects

Impurity-related effects often appear as unpredictable viscosity build, abnormal thixotropy loss, or time-dependent drift during holding and recirculation. Inline rheology signals show increased noise or sporadic deviations unrelated to process changes. Batch-to-batch consistency deteriorates despite stable formulation and processing conditions.

In concrete applications, VMAs affected by residual monomers exhibit poor robustness, sensitivity to cement alkalis, and inconsistent performance under extended mixing or pumping. These effects frequently intensify with storage time, reflecting ongoing interaction between impurities and the polymer network.

Analytical Detection and Industry 4.0 Correlation

Detection of residual monomers requires targeted analytical methods such as GC-MS, HPLC, or headspace analysis, which are not routinely applied in incoming QC. Proxy indicators include abnormal conductivity, unexpected pH drift, or accelerated viscosity decay under controlled aging tests. Rheological stress testing under elevated temperature and shear provides functional evidence of impurity sensitivity.

Industry 4.0 platforms correlate impurity indicators with manufacturing performance and field complaints, enabling identification of supplier or lot-specific risk. Advanced analytics distinguish impurity-driven instability from shear or thermal degradation, supporting focused corrective action.

Table 19.3.6: Residual Monomers and Impurities and Their Effects on VMA Rheology

Residual Species	Origin	Manufacturing Signal	Functional Impact
Unreacted monomers	Incomplete polymerization	pH drift	Charge imbalance
Hydrophobic residues	Comonomer excess	Thixotropy loss	Network disruption
Oxidative byproducts	Harsh processing	Gradual viscosity decay	MWD degradation
Residual salts	Insufficient washing	Inline variability	Ionic sensitivity
Volatile impurities	Storage aging	Signal noise	Control instability

Control Strategies and Technology Transfer Considerations

Mitigation requires tightening impurity specifications and incorporating impurity-sensitive performance tests into raw material qualification. Aging and stress tests should be standardized to reveal latent instability associated with residual species. Supplier audits focusing on purification and post-treatment practices provide additional risk reduction.

Industry 4.0-enabled material genealogy systems link impurity profiles with processing history and application performance, enabling rapid identification of recurring issues. For technology transfer, explicit communication of impurity limits, analytical methods, and acceptable aging behavior is essential. Failure to transfer these constraints often results in unexplained rheology drift when sourcing or production location changes, despite nominal compliance with raw material specifications.

19.3.7 Storage Aging of Polymer Feedstocks

Chemical and Physical Drivers of Storage Aging

Polymer feedstocks undergo continuous physicochemical evolution during storage, even under nominally controlled conditions. Exposure to oxygen, moisture, residual heat, and trace impurities initiates slow degradation pathways that alter molecular weight distribution, degree of ionization, and associative functionality. In aqueous polymer dispersions, hydrolysis, oxidation, and counterion exchange proceed gradually, while dry polymer powders experience moisture uptake, partial agglomeration, and surface oxidation.

From a polymer chemistry standpoint, storage aging disproportionately affects the most rheologically active fractions, particularly high-molecular-weight chains and labile associative groups. These changes occur without obvious visual or compositional indicators, making aging a silent contributor to VMA instability.

Mechanisms of Rheological Drift During Storage

Oxidative aging promotes backbone scission in polysaccharides and weakens associative junctions in acrylic polymers, narrowing effective molecular weight distribution. Moisture-driven hydrolysis alters charge density and hydration behavior, while CO_2 absorption shifts pH and counterion balance. In liquid feedstocks, slow phase separation or microgel formation modifies dissolution kinetics during subsequent processing.

These mechanisms reduce viscosity build rate, thixotropic recovery, and shear resistance once the polymer is formulated into a VMA. Aging effects accumulate with storage duration and are exacerbated by elevated temperature, headspace oxygen, and repeated container opening.

Manufacturing and Field Performance Consequences

Storage-aged polymers introduce time-dependent variability that manifests as gradual viscosity loss, delayed hydration, or inconsistent neutralization behavior during manufacturing. Inline rheology reveals lot-age-dependent trends rather than random scatter, often misattributed to process drift. Batch-to-batch consistency deteriorates despite unchanged formulations and operating conditions.

In concrete applications, VMAs produced from aged feedstocks exhibit reduced robustness, shorter slump retention, and increased sensitivity to mixing energy and temperature. Performance complaints frequently correlate with storage duration rather than supplier lot, complicating root-cause identification without explicit age tracking.

Monitoring, Analytics, and Industry 4.0 Enablement

Effective management of storage aging requires systematic tracking of feedstock age, storage conditions, and exposure history. Accelerated aging tests combined with rheological fingerprinting provide predictive insight into shelf-life limitations. Simple indicators such as peroxide value, pH drift, or moisture uptake serve as early warning signals when calibrated against performance outcomes.

Industry 4.0 systems integrate warehouse data, environmental sensors, and manufacturing performance metrics into material genealogy models. Predictive analytics identify aging thresholds beyond which polymer performance risk increases, enabling proactive inventory rotation or conditional requalification.

Table 19.3.7: Storage Aging Effects on Polymer Feedstock Performance

Aging Driver	Storage Condition	Manufacturing Signal	Functional Impact
Oxidative exposure	Air headspace	Gradual viscosity decay	Network weakening
Moisture uptake	Humid storage	Delayed hydration	Poor build rate
Thermal aging	Elevated temperature	Time-dependent drift	MWD narrowing
CO_2 absorption	Open containers	pH shift	Ionic imbalance
Extended storage	Long shelf life	Lot-age trend	Reduced robustness

Control Strategies and Technology Transfer Implications

Mitigation requires defining storage conditions and maximum allowable age as critical raw material attributes. Controlled atmosphere packaging, temperature limits, and moisture barriers reduce aging kinetics. First-in–first-out inventory discipline must be enforced with digital traceability rather than manual tracking.

Industry 4.0-enabled inventory management links feedstock age and storage history with rheological performance, supporting data-driven shelf-life determination. For technology transfer, explicit transfer of storage specifications, aging sensitivity, and requalification criteria is essential. Omission of these parameters is a common root cause of rheology variability when production is relocated or inventory practices differ, despite identical formulations.

19.3.8 Specification Window Versus Functional Window Misalignment

Conceptual Distinction Between Specification and Functional Windows

Industrial polymer specifications define acceptance limits for a limited set of measurable parameters such as solids content, pH, viscosity at defined shear, and residual monomer level. These limits describe a specification window that ensures material consistency from a transactional and contractual perspective. In contrast, the functional window represents the multidimensional performance space within which a polymer delivers stable rheology, robustness to formulation variables, and predictable field behavior in concrete systems. The two windows overlap only partially, particularly for associative polymers and modified polysaccharides whose performance depends on subtle molecular and colloidal interactions.

From a polymer chemistry standpoint, rheological function emerges from chain architecture, molecular weight distribution tails, degree of neutralization, and associative junction dynamics. Many of these parameters are weakly constrained or entirely invisible within conventional specifications, allowing materials to pass QC while operating at the edge or outside the true functional window.

Molecular Origins of Misalignment

Misalignment arises when specification parameters fail to capture sensitivity to second-order molecular attributes. Two materials with identical Brookfield viscosity and solids may differ markedly in high-molecular-weight fraction, hydrophobe distribution, or counterion population. These differences translate directly into yield stress development, thixotropic recovery, and shear stability once formulated into VMAs.

In acrylic VMAs, slight shifts in neutralization profile or hydrophobe content can preserve specification viscosity while fundamentally altering associative network resilience. In polysaccharide-based systems, controlled degradation can maintain average molecular weight yet erode the long-chain fraction that dominates low-shear viscosity. The result is functional divergence hidden beneath apparent compliance.

Manufacturing and Quality Control Consequences

When specification and functional windows are misaligned, manufacturing performance becomes erratic despite incoming materials meeting all stated requirements. Operators observe unexplained variability in viscosity build rate, inline rheology stability, or aging behavior. Corrective actions often focus on process adjustments rather than raw material fitness, compounding variability.

Quality control laboratories inadvertently reinforce this gap by optimizing tests for repeatability and speed rather than functional relevance. Single-point viscosity measurements at moderate shear fail to detect susceptibility to shear fatigue, ionic challenge, or extended mixing. As a result, off-functional materials are released into production with high confidence.

Impact on Concrete Performance and Field Robustness

In concrete applications, misalignment manifests as reduced robustness to cement chemistry, temperature, and mixing energy. VMAs produced from edge-of-window polymers may perform adequately under standard lab conditions yet fail under extended pumping, delayed placement, or high-alkali environments. Field complaints frequently involve loss of slump retention, segregation control failure, or inconsistent finishability.

These failures are difficult to trace because post-event analysis confirms compliance with raw material specifications. Without functional benchmarks, the root cause remains obscured, leading to repeated incidents across projects and regions.

Industry 4.0 Diagnostics and Window Mapping

Industry 4.0 frameworks enable quantitative mapping between specification space and functional performance. By correlating raw material data, inline rheology, process historian signals, and field outcomes, manufacturers can identify regions within the specification window associated with elevated risk. Multivariate models reveal which combinations of "in-spec" parameters predict instability.

Digital twins of formulation and manufacturing processes allow virtual stress testing of materials across the full functional window. These tools shift quality assurance from binary pass–fail logic toward probabilistic performance assurance, aligning material acceptance with real-world demands.

Table 19.3.8: Specification Window Versus Functional Window Characteristics

Aspect	Specification Window	Functional Window
Primary purpose	Commercial acceptance	Performance assurance
Typical parameters	pH, solids, viscosity	Yield stress, thixotropy, shear stability
Sensitivity to impurities	Low	High
Predictive value for field	Limited	High
QC decision logic	Pass or fail	Risk-weighted

Control Strategies and Technology Transfer Alignment

Closing the gap requires redefining specifications to include functionally discriminating tests such as low-shear viscosity, shear recovery, ionic tolerance, and accelerated aging response. These tests need not replace conventional QC but should act as secondary release or monitoring criteria for critical applications.

For technology transfer, explicit communication of the functional window is essential. Receiving sites must understand not only numeric limits but also which regions within those limits are preferred or risky. Industry 4.0-enabled dashboards that visualize specification compliance alongside functional risk indicators provide a practical bridge between laboratory data, manufacturing control, and field performance.

19.4 Stability, Microbial, and Shelf-Life Failures

Stability and shelf-life failures in liquid VMA products reflect slow, cumulative degradation processes rather than acute manufacturing errors. Polysaccharide-based and partially bio-derived VMAs are particularly vulnerable to microbial attack, resulting in viscosity loss, odor formation, and gas evolution during storage. In many cases, these failures stem from inadequate preservation strategies or inconsistent hygiene practices rather than inherent polymer instability.

Physicochemical instability, including phase separation, sedimentation, or post-neutralization viscosity creep, further complicates storage performance. These phenomena are influenced by solids content, ionic balance, and container interactions over time. Robust troubleshooting combines accelerated aging protocols with mechanistic interpretation, enabling predictive shelf-life modeling and rational selection of stabilizers and biocides suitable for alkaline environments.

19.4.1 Microbial Degradation of Polysaccharide VMAs

Susceptibility of Polysaccharide Structures to Microbial Attack

Polysaccharide-based VMAs derived from cellulose ethers, starch derivatives, guar, or modified natural gums inherently contain glycosidic linkages and residual nutrients that are accessible to microbial metabolism. Even after chemical modification, etherification, or partial oxidation, these polymers retain structural motifs recognizable by bacteria and fungi. In aqueous, mildly alkaline environments typical of liquid VMA concentrates, microbial colonization can initiate rapidly when preservation is insufficient.

From a polymer chemistry perspective, microbial susceptibility is governed by backbone accessibility, degree of substitution, and molecular weight distribution. High-molecular-weight fractions that dominate low-shear viscosity are preferentially targeted by enzymatic cleavage, making rheological performance disproportionately sensitive to early-stage biodegradation even at low microbial load.

Mechanisms of Biodegradation and Rheological Collapse

Microorganisms secrete specific enzymes such as cellulases, amylases, and hemicellulases that cleave polysaccharide chains into lower-molecular-weight fragments. Chain scission reduces hydrodynamic volume and disrupts transient entanglements responsible for viscosity and yield stress development. Unlike oxidative degradation, microbial attack proceeds heterogeneously, creating localized regions of severe molecular weight loss within an otherwise intact matrix.

Secondary metabolic processes generate organic acids, alcohols, and gases such as CO_2, shifting pH and ionic balance. These changes further destabilize polymer hydration and associative behavior, accelerating viscosity decay beyond what would be expected from molecular weight loss alone. Rheological collapse often appears sudden after a latent induction period, misleading operators into attributing failure to handling or formulation errors.

Manufacturing and Storage Manifestations

Microbial degradation manifests during storage as progressive viscosity loss, stringiness reduction, and loss of thixotropic recovery. Early indicators include faint odor development, microbubble formation, or slight turbidity increase, frequently overlooked during routine inspection. Inline viscosity monitoring during reprocessing or dilution reveals unstable signals and rapid decay under shear.

Batch-to-batch variability increases when partially degraded material is blended with fresh production, contaminating otherwise stable lots. Once established, microbial populations persist in tanks, transfer lines, and valves, creating recurring failures despite compliant raw materials and formulations.

Impact on Concrete Performance

VMAs affected by microbial degradation exhibit delayed viscosity build, poor segregation control, and reduced robustness to cement chemistry. Slump retention becomes erratic, particularly under extended mixing or pumping, where weakened polymer networks fail to resist shear-induced dilution. Field performance deteriorates unpredictably, with complaints often clustering toward the end of product shelf life.

These failures are rarely reproducible in laboratory trials using fresh material, complicating root-cause analysis and eroding confidence in polysaccharide-based systems when microbial control is inadequate.

Detection, Monitoring, and Industry 4.0 Integration

Conventional QC tests lack sensitivity to early microbial activity. Viable cell counts, ATP assays, and selective plating provide direct evidence but are seldom implemented in admixture plants. Proxy indicators include unexplained viscosity drift during storage, rising conductivity, and pH decline in nominally buffered systems.

Industry 4.0 architectures integrate storage time, temperature, tank turnover, and cleaning-in-place data with rheological performance trends. Pattern recognition algorithms detect degradation signatures consistent with biological activity rather than chemical aging, enabling targeted sanitation or disposal decisions before widespread contamination occurs.

Control Strategies and Technology Transfer Considerations

Effective mitigation requires a multilayer approach combining hygiene design, validated biocide systems, and controlled storage conditions. Biocide selection must account for alkaline stability, polymer compatibility, and resistance development, with periodic efficacy verification under real storage conditions. Dead zones in tanks and low-flow piping must be eliminated through equipment design and cleaning protocols.

For technology transfer, explicit communication of microbial risk, sanitation practices, and preservative maintenance is critical. Facilities receiving transferred formulations often underestimate biological degradation pathways, leading to premature shelf-life failures despite strict adherence to formulation recipes. Digital documentation of microbial control strategies and performance limits is essential to preserve functional equivalence across production sites.

19.4.2 Biocide Compatibility and Efficacy Limits

Chemical Compatibility Constraints in VMA Systems

Biocides used in polysaccharide-based VMA formulations must remain chemically stable and functionally active in highly alkaline, high-ionic-strength aqueous environments. Many conventional preservatives are optimized for neutral or mildly acidic conditions and undergo hydrolysis, ionization loss, or nucleophilic attack at pH values typical of VMA concentrates. As a result, nominally adequate biocide dosages may deliver only transient protection.

From a polymer chemistry perspective, biocide–polymer interactions impose additional constraints. Cationic biocides can associate with anionic polysaccharide sites, reducing bioavailability while altering polymer hydration and rheology. Nonionic or hydrophobic biocides may partition into associative domains, locally plasticizing the network and subtly modifying viscosity response even before microbial control fails.

Efficacy Limits Under Industrial Storage Conditions

Biocide efficacy is not static and decays over time due to chemical consumption, volatilization, adsorption to container surfaces, or microbial adaptation. In VMA storage, repeated oxygen ingress, temperature fluctuations, and dilution events accelerate biocide depletion. Biofilms formed on tank walls and transfer lines provide protected niches where microorganisms survive sublethal biocide exposure and reseed the bulk product.

These effects impose a practical upper limit on achievable shelf life, regardless of initial biocide loading. Increasing dosage beyond this limit often yields diminishing returns and may trigger regulatory, odor, or compatibility issues without extending effective protection.

Rheological Side Effects of Biocide Systems

Several biocides exert secondary effects on polysaccharide rheology. Aldehyde-based systems promote slow crosslinking or chain stiffening, leading to post-addition viscosity drift. Isothiazolinones can interact with reducing end groups or residual impurities, subtly altering molecular weight distribution over extended storage. Oxidizing biocides accelerate backbone scission when overdosed or improperly buffered.

These interactions create a narrow operating window in which microbial control is achieved without compromising rheological integrity. Exceeding this window often results in apparent chemical instability misdiagnosed as polymer aging or formulation error.

Manufacturing and Quality Control Implications

Inconsistent biocide performance manifests as sporadic shelf-life failures across nominally identical batches. QC release testing, typically performed shortly after production, fails to capture time-dependent loss of efficacy. Manufacturing teams respond by escalating dosage or switching products without systematic evaluation, increasing variability and cost.

Lack of standardized biocide challenge testing under alkaline, polymer-rich conditions further obscures performance limits. Supplier datasheets rarely reflect real VMA environments, leading to overreliance on theoretical compatibility claims.

Industry 4.0-Based Performance Mapping

Industry 4.0 platforms enable correlation of biocide type, dosage, storage conditions, and observed rheological decay. By integrating microbiological indicators, rheology trends, and environmental data, manufacturers can identify depletion kinetics and failure thresholds for specific preservative systems. Digital twins simulate storage scenarios, supporting predictive shelf-life assignment rather than fixed-date expiration.

These tools shift biocide selection from empirical trial to data-driven optimization, balancing efficacy, compatibility, and regulatory constraints.

Table 19.4.2: Biocide Classes and Performance Constraints in Polysaccharide VMAs

Biocide Class	Primary Limitation	Manufacturing Signal	Functional Impact
Isothiazolinones	Alkaline instability	Delayed viscosity loss	Limited shelf life
Aldehydes	Polymer interaction	Viscosity creep	Rheology drift
Quaternary ammonium	Polymer binding	Reduced efficacy	Microbial breakthrough
Oxidizing agents	Chain scission risk	Rapid viscosity decay	Performance collapse

Organic acids	pH dependence	Ineffective preservation	Odor formation

Control Strategies and Technology Transfer Considerations

Effective control requires validation of biocide systems under representative formulation, pH, and storage conditions rather than reliance on generic recommendations. Rotational or combination biocide strategies reduce resistance development and extend protection without excessive dosing. Hygiene and cleaning protocols must be treated as integral components of preservation strategy rather than secondary safeguards.

For technology transfer, explicit documentation of biocide rationale, efficacy limits, and monitoring indicators is essential. Receiving sites must replicate not only dosage but also storage practices, sanitation frequency, and environmental controls. Failure to transfer these parameters commonly results in premature microbial failure despite strict formulation compliance.

19.4.3 Phase Separation and Sedimentation Mechanisms

Colloidal and Thermodynamic Drivers of Instability

Phase separation and sedimentation in liquid VMA products arise from slow loss of colloidal and thermodynamic balance rather than gross formulation errors. Polysaccharide-based and associative polymer VMAs exist as highly hydrated, metastable systems in which polymer chains, counterions, and low-molecular-weight components coexist under finely balanced osmotic and electrostatic forces. Minor perturbations in ionic strength, pH, or polymer conformation shift this balance toward demixing.

From a polymer chemistry viewpoint, instability is often initiated by gradual contraction of the polymer coil due to counterion exchange or progressive dehydration. As hydrodynamic volume decreases, excluded volume interactions weaken, allowing denser polymer-rich domains to form and settle under gravity.

Mechanisms of Phase Separation in Liquid VMAs

Phase separation typically proceeds through slow polymer-rich phase formation rather than sharp liquid–liquid demixing. In associative acrylic systems, hydrophobic junctions aggregate over time into microdomains, locally increasing density and reducing network homogeneity. In polysaccharide VMAs, partial de-substitution, oxidative aging, or microbial byproducts reduce solvation efficiency, promoting chain–chain association.

Temperature fluctuations accelerate these processes by repeatedly disrupting and reforming hydration shells. Each cycle slightly favors the most thermodynamically stable, compact configuration, cumulatively driving macroscopic separation without obvious compositional change.

Sedimentation and Structural Collapse

Sedimentation is frequently linked to the presence of undissolved polymer fragments, microgels, or agglomerates formed during storage or incomplete hydration. These entities act as nuclei for further polymer association, growing in size until gravitational forces exceed Brownian stabilization. High-solids formulations intensify sedimentation risk due to elevated collision frequency and reduced free water.

Once sediment forms, redispersion is often incomplete. Shear applied during agitation breaks weak flocs but fails to restore original molecular conformation, resulting in permanent loss of viscosity and thixotropic recovery.

Manufacturing and Handling Manifestations

In production and storage, phase separation appears as clear supernatant formation, bottom-layer thickening, or gel-like deposits on tank walls. Inline density and viscosity sensors detect slow drift rather than abrupt deviation, frequently misinterpreted as sensor noise. Reblending restores apparent homogeneity but not original rheological performance.

During transfer and dosing, sedimented material creates inconsistent feed concentration, leading to erratic VMA addition rates and unstable concrete rheology. These effects intensify toward the end of container life, generating sporadic field complaints.

Impact on Concrete Performance

VMAs affected by phase separation deliver inconsistent polymer availability during batching. Early portions of a container may underperform, while later portions overdose the system. Concrete exhibits variable segregation resistance, finishability, and air stability within the same project.

These failures are difficult to reproduce under laboratory conditions using freshly homogenized samples, masking the underlying storage-driven instability.

Monitoring and Industry 4.0 Approaches

Detection of phase separation requires trend-based rather than snapshot monitoring. Low-shear viscosity profiling, sediment height tracking, and density gradient measurements provide early indicators. Coupling these data with storage temperature history and turnover rates enables predictive assessment of separation risk.

Industry 4.0 systems integrate sensor data from storage tanks, agitators, and dosing units with rheological outcomes. Machine learning models identify patterns associated with incipient separation, enabling corrective action before irreversible sedimentation occurs.

Table 19.4.3: Phase Separation and Sedimentation Drivers in Liquid VMAs

Mechanism	Initiating Factor	Manufacturing Signal	Functional Impact
Polymer coil contraction	Ionic drift	Viscosity decay	Reduced hydration
Associative clustering	Hydrophobe aggregation	Density increase	Network inhomogeneity
Microgel formation	Aging or impurities	Bottom deposits	Incomplete redispersion
Thermal cycling	Temperature swings	Layer formation	Performance variability
Incomplete hydration	Poor dispersion	Persistent sediment	Permanent viscosity loss

Control Strategies and Technology Transfer Considerations

Mitigation requires formulation design with adequate colloidal stability margin, including optimized counterion selection, controlled solids content, and hydration kinetics tailored to storage conditions. Gentle, continuous agitation during storage reduces sediment formation but cannot compensate for intrinsic instability.

For technology transfer, explicit definition of acceptable separation behavior, redispersion limits, and agitation requirements is essential. Receiving sites often assume that visual homogeneity equates to functional integrity, leading to misdiagnosis of performance drift. Digital documentation of separation sensitivity and monitoring thresholds ensures consistent handling across production locations.

19.4.4 Post-Neutralization Viscosity Creep

Chemical Origin of Post-Neutralization Instability

Post-neutralization viscosity creep refers to the slow, time-dependent increase in viscosity observed after alkaline adjustment of polymer-based VMA concentrates. This phenomenon is most pronounced in polysaccharide derivatives and weakly acidic associative acrylic polymers where neutralization alters chain ionization, hydration, and intermolecular association. Although neutralization is intended to stabilize solubility and performance, it frequently initiates secondary equilibration processes that continue for days or weeks.

From a polymer chemistry perspective, neutralization converts latent acidic sites into charged moieties, expanding polymer coils and increasing electrostatic repulsion. However, full conformational equilibration is kinetically delayed, particularly in high-solids systems where chain mobility is restricted. The result is a gradual viscosity rise that occurs long after manufacturing release.

Molecular and Colloidal Mechanisms

Viscosity creep arises from progressive hydration and delayed ion diffusion into dense polymer domains. In polysaccharide VMAs, alkaline conditions promote slow de-aggregation of partially collapsed chains, increasing effective hydrodynamic volume over time. In associative acrylic systems, neutralization enhances hydrophobe exposure, enabling gradual formation of additional junction zones and network reinforcement.

Counterion exchange further contributes to creep. Replacement of monovalent ions with divalent species from water or raw materials reduces charge screening efficiency, strengthening interchain interactions. These molecular rearrangements proceed silently within specification limits, masking the onset of instability.

Manufacturing Manifestations and Process Control Challenges

In production, post-neutralization creep appears as acceptable initial viscosity followed by steady upward drift during storage or recirculation. Inline rheometers register monotonic increase unrelated to temperature or solids variation, often prompting unnecessary dilution or process intervention. Once diluted, the system may re-creep, compounding variability.

Neutralization performed in-line or late in the process amplifies creep risk by limiting equilibration time before packaging. Differences in mixing intensity, residence time, and alkali addition rate create site-specific behavior, complicating technology transfer.

Consequences for Concrete Performance

VMAs exhibiting viscosity creep show inconsistent dosing efficiency and variable thickening power over shelf life. Early use may underperform, while aged material overdoses the concrete, increasing yield stress and impairing workability. Thixotropic recovery may improve initially but becomes overly rigid at rest, reducing pumpability and finishability.

Field issues often emerge weeks after production, disconnecting cause and effect in post-project analysis. Laboratory qualification using freshly neutralized samples fails to predict these outcomes.

Detection and Industry 4.0-Based Prediction

Single-point QC viscosity measurements are insufficient to capture creep behavior. Time-resolved viscosity profiling under controlled temperature and low shear provides functional insight. Monitoring conductivity and ion balance supports identification of counterion-driven effects.

Industry 4.0 systems correlate neutralization conditions, storage time, and viscosity evolution across batches. Predictive models identify formulations and process conditions prone to creep, enabling pre-emptive adjustment of neutralization strategy or release criteria.

Table 19.4.4: Drivers and Manifestations of Post-Neutralization Viscosity Creep

Driver	Initiating Condition	Manufacturing Signal	Functional Impact
Delayed hydration	High solids	Gradual viscosity rise	Overthickening
Enhanced association	Increased ionization	Thixotropy increase	Pumpability loss
Counterion exchange	Divalent ion ingress	Conductivity drift	Rheology instability
Late neutralization	Short equilibration	Post-release creep	Shelf-life reduction
Inhomogeneous mixing	Local pH gradients	Batch variability	Inconsistent dosing

Control Strategies and Technology Transfer Considerations

Mitigation requires designing neutralization protocols that allow full equilibration prior to packaging. Controlled alkali addition rate, sufficient mixing energy, and defined aging periods reduce latent creep potential. Buffer systems or partial pre-neutralization strategies stabilize charge development without excessive network growth.

For technology transfer, explicit specification of neutralization sequence, target ion profile, and acceptable viscosity evolution over time is essential. Sites receiving transferred formulations frequently replicate final viscosity targets without reproducing equilibration history, leading to delayed instability. Digital capture of neutralization parameters and viscosity-time fingerprints ensures consistent performance across manufacturing locations.

19.4.5 Container and Packaging Interaction Effects

Material–Formulation Interactions at the Container Interface

Liquid VMA products remain in continuous contact with container and packaging materials throughout storage, transport, and handling. Although packaging is often treated as inert, polymer chemistry and surface science demonstrate that sustained interfacial interaction can significantly influence product stability. High-alkaline, polymer-rich formulations interact with container walls through adsorption, ion exchange, and extractable migration processes that evolve over time.

From a molecular standpoint, polysaccharide chains and associative acrylic polymers exhibit affinity for polar or charged surfaces. Adsorption at the container interface selectively removes high-molecular-weight or highly associative fractions from the bulk, progressively altering rheological balance without measurable change in solids content.

Adsorption and Depletion Mechanisms

Polymer adsorption is driven by hydrogen bonding, electrostatic attraction, and hydrophobic interactions depending on container composition. Polyethylene and polypropylene containers present low-energy hydrophobic surfaces that preferentially attract associative segments, while metal or coated steel containers introduce charged sites that bind ionic polymer domains.

This selective depletion reduces effective molecular weight distribution and associative junction density in the bulk phase. Over extended storage, adsorption layers thicken and partially reorganize, becoming difficult to desorb even under agitation, leading to irreversible rheological loss.

Migration of Extractables and Additives

Packaging materials contain stabilizers, plasticizers, slip agents, and residual monomers that can migrate into VMA formulations. Under alkaline conditions, these species dissolve or hydrolyze, introducing low-molecular-weight impurities that modify ionic strength and polymer solvation. Even trace levels can perturb associative equilibria, causing viscosity drift or loss of thixotropic recovery.

In flexible packaging or intermediate bulk containers, multilayer structures and liners present additional sources of extractables. Temperature elevation during transport accelerates migration kinetics, amplifying interaction effects late in the shelf-life period.

Gas Permeation and Headspace Effects

Container permeability to oxygen and carbon dioxide influences long-term stability. Oxygen ingress promotes oxidative degradation of polysaccharides and biocide depletion, while CO_2 absorption lowers pH and alters counterion balance. Rigid plastic containers with high surface-area-to-volume ratios are particularly susceptible, especially when partially emptied and resealed.

These gas exchange processes operate independently of formulation quality and are rarely considered during product development, yet they materially affect storage performance.

Manufacturing and Field Manifestations

Container-induced instability appears as gradual viscosity loss, unexpected pH drift, or inconsistent behavior between packaging formats. Identical formulations packaged in drums, IBCs, or bulk tanks may exhibit different shelf lives and rheological profiles. Re-homogenization restores appearance but not original performance.

Field issues often correlate with specific packaging types or reuse cycles rather than formulation changes. Without explicit packaging evaluation, root-cause analysis incorrectly focuses on polymer quality or processing conditions.

Industry 4.0 Monitoring and Traceability

Industry 4.0 systems enable correlation of packaging type, container age, reuse history, and storage environment with rheological performance. Digital tracking of container lots and sensor-based monitoring of headspace conditions support identification of high-risk packaging configurations. Data analytics distinguish container-driven effects from intrinsic chemical aging.

These tools facilitate evidence-based decisions on container selection, liner materials, and maximum reuse cycles.

Table 19.4.5: Container and Packaging Interaction Effects on VMA Stability

Interaction Mechanism	Container Factor	Manufacturing Signal	Functional Impact
Polymer adsorption	Surface chemistry	Viscosity decline	Network depletion
Additive migration	Plastic materials	pH or conductivity drift	Rheology instability
Oxygen ingress	Permeable walls	Accelerated aging	MWD degradation

CO_2 absorption	Headspace exposure	Gradual pH drop	Ionic imbalance
Container reuse	Residual films	Batch inconsistency	Performance variability

Control Strategies and Technology Transfer Considerations

Mitigation requires qualification of packaging materials as integral formulation components. Compatibility testing must include long-term storage trials under representative temperature and headspace conditions. Liners and coatings should be selected for minimal extractables and low gas permeability under alkaline exposure.

For technology transfer, explicit specification of approved container types, maximum reuse cycles, and handling practices is essential. Transferring formulations without packaging constraints frequently leads to unexplained shelf-life reduction and performance drift. Digital documentation linking container selection to functional performance preserves consistency across manufacturing and distribution networks.

19.4.6 Temperature Cycling During Storage and Transport

Thermal Sensitivity of Polymer-Based VMA Systems

Liquid VMA formulations are commonly exposed to repeated temperature fluctuations during storage, transport, and intermediate handling. Day–night warehouse cycles, seasonal variation, and uncontrolled transport conditions impose thermal stress even when average temperatures remain within nominal limits. Polymer-based VMAs respond nonlinearly to such cycling because hydration, association, and ionic equilibria are temperature dependent and kinetically constrained.

From a polymer chemistry standpoint, temperature cycling does not simply accelerate aging uniformly. Instead, it repeatedly perturbs polymer conformation and solvation state, driving the system toward progressively more thermodynamically stable but functionally inferior configurations.

Molecular and Colloidal Mechanisms Activated by Cycling

Heating phases promote polymer coil contraction, dehydration of associative domains, and increased mobility of counterions and low-molecular-weight species. Cooling phases rehydrate chains but do not fully restore the original conformational state, particularly in high-solids or partially associated systems. Each cycle therefore leaves a small residual structural change.

In polysaccharide VMAs, repeated thermal expansion and contraction weaken hydrogen-bonded hydration shells and facilitate interchain association. In associative acrylic systems, hydrophobic junctions reorganize and coarsen over cycles, increasing heterogeneity and reducing shear reversibility. When freezing thresholds are approached, even transiently, microphase damage becomes irreversible.

Manifestations During Manufacturing and Storage

Temperature cycling manifests as gradual viscosity drift, increased thixotropy, or delayed dissolution behavior upon dilution. Inline rheology records periodic fluctuations superimposed on a long-term trend, often misinterpreted as sensor instability. Visual inspection rarely reveals damage until advanced stages, when haze, microgels, or slight sediment appear.

Drums or IBCs stored near external walls or exposed to direct sunlight exhibit more severe degradation than centrally stored units, creating within-lot variability that complicates quality assurance.

Consequences for Concrete Performance

VMAs exposed to thermal cycling show reduced robustness under variable jobsite temperatures. Slump retention becomes unpredictable, and segregation resistance degrades under extended mixing or pumping. Performance inconsistencies often appear only under marginal conditions, leading to intermittent field complaints that are difficult to reproduce under controlled laboratory testing.

These effects are amplified in cold or hot climates where transport cycles are more extreme, particularly when combined with partial container emptying and resealing.

Monitoring and Industry 4.0-Based Mitigation

Effective management requires continuous temperature history tracking rather than reliance on average values. Low-cost temperature loggers and integrated warehouse sensors provide cycle count, amplitude, and duration data. Correlating these metrics with rheological evolution enables identification of critical cycling thresholds.

Industry 4.0 analytics integrate thermal history with storage time, packaging type, and performance data to predict degradation risk. Predictive alerts support inventory prioritization or conditional requalification of thermally stressed material.

Table 19.4.6: Effects of Temperature Cycling on VMA Stability

Thermal Driver	Exposure Scenario	Manufacturing Signal	Functional Impact
Daily cycling	Uncontrolled storage	Viscosity drift	Network fatigue
Seasonal swings	Long-distance transport	Thixotropy increase	Reduced robustness
Local overheating	Sun exposure	Microgel formation	Incomplete hydration
Near-freezing events	Winter logistics	Permanent viscosity loss	Performance collapse
Repeated cycling	Extended storage	Lot variability	Inconsistent dosing

Control Strategies and Technology Transfer Considerations

Mitigation requires defining acceptable temperature ranges and maximum cycling intensity as critical quality attributes. Insulated storage, controlled transport, and avoidance of extreme conditions reduce cumulative damage. Formulation-level mitigation through thermal stabilizers or hydration modifiers provides limited protection and cannot substitute for environmental control.

For technology transfer, explicit transfer of thermal sensitivity data, acceptable exposure limits, and monitoring practices is essential. Receiving sites often replicate formulation and process conditions while overlooking environmental differences, leading to premature instability. Digital thermal history documentation linked to batch genealogy ensures consistent shelf-life management across regions.

19.4.7 Accelerated Aging Test Correlation Failures

Rationale and Structural Limitations of Accelerated Aging

Accelerated aging protocols are widely applied to estimate shelf life of liquid VMA products by exposing formulations to elevated temperature, intensified shear, or oxidative environments. While these tests provide rapid comparative screening, they frequently fail to reproduce degradation pathways active under real storage and transport conditions. The fundamental limitation arises from the nonlinearity of polymer aging mechanisms, which cannot be scaled reliably by simple time–temperature superposition.

From a polymer chemistry perspective, many degradation processes in polysaccharide and associative polymer VMAs are kinetically gated rather than purely thermally activated. Enzymatic activity, counterion exchange, adsorption equilibria, and slow conformational rearrangements respond differently to acceleration than chemical scission reactions, leading to distorted failure modes.

Divergence of Mechanisms Under Accelerated Conditions

Elevated-temperature aging amplifies hydrolysis, oxidation, and dehydration but suppresses microbial growth and alters biocide efficacy. As a result, microbial degradation observed during ambient storage is often absent in accelerated tests, while chemically driven viscosity loss is overrepresented. High-shear protocols exaggerate mechanical degradation and associative junction fatigue, which may not dominate under quiescent storage.

In polysaccharide VMAs, accelerated heat aging induces backbone scission and solubility loss that do not mirror gradual enzymatic cleavage observed at lower temperatures. In associative acrylic systems, thermal stress shifts hydrophobe association equilibria into regimes rarely accessed during normal storage, producing artifacts rather than predictive insight.

Manufacturing Consequences of Misleading Correlation

When accelerated tests poorly correlate with real-time aging, shelf-life assignments become unreliable. Materials passing aggressive accelerated protocols may fail prematurely in the field, while conservatively rejected formulations may perform adequately in practice. Manufacturing teams lose confidence in qualification data and compensate by increasing safety margins, raising cost and complexity.

Quality systems that rely on single accelerated indicators propagate false negatives and false positives across production sites. Technology transfer magnifies this issue when receiving facilities validate shelf life using different acceleration protocols, generating inconsistent conclusions for identical formulations.

Impact on Concrete Performance Prediction

Accelerated aging correlation failures translate directly into field risk. VMAs qualified through non-representative aging exhibit unexpected viscosity loss, creep, or instability late in shelf life, undermining performance guarantees. Conversely, overconservative aging criteria restrict usable inventory and create artificial supply constraints without improving concrete robustness.

These mismatches erode trust between admixture suppliers and concrete producers, particularly when post-failure analysis confirms compliance with approved aging protocols.

Industry 4.0-Based Correlation Improvement

Industry 4.0 approaches enable reconciliation of accelerated and real-time aging by integrating multivariate data rather than relying on single stress factors. Continuous capture of storage temperature profiles, microbial indicators, rheological evolution, and field performance supports development of hybrid aging models. Machine learning tools identify which accelerated conditions best approximate specific real-world degradation pathways.

Digital twins of storage environments allow virtual aging under realistic cycling, headspace, and contamination scenarios. These models refine acceleration factors and improve predictive value without excessive physical testing.

Table 19.4.7: Causes of Accelerated Aging Test Misalignment

Accelerated Stress	Overrepresented Effect	Missing Mechanism	Consequence
Elevated temperature	Chemical scission	Microbial activity	False shelf-life extension
High shear	Mechanical degradation	Quiescent equilibration	Overconservative rejection
Oxidative exposure	Backbone oxidation	Ion exchange	Non-representative failure

Short-duration tests	Rapid responses	Slow adsorption	Missed late-stage instability
Single-factor protocols	Isolated mechanisms	Coupled effects	Poor field correlation

Control Strategies and Technology Transfer Considerations

Improved correlation requires combining moderate acceleration with extended duration and inclusion of multiple stressors. Aging protocols should reflect realistic temperature cycling, headspace conditions, and microbial challenge where relevant. Performance should be evaluated through rheological fingerprints rather than single viscosity endpoints.

For technology transfer, aging protocols and acceptance criteria must be transferred as functional methodologies, not just numeric limits. Receiving sites frequently substitute locally convenient tests, breaking correlation and introducing hidden risk. Digital archiving of aging response curves and correlation models ensures consistent shelf-life assessment across organizations and regions.

19.4.8 Odor, Gas Formation, and Visual Defects

Chemical and Biological Origins of Sensory Defects

Odor development, gas formation, and visual defects in liquid VMA products represent late-stage manifestations of underlying chemical or biological instability rather than isolated quality issues. These phenomena are most frequently associated with polysaccharide-based or partially bio-derived systems where microbial metabolism, oxidative reactions, or additive decomposition generate volatile and chromophoric byproducts. Although often treated as cosmetic defects, they provide critical diagnostic information on degradation pathways active during storage.

From a polymer chemistry standpoint, sensory defects emerge when low-molecular-weight species accumulate to levels that exceed solubility or volatility thresholds. These species are typically absent at production release and evolve gradually under storage conditions, escaping routine QC detection.

Odor Formation Mechanisms

Odors arise primarily from microbial metabolism producing short-chain organic acids, alcohols, amines, and sulfur-containing compounds. In preserved systems, sublethal biocide exposure shifts microbial populations toward odor-producing strains rather than eliminating growth. Chemical contributors include aldehydes generated by oxidative cleavage of polysaccharides and residual monomers undergoing slow side reactions.

Odor intensity often increases sharply after an induction period, coinciding with rapid viscosity loss or pH drift. Once established, odor cannot be reversed through filtration or reblending, indicating fundamental chemical change rather than contamination alone.

Gas Evolution and Container Pressurization

Gas formation results from microbial respiration producing CO_2 or, less commonly, hydrogen and methane under anaerobic microenvironments. Chemical gas evolution may occur through carbonate formation when alkaline VMAs absorb atmospheric CO_2, particularly in partially filled containers. In severe cases, pressure buildup deforms containers or triggers venting events, creating safety and handling concerns.

Gas evolution disrupts rheological measurements by introducing microbubbles, leading to apparent viscosity fluctuations and noisy inline signals. These artifacts complicate process control and mask concurrent degradation.

Visual Defects as Structural Indicators

Visual defects include haze, turbidity, phase separation, sediment, surface films, and color change. Haze and turbidity indicate microgel formation, microbial biomass, or polymer aggregation. Surface films often consist of biofilms or oxidized polymer layers formed at the air–liquid interface. Color changes, typically yellowing or browning, signal oxidative degradation or Maillard-type reactions involving residual sugars and amines.

These visual cues correlate strongly with loss of functional performance, even when bulk viscosity remains temporarily within specification limits.

Manufacturing and Field Implications

Sensory defects trigger customer rejection and regulatory scrutiny regardless of measured performance. Manufacturing responses such as dilution, deodorization, or blending fail to restore polymer integrity and often exacerbate instability. Reuse of containers or equipment contaminated with odor-generating residues perpetuates recurring failures across batches.

In concrete applications, VMAs exhibiting sensory defects deliver inconsistent performance and increased air entrainment variability. Gas-laden products dose inaccurately, while degraded polymers fail to provide reliable segregation control.

Industry 4.0 Detection and Preventive Analytics

Industry 4.0 systems enhance early detection by correlating subtle rheological drift, pH changes, and storage history with eventual sensory failure. Gas sensors, pressure monitoring, and machine vision systems detect early visual anomalies before customer-visible defects appear. Predictive analytics distinguish microbial-driven gas evolution from chemical CO_2 absorption based on pattern recognition.

These capabilities enable proactive disposal or requalification decisions, reducing downstream risk and reputational impact.

Table 19.4.8: Odor, Gas, and Visual Defects and Their Diagnostic Value

Defect Type	Primary Origin	Manufacturing Signal	Functional Implication
Sour or sulfur odor	Microbial metabolism	pH decline	Polymer degradation
Alcoholic odor	Oxidative byproducts	Conductivity drift	Rheology instability
Gas bubbles	Microbial respiration	Pressure buildup	Measurement artifacts
Haze or turbidity	Microgels or biomass	Light scattering	Incomplete hydration
Color change	Oxidation reactions	Yellowing	Performance loss

Control Strategies and Technology Transfer Considerations

Prevention requires eliminating root causes rather than treating symptoms. Effective microbial control, oxygen management, and packaging compatibility are prerequisites. Sensory defect thresholds should be defined as release-critical attributes, not customer complaints. Routine sensory inspection combined with objective measurements such as turbidity and gas evolution rate improves early intervention.

For technology transfer, explicit communication that odor and visual changes represent functional failure is essential. Receiving sites often tolerate early sensory deviations, assuming performance remains acceptable, leading to delayed corrective action. Digital capture of sensory indicators and associated performance loss ensures consistent decision-making across sites.

19.5 Integrated Troubleshooting Framework for Industry 4.0

Modern VMA manufacturing increasingly depends on integrated troubleshooting frameworks that unify formulation knowledge with digital process intelligence. Isolated corrective actions based on single-parameter deviations are no longer sufficient in high-throughput plants operating under narrow rheological tolerances. Industry 4.0 architectures enable continuous aggregation of raw material data, process sensor outputs, and quality control results into coherent datasets suitable for multivariate analysis.

By leveraging inline rheology, advanced SPC, and predictive analytics, manufacturers can transition from reactive troubleshooting to proactive process stabilization. Digital twins and knowledge management systems further support technology transfer across sites by formalizing empirical know-how into reproducible decision pathways. Such integrated frameworks reduce variability, accelerate scale-up, and enhance resilience against both raw material and process disturbances.

19.5.1 Data Integration Across Raw Materials and Processing

Raw Material Fingerprinting as a Foundation

Effective data integration begins with detailed fingerprinting of critical raw materials, including polymers, preservatives, solvents, and water sources. Parameters such as molecular weight distribution, degree of substitution, ionic content, and trace impurity profiles are digitized and linked to supplier lot numbers. Variability in polysaccharide-derived VMAs, in particular, requires higher-resolution input data due to sensitivity to seasonal and biological factors. When integrated with process data, raw material fingerprints enable rapid discrimination between formulation-driven and process-induced deviations.

Process Data Synchronization and Contextualization

Processing variables such as addition sequence, shear history, residence time, and thermal exposure exert strong influence on polymer conformation and solution structure. Industry 4.0 systems contextualize these variables by synchronizing them with raw material attributes and batch timelines. For example, a viscosity drop can be algorithmically associated with elevated shear during dilution only when specific molecular weight fractions are present. This contextual linkage prevents misdiagnosis and avoids corrective actions that exacerbate instability, such as indiscriminate polymer overdosing.

Multivariate Analytics for Root-Cause Resolution

Once integrated, datasets are subjected to multivariate statistical tools including Principal Component Analysis and Partial Least Squares regression. These methods reveal latent relationships between raw material variability, process excursions, and quality outcomes that are invisible to conventional analysis. Predictive models rank probable root causes and recommend prioritized interventions, reducing troubleshooting time and operator subjectivity. Over time, model accuracy improves through continuous learning, reinforcing process robustness under evolving operating conditions.

Technology Transfer and Cross-Site Consistency

Integrated data frameworks are particularly valuable during technology transfer, where differences in equipment design, water chemistry, and operational culture introduce hidden risks. Standardized data structures and shared analytics dashboards allow receiving sites to benchmark performance against reference plants in real time. Deviations are interpreted within a common troubleshooting logic rather than local heuristics, ensuring consistent product quality and faster stabilization after startup.

Table 19.5.1: Key Data Streams and Their Troubleshooting Value in Industry 4.0 VMA Manufacturing

Data Stream	Typical Source	Diagnostic Contribution	Troubleshooting Value
Polymer fingerprint	Supplier COA, lab analysis	Molecular variability	Root-cause isolation
Inline rheology	Process sensors	Early viscosity drift	Proactive correction
pH and conductivity	Inline probes	Degradation indicators	Stability assessment
Shear and residence time	Process historians	Structural damage risk	Process optimization
Storage history	Digital logs	Aging effects	Shelf-life prediction

Strategic Implications for Manufacturers

Integrated troubleshooting frameworks transform data from passive records into active process control assets. Manufacturers adopting these systems achieve lower variability, faster response to disturbances, and improved confidence during formulation changes or raw material substitutions. As regulatory pressure and sustainability constraints drive further complexity into VMA formulations, such frameworks become not optional enhancements but core operational capabilities.

19.5.2 Inline Sensor Selection and Validation

Sensor Strategy Aligned with Polymer-Specific Failure Modes

Inline sensor selection for Viscosity-Modifying Admixtures manufacturing must be driven by polymer-specific degradation, hydration, and instability mechanisms rather than generic process monitoring checklists. VMAs exhibit non-Newtonian rheology, time-dependent structuring, and sensitivity to shear and dilution history, requiring sensors capable of resolving subtle, continuous changes under real process conditions. Sensors selected solely for robustness or cost frequently lack the resolution or response dynamics needed to detect early-stage molecular or microbial degradation. An effective strategy maps each critical sensor to a known failure pathway, ensuring that measured signals carry diagnostic value rather than redundant confirmation of downstream QC results.

Inline Rheology and Flow Characterization

Inline rheometers provide the most direct insight into VMA structural integrity by capturing viscosity evolution under controlled shear rates. For polymeric VMAs, multi-shear or oscillatory-capable instruments outperform single-point viscometers by distinguishing between reversible structuring and irreversible chain scission. Validation requires correlation against laboratory rheology using representative shear protocols and hydration times, with acceptance criteria defined on trend fidelity rather than absolute value matching. Drift analysis over extended operation is essential, as fouling or microbubble entrapment can introduce false viscosity shifts that mimic degradation.

Chemical and Electrochemical Sensors for Stability Monitoring

pH and conductivity sensors serve as early indicators of chemical instability, preservative depletion, or microbial activity. In polysaccharide-based systems, gradual pH decline combined with rising conductivity often precedes odor formation and viscosity loss. Sensor materials must be compatible with high ionic strength, biocide-containing formulations to avoid accelerated aging or signal drift. Validation protocols should include exposure to worst-case formulation envelopes and cleaning cycles to ensure long-term reliability under industrial conditions.

Gas, Optical, and Turbidity Detection

Gas evolution sensors and pressure transducers detect microbial respiration or chemical CO_2 absorption before visible container deformation occurs. Optical sensors, including turbidity meters and inline cameras, provide structural information on haze formation, microgels, or biofilm development. These sensors are particularly valuable in Industry 4.0 environments where machine vision algorithms flag deviations invisible to operators during routine inspections. Validation focuses on sensitivity thresholds linked to functional performance loss rather than aesthetic appearance alone.

Sensor Validation, Calibration, and Data Integrity

Inline sensor validation extends beyond initial calibration to include lifecycle performance under real manufacturing stress. Cross-validation with laboratory methods establishes baseline confidence, while statistical comparison of trends ensures ongoing accuracy. Redundancy strategies, such as pairing rheology with flow and density measurements, improve fault tolerance and reduce false alarms.

Data integrity safeguards, including timestamp synchronization and automated plausibility checks, prevent erroneous troubleshooting decisions driven by sensor malfunction rather than true process change.

Integration into Predictive and Prescriptive Analytics

Validated sensors form the foundation of predictive analytics and digital twins by supplying high-quality, context-rich data. Sensor outputs are weighted based on historical reliability and diagnostic relevance within multivariate models. Over time, machine learning systems refine sensor importance rankings, identifying which measurements provide the earliest and most reliable warning for specific failure modes. This adaptive prioritization improves both responsiveness and operator trust in automated recommendations.

Table 19.5.2: Inline Sensor Types, Validation Focus, and Diagnostic Role in VMA Manufacturing

Sensor Type	Primary Measurement	Key Validation Focus	Diagnostic Role
Inline rheometer	Viscosity vs shear	Trend correlation	Structural integrity
Flow and density	Mass balance	Drift stability	Dosing accuracy
pH probe	Acid–base status	Chemical resistance	Degradation onset
Conductivity	Ionic strength	Fouling tolerance	Microbial activity
Gas or pressure sensor	CO_2 evolution	Sensitivity threshold	Storage instability
Optical or turbidity	Light scattering	False positive control	Microstructure changes

Operational and Technology Transfer Considerations

Sensor strategies must be standardized across sites to ensure comparable data and transferable analytics. Local substitution of sensor models or installation geometries undermines cross-plant learning and weakens predictive models. During technology transfer, sensor validation data should be included alongside formulation and process documentation, emphasizing known limitations and failure signatures. Training operators to interpret sensor trends within a polymer chemistry framework, rather than reacting to alarms in isolation, remains critical for realizing the full value of Industry 4.0 systems.

19.5.3 Digital Twin Development for VMA Manufacturing

Digital Twins as Polymer-Centric Process Models

Digital twins for Viscosity-Modifying Admixture manufacturing extend beyond equipment simulation to represent the evolving physicochemical state of polymer solutions throughout production and storage. Unlike static mass and energy balances, effective twins incorporate polymer hydration kinetics, molecular weight distribution sensitivity, associative interactions, and degradation pathways as time-dependent variables. These models reflect the reality that VMA performance is governed by structural evolution rather than nominal composition alone. When correctly implemented, digital twins act as continuously updated mirrors of the real process, enabling early identification of instability trajectories before specification limits are crossed.

Model Architecture and Data Foundations

The core architecture combines first-principles polymer science with data-driven learning layers. Mechanistic sub-models describe dissolution, shear-induced chain alignment, adsorption equilibria, and preservative efficacy, while machine learning components capture nonlinear interactions and site-specific behavior. High-quality input data from validated inline sensors are essential, as model fidelity degrades rapidly when fed with noisy or poorly contextualized signals. Time alignment of raw material fingerprints, processing events, and storage conditions ensures that model outputs remain causally interpretable rather than statistically coincidental.

Calibration, Validation, and Continuous Learning

Digital twin calibration begins during pilot or early commercial production, using controlled perturbations to map model sensitivity to formulation and process variables. Validation focuses on the ability to predict trend direction, onset timing of defects, and relative severity rather than absolute numerical accuracy. As production continues, the twin evolves through continuous learning, incorporating new failure modes, raw material changes, and seasonal effects. Governance protocols are required to prevent uncontrolled model drift, particularly when machine learning layers update autonomously.

Use Cases in Manufacturing and Troubleshooting

In routine manufacturing, digital twins support proactive setpoint optimization, dilution strategy selection, and preservative dosing adjustments based on predicted stability windows. During troubleshooting, simulated scenarios test corrective actions virtually, avoiding costly trial-and-error on live batches. For example, the twin can evaluate whether a viscosity drop originates from shear history or early microbial activity, guiding targeted intervention. Such capabilities significantly reduce downtime and material waste in high-volume plants.

Scale-Up and Technology Transfer Enablement

Digital twins play a critical role during scale-up and cross-site technology transfer by decoupling formulation knowledge from specific equipment configurations. By parameterizing vessel geometry, mixing energy, and residence time, the same twin framework can be deployed across plants with different hardware. Receiving sites benefit from predictive insights grounded in reference plant behavior, shortening stabilization periods and reducing dependence on local experiential knowledge. This standardization is particularly valuable for global VMA platforms serving diverse cement markets.

Integration with Industry 4.0 Decision Systems

Within Industry 4.0 ecosystems, digital twins interface with Manufacturing Execution Systems and advanced analytics dashboards. Outputs are translated into prescriptive recommendations, such as adjustment ranges or risk flags, rather than raw model variables. Human-in-the-loop design remains essential, ensuring that polymer chemists and process engineers retain authority over critical decisions. Over time, confidence in twin-guided actions increases as predictive accuracy is demonstrated across multiple campaigns and sites.

Table 19.5.3: Digital Twin Components and Their Functional Contribution in VMA Manufacturing

Twin Component	Primary Inputs	Functional Role	Industrial Benefit

Mechanistic polymer model	MW distribution, shear, pH	Structural evolution	Root-cause clarity
Sensor data layer	Inline rheology, chemistry	Real-time state update	Early warning
Machine learning layer	Historical batches	Nonlinear pattern capture	Prediction accuracy
Scenario simulator	Process variables	Virtual trials	Risk reduction
MES interface	Production schedules	Prescriptive output	Operational efficiency

Implementation Risks and Mitigation

Common failure modes include overfitting to limited datasets, neglect of polymer chemistry constraints, and poor change management at the plant level. Mitigation requires phased deployment, transparent model logic, and continuous collaboration between data scientists and formulation experts. Digital twins should be treated as living assets, requiring maintenance, validation, and periodic recalibration as formulations and operating environments evolve.

19.5.4 Multivariate Root-Cause Analysis

Limitations of Univariate Troubleshooting in VMA Systems

Viscosity-Modifying Admixtures exhibit tightly coupled dependencies between polymer structure, processing history, and environmental exposure, rendering univariate troubleshooting ineffective for most industrial failures. Single-parameter alarms, such as viscosity out-of-spec or pH drift, capture symptoms rather than mechanisms and frequently lead to corrective actions that mask or amplify underlying degradation. Multivariate Root-Cause Analysis addresses this limitation by evaluating correlated variable behavior across the full manufacturing and storage timeline, enabling mechanistic interpretation aligned with polymer chemistry realities.

Data Conditioning and Variable Selection

Effective multivariate analysis begins with rigorous data conditioning, including sensor validation, noise filtering, and contextual tagging of process events. Variables selected must represent causal drivers rather than redundant outcomes, with preference given to parameters reflecting polymer conformation, hydration kinetics, and degradation potential. Inclusion of raw material fingerprints, shear exposure indices, and storage conditions enhances model sensitivity to early-stage instability. Poor variable discipline inflates model complexity without improving diagnostic power, undermining operator confidence.

Principal Component and Latent Structure Analysis

Principal Component Analysis is widely applied to compress high-dimensional VMA datasets into a limited number of latent variables capturing dominant variance patterns. In stable processes, batch trajectories cluster tightly within defined confidence regions. Deviations signal abnormal interactions such as unexpected adsorption behavior or preservative depletion. Loadings analysis links these deviations to specific variables, providing actionable insight rather than abstract statistical flags. PCA also supports early fault detection by identifying subtle drift long before specification limits are reached.

Regression-Based Attribution and Predictive Diagnosis

Partial Least Squares regression extends root-cause analysis by explicitly linking process and raw material variables to quality attributes such as viscosity retention, segregation resistance, and shelf stability. Regression coefficients quantify the relative contribution of each factor, supporting prioritized corrective actions. Predictive diagnosis evaluates hypothetical adjustments virtually, estimating their impact on quality outcomes before physical implementation. This capability reduces trial batches and accelerates convergence toward stable operating windows.

Dynamic and Time-Resolved Analysis

Many VMA failures evolve over time rather than appearing abruptly, requiring dynamic multivariate methods that capture temporal behavior. Multivariate Statistical Process Control and time-series modeling track trajectory deviations across production, cooling, and storage phases. Time alignment reveals cause-and-effect relationships, such as delayed viscosity loss following excessive shear or oxygen exposure. These insights are particularly valuable for diagnosing late-emerging defects that escape conventional release testing.

Operationalization within Industry 4.0 Environments

For industrial impact, multivariate root-cause outputs must be translated into interpretable diagnostics and decision support. Industry 4.0 platforms embed multivariate models within dashboards that visualize deviation scores, variable contributions, and confidence levels. Alarm logic shifts from threshold breaches to pattern recognition, reducing nuisance alerts and operator fatigue. Continuous model retraining ensures relevance as formulations, suppliers, and operating conditions evolve.

Table 19.5.4: Multivariate Tools and Their Diagnostic Application in VMA Manufacturing

Tool	Primary Input Data	Diagnostic Strength	Typical Use Case
PCA	Process and sensor data	Drift detection	Early instability
PLS regression	Raw materials and QC	Causal attribution	Corrective planning
MSPC	Time-resolved signals	Trajectory analysis	Storage failures
Contribution plots	Latent variables	Variable ranking	Root-cause clarity
Scenario prediction	Model coefficients	Virtual testing	Risk mitigation

Technology Transfer and Knowledge Retention

Multivariate root-cause frameworks enable consistent troubleshooting logic across sites by replacing anecdotal explanations with data-supported diagnostics. During technology transfer, historical deviation patterns and associated corrective actions are embedded within the analytics platform, preserving institutional knowledge. Receiving sites benefit from predefined interpretation rules, reducing dependence on local trial-and-error and accelerating stabilization under unfamiliar operating conditions.

19.5.5 Predictive Maintenance Linked to Rheology Signals

Rheology as an Early Indicator of Equipment Health

In Viscosity-Modifying Admixture manufacturing, rheological signals reflect not only polymer state but also the mechanical condition of processing equipment. Subtle, repeatable deviations in viscosity profiles often originate from changes in shear field uniformity, dead zones, or surface fouling rather than formulation drift. Predictive maintenance strategies that leverage inline rheology convert these deviations into actionable indicators of equipment degradation. This approach aligns maintenance planning with functional performance risk rather than calendar-based schedules.

Shear Signature Analysis and Mechanical Degradation

Each mixing and transfer system imposes a characteristic shear signature on polymer solutions, which remains stable under normal operation. Progressive wear of impellers, pump clearances, or static mixer elements alters this signature, producing consistent rheological shifts at constant formulation. Multivariate comparison of current rheology trends against historical baselines distinguishes mechanical causes from chemical degradation. Early detection prevents misinterpretation of viscosity loss as polymer instability, avoiding unnecessary reformulation or additive overdosing.

Fouling, Deposits, and Cleanability Assessment

Polymer deposition on vessel walls, sensors, or heat exchange surfaces introduces localized shear gradients and residence time anomalies. These effects manifest as increased viscosity variability, hysteresis between up and downstream measurements, or delayed response to dilution. Rheology-linked predictive maintenance flags such patterns before fouling becomes visually apparent or disrupts flow. Cleaning effectiveness can be quantitatively verified by restoration of baseline rheological behavior, replacing subjective visual inspection with performance-based criteria.

Integration with Maintenance and Asset Management Systems

Industry 4.0 environments integrate rheology-derived health indicators into Computerized Maintenance Management Systems. Algorithms translate trend deviations into maintenance priority scores, balancing risk of performance failure against operational cost. Maintenance actions are scheduled during planned downtime, minimizing production disruption. Over time, correlation between rheological anomalies and confirmed mechanical faults refines predictive accuracy and builds confidence in data-driven maintenance decisions.

Distinguishing Process Drift from Equipment Failure

A critical advantage of rheology-based predictive maintenance lies in its ability to separate process-induced variability from equipment-related issues. When rheological changes occur without corresponding shifts in raw material fingerprints or chemical sensor data, mechanical causes are implicated. Conversely, concurrent changes across rheology, pH, and conductivity point toward polymer degradation. This discrimination reduces unnecessary maintenance interventions and focuses resources where they deliver measurable benefit.

Table 19.5.5: Rheology-Linked Predictive Maintenance Indicators in VMA Manufacturing

Rheological Signal	Likely Equipment Issue	Diagnostic Confidence	Maintenance Action
Gradual viscosity drift	Impeller wear	High	Planned replacement
Increased signal noise	Pump cavitation	Medium	Inspection and tuning
Hysteresis effects	Surface fouling	High	Targeted cleaning
Delayed response	Dead zones	Medium	Process redesign
Upstream–downstream mismatch	Sensor fouling	High	Sensor maintenance

Operational Benefits and Technology Transfer Impact

Predictive maintenance anchored in rheology reduces unplanned downtime, stabilizes product quality, and extends equipment life. For multi-site operations, standardized rheological health metrics support consistent maintenance strategies and shared learning across plants. During technology transfer, baseline shear signatures and maintenance thresholds accompany process documentation, enabling receiving sites to establish predictive maintenance programs rapidly without prolonged empirical tuning.

19.5.6 Automated SPC and Alarm Management

Transition from Threshold Control to Pattern Recognition

Automated Statistical Process Control in Viscosity-Modifying Admixture manufacturing must move beyond fixed upper and lower limits applied to isolated variables. Polymer-based systems exhibit natural variability driven by hydration kinetics, shear history, and temperature dependence, which can trigger false alarms under conventional SPC schemes. Industry 4.0-enabled SPC emphasizes pattern recognition across correlated signals, identifying abnormal trajectories rather than momentary excursions. This approach aligns alarm generation with true risk to polymer integrity and functional performance.

Design of Multivariate SPC Architectures

Multivariate SPC frameworks integrate inline rheology, flow, pH, conductivity, temperature, and pressure into unified control models. Control statistics such as Hotelling's T^2 and Squared Prediction Error capture deviations from established multivariate operating envelopes. These metrics are more sensitive to subtle instability pathways, including early polymer degradation or preservative exhaustion, while remaining robust against benign process noise. Model tuning requires historical data representing both stable operation and known failure modes to ensure discriminative power.

Adaptive Control Limits and Context Awareness

Automated SPC systems benefit from adaptive control limits that adjust based on contextual variables such as formulation grade, batch size, or ambient conditions. For example, acceptable viscosity variability during initial polymer hydration differs from that during final dilution. Context-aware SPC reduces nuisance alarms and prevents operator desensitization. Control logic incorporates process phase recognition, enabling different alarm sensitivities during charging, mixing, cooling, and storage transfer.

Alarm Rationalization and Hierarchical Response

Alarm management is most effective when alerts are prioritized according to functional risk rather than sensor deviation magnitude alone. Hierarchical alarm structures classify events into advisory, warning, and critical levels, each linked to predefined response actions. Low-level advisories prompt observation or data review, while higher-level alarms trigger targeted intervention or batch hold. Such rationalization ensures that operator attention is reserved for events with genuine impact on product performance or safety.

Human-in-the-Loop Decision Support

Despite high automation, human expertise remains central to effective SPC deployment. Dashboards translate multivariate statistics into intuitive visualizations, highlighting contributing variables and confidence levels. Operators and process engineers interpret these insights within a polymer chemistry framework, validating automated recommendations before execution. This human-in-the-loop design maintains accountability and facilitates continuous improvement through feedback into the control models.

Table 19.5.6: Automated SPC Elements and Their Functional Role in VMA Manufacturing

SPC Element	Input Data	Primary Function	Industrial Benefit
Multivariate control charts	Sensor networks	Pattern deviation detection	Early fault identification
Adaptive limits	Process context	Noise reduction	Fewer false alarms
Alarm prioritization	Risk models	Response focus	Operator efficiency
Contribution analysis	Latent variables	Root-cause insight	Faster resolution
Dashboard visualization	SPC outputs	Decision support	Knowledge retention

Integration with Continuous Improvement and Technology Transfer

Automated SPC systems accumulate a rich history of deviation patterns and responses, forming a data-backed foundation for continuous improvement. During technology transfer, SPC models and alarm logic are deployed alongside formulations, ensuring consistent interpretation of process behavior across sites. Differences in equipment or operating culture are accommodated through controlled model adaptation rather than ad hoc alarm suppression. This disciplined approach sustains quality, accelerates stabilization, and strengthens organizational learning.

19.5.7 Knowledge Capture for Multi-Site Technology Transfer

From Tacit Expertise to Structured Industrial Knowledge

Successful technology transfer in Viscosity-Modifying Admixture manufacturing depends on converting tacit formulation and process know-how into structured, transferable knowledge assets. Much of the critical understanding around polymer hydration behavior, sensitivity to shear, and early degradation signals resides informally with experienced operators and formulation chemists. When this expertise remains undocumented, receiving sites rely on trial-and-error learning, increasing variability and startup risk. Industry 4.0 frameworks enable systematic capture of this knowledge by linking expert decisions to validated process data and outcomes.

Digital Documentation of Formulation–Process Interactions

Knowledge capture extends beyond static recipes to include dynamic interactions between formulation variables and processing conditions. Digital platforms record how changes in polymer molecular weight distribution, preservative system, or dilution sequence influence rheological evolution and stability margins. Contextual annotations, such as rationale for specific setpoints or known incompatibilities, transform historical batch records into actionable guidance. This depth of documentation supports informed decision-making at receiving sites facing similar but not identical operating environments.

Encoding Troubleshooting Logic and Failure Signatures

Recurring failure modes in VMA manufacturing often present with characteristic multivariate signatures rather than unique single-variable alarms. Capturing these signatures, along with validated corrective actions, creates a reusable troubleshooting library. Industry 4.0 systems encode this logic within analytics engines, enabling automated pattern recognition and recommendation generation. Over time, the library evolves as new raw materials, equipment, or regulatory constraints introduce novel interactions, preserving organizational learning.

Standardization of Data Models and Interpretation Rules

Cross-site technology transfer requires consistent data structures, sensor definitions, and interpretation rules. Without standardization, similar process behaviors appear different in analytics outputs, undermining confidence in shared models. Establishing common ontologies for raw materials, process steps, and quality attributes ensures that insights generated at one site remain valid elsewhere. Controlled flexibility allows local adaptation while preserving comparability of key indicators across the manufacturing network.

Human Capital and Training Integration

Digital knowledge capture complements, rather than replaces, human expertise. Training programs integrate data-driven insights with polymer chemistry fundamentals, enabling personnel at receiving sites to interpret analytics outputs critically. Simulation environments based on digital twins allow new teams to explore process sensitivities and troubleshooting scenarios safely. This blended approach accelerates competence development and reduces dependence on prolonged on-site support from originating plants.

Table 19.5.7: Knowledge Capture Elements Supporting Multi-Site VMA Technology Transfer

Knowledge Asset	Source	Capture Method	Transfer Benefit
Formulation rationale	R&D and production	Digital records	Consistent replication
Process sensitivity maps	Manufacturing data	Analytics models	Faster stabilization
Failure signatures	SPC and RCA history	Pattern libraries	Early detection

Corrective action logic	Expert decisions	Rule encoding	Reduced trial-and-error
Training simulations	Digital twins	Virtual scenarios	Skill acceleration

Strategic Impact on Global Manufacturing Networks

Structured knowledge capture reduces variability across sites, shortens ramp-up periods, and protects intellectual capital as personnel change. Manufacturers achieve greater resilience to raw material shifts and regulatory pressures by disseminating validated operating principles rather than isolated recipes. In a competitive environment characterized by complex polymer systems and demanding performance requirements, robust knowledge capture becomes a strategic enabler of scalable, high-quality VMA production.

20. Troubleshooting VMA Application Issues

Viscosity-modifying admixtures frequently meet formal product specifications while still exhibiting inconsistent behavior in concrete production and placement. Troubleshooting therefore requires moving beyond formulation intent and certificate data toward a systems-level understanding that links polymer chemistry, process variability, and application conditions. Effective resolution integrates manufacturing discipline, site practices, and digitally enabled feedback to ensure that rheological control achieved in design is reliably reproduced in the field.

20.1 From Manufacturing Conformance to Field Performance Reliability

Consistency in VMA manufacturing does not automatically translate into robustness under real construction environments. Variations in batching accuracy, material logistics, and concrete production workflows can amplify small formulation sensitivities. This perspective frames troubleshooting as a continuum that spans polymer production, quality assurance, and site execution rather than an isolated product issue.

20.1.1 Specification Compliance versus Functional Performance

Compliance with viscosity, solids content, and pH specifications confirms manufacturing control but does not guarantee functional equivalence in concrete systems. VMAs interact dynamically with cement phases, fillers, and other admixtures, often in ways not captured by standard acceptance tests. Troubleshooting must therefore prioritize performance-based indicators such as segregation resistance, yield stress evolution, and placement stability.

Mechanistic Gap between Certificate Values and In-Concrete Behavior

Specification parameters such as viscosity, solids content, and pH are designed to verify manufacturing consistency and batch-to-batch reproducibility. These metrics describe the admixture in isolation and under controlled laboratory conditions, yet they do not capture the dynamic, multicomponent environment encountered in concrete. Polymeric VMAs function through weak adsorption, solution structuring, and time-dependent conformational adjustment, all of which are strongly influenced by cement mineralogy, fines loading, and competitive admixture chemistry. As a result, two VMA batches that are fully compliant on paper may exhibit divergent rheological impact once introduced into real mix designs.

At the molecular level, functional performance is governed by polymer availability at the solid–liquid interface, chain mobility under shear, and persistence of associative interactions during hydration. These attributes are largely invisible to standard acceptance tests but directly control segregation resistance, yield stress evolution, and stability during placement.

Limitations of Conventional Acceptance Testing

Routine quality control methods typically rely on single-point viscosity measurements, density checks, and pH confirmation. While suitable for detecting gross manufacturing deviations, these tests lack sensitivity to adsorption kinetics, shear-thinning behavior, and time-dependent rheological drift. VMAs that pass all acceptance criteria may still display delayed activation, excessive structuring, or rapid performance decay when exposed to realistic mixing energy and temperature histories.

Laboratory qualification protocols further simplify conditions by using standardized cement, fixed water chemistry, and short observation periods. Such simplifications mask interaction effects that become dominant in production, particularly in mixes containing high SCM levels, recycled water, or modern PCEs with strong surface affinity.

Performance-Based Indicators for Effective Troubleshooting

Troubleshooting oriented toward functional reliability requires replacing specification-centric thinking with performance-driven diagnostics. Key indicators include segregation resistance under variable fines content, yield stress development under representative shear profiles, and stability during pumping and finishing windows. Time-resolved rheological measurements and comparative mix trials provide far greater insight than static certificate values.

Industry 4.0 systems increasingly enable this transition by correlating plant dosing records, mixer torque signals, and field performance observations. Such data integration allows identification of performance trends that systematically deviate despite nominal specification compliance.

Table 20.1.1: Specification Compliance versus Functional Performance Disconnect

Aspect	Specification Focus	Hidden Variable	Concrete Performance Impact
Viscosity	Single-point value	Shear dependency	Pumpability variation
Solids content	Mass balance	Effective polymer availability	Inconsistent stabilization
pH	Chemical stability	Adsorption kinetics	Slump instability
Certificate conformity	Batch uniformity	Interaction with SCMs	Segregation risk
Lab qualification	Controlled conditions	Field variability	Placement sensitivity

Corrective and Preventive Strategies

Reducing the gap between specification compliance and functional performance requires redefining acceptance criteria to include application-relevant indicators. Incorporating rheological envelopes, time-dependent stability indices, and interaction screening with key admixtures strengthens predictive capability. From an Industry 4.0 perspective, linking QA databases with plant and site performance data enables continuous recalibration of operating windows rather than reliance on static limits.

For technology transfer, clear communication of what specifications do and do not guarantee is essential. Training technical service teams to interpret certificates in conjunction with mix design context prevents misdiagnosis and supports more effective, data-backed troubleshooting in the field.

20.1.2 Limitations of Laboratory Qualification for Field Conditions

Laboratory qualification protocols are inherently simplified, using controlled water chemistry, fixed mixing energy, and short observation windows. Field conditions introduce extended mixing, temperature gradients, and recycled water variability that can alter polymer conformation and adsorption behavior. Understanding these limitations is essential to interpreting test results and avoiding false confidence in lab-derived performance claims.

Structural Simplification of Laboratory Qualification Protocols

Laboratory qualification of VMAs is intentionally designed to isolate formulation performance under controlled and repeatable conditions. Standardized cement types, potable water, fixed water-to-binder ratios, and defined mixing energy are employed to reduce noise and enable comparative screening. While effective for formulation ranking, such protocols suppress many of the interacting variables that dominate real concrete production. As a consequence, laboratory results often reflect idealized polymer behavior rather than the response of VMAs under industrial and site-specific constraints.
At the polymer scale, this simplification stabilizes solvation and adsorption equilibria that are rarely preserved during plant-scale batching or extended transport.

Field-Induced Deviations Affecting Polymer Functionality

Concrete production and placement expose VMAs to prolonged and heterogeneous shear histories, temperature gradients between materials and ambient conditions, and variable residence times prior to discharge. Recycled and wash water introduce dissolved ions, fine particulates, and organic residues that modify polymer charge screening and hydration kinetics. These factors collectively influence polymer conformation, adsorption competition, and associative network formation, frequently shifting the effective operating window outside that observed in laboratory trials.

Time-dependent effects are particularly critical, as VMAs that appear stable within short lab observation periods may exhibit delayed thickening, performance decay, or over-structuring during transport and waiting on site.

Mismatch between Laboratory Metrics and Application-Relevant Performance

Most qualification programs emphasize early-age workability, visual stability, or simple segregation indices measured within minutes of mixing. Such metrics provide limited insight into pumpability, slump retention stability, or finishability after extended holding times. Moreover, laboratory mixing energy is typically insufficient to replicate the cumulative shear imparted by truck mixers, transfer pumps, and placement equipment. The resulting mismatch leads to systematic overestimation of robustness when laboratory data are extrapolated directly to field performance.

Table 20.1.2: Key Differences between Laboratory Qualification and Field Conditions

Parameter	Laboratory Qualification	Field Reality	Impact on VMA Performance
Water quality	Potable, controlled	Recycled, variable ions	Altered adsorption behavior
Mixing energy	Fixed, low residence	Extended, cumulative shear	Polymer degradation or drift
Temperature	Stable	Gradients and seasonal swings	Shifted hydration kinetics
Observation time	Short	Extended transport and waiting	Delayed instability
System complexity	Simplified	Multiple interacting admixtures	Competitive adsorption effects

Implications for Interpretation and Troubleshooting

Recognizing the inherent limitations of laboratory qualification is essential for correct interpretation of test outcomes. Laboratory data should be viewed as boundary conditions rather than predictive guarantees of field behavior. Robust troubleshooting integrates lab results with plant trials, time-resolved rheology, and monitored field performance to validate functional stability across realistic operating envelopes.

From an Industry 4.0 perspective, digital comparison of laboratory baselines with plant and site data enables early identification of divergence and reduces reliance on static qualification protocols. For technology transfer, explicitly communicating the scope and limitations of laboratory testing prevents false confidence and supports more resilient VMA deployment strategies.

20.1.3 Impact of Transport, Storage, and Dosing Accuracy

VMAs are sensitive to thermal history, prolonged storage, and shear exposure during pumping and recirculation. Inaccurate dosing systems or delayed addition points can shift effective polymer concentration outside its optimal window. Troubleshooting at this level requires coordination between logistics, plant maintenance, and admixture supply practices.

Influence of Logistics and Storage on Polymer Integrity

Viscosity-modifying admixtures are not chemically static during transport and storage, particularly when supplied as aqueous solutions or high-solids concentrates. Extended storage under elevated or fluctuating temperatures alters polymer solvation state, promotes partial associative rearrangement, and may accelerate hydrolytic side reactions in susceptible backbones. Repeated thermal cycling during transport further affects solution homogeneity, leading to localized viscosity gradients and reduced dosing reproducibility. These effects are rarely captured by shelf-life specifications yet directly influence functional polymer availability at the point of use.

From a molecular standpoint, changes in hydration shell structure and chain entanglement density modify the effective thickening efficiency once the VMA is introduced into concrete.

Shear Exposure during Transfer, Pumping, and Recirculation

VMAs are routinely exposed to mechanical shear during tanker unloading, recirculation in storage tanks, and high-pressure dosing lines. Prolonged or excessive shear can disrupt weak associative networks or induce partial chain scission in high-molecular-weight systems, particularly for cellulose ethers and associative synthetic polymers. Even in the absence of measurable molecular weight reduction, repeated shear alters polymer conformation and delays viscosity recovery after dosing.

Such shear history effects accumulate invisibly, resulting in batches that meet incoming inspection criteria but underperform in stabilizing fresh concrete.

Dosing Accuracy and Addition Point Sensitivity

Accurate metering is critical due to the narrow effective dosage window of most VMAs. Calibration drift, air entrainment in dosing lines, or pulsation in low-flow pumps introduce systematic under- or over-dosing that shifts the polymer concentration away from its optimal functional range. Delayed addition or incorrect sequencing relative to water and other admixtures further modifies adsorption dynamics and reduces efficiency. Small absolute dosing errors translate into disproportionate changes in yield stress and segregation resistance.
These sensitivities are amplified in modern low-water and high-fines concrete systems, where VMA demand is sharply defined.

Table 20.1.3: Logistics and Dosing-Related Risk Factors Affecting VMA Performance

Risk Factor	Primary Cause	Operational Indicator	Concrete Performance Impact
Thermal aging	Prolonged hot storage	Gradual viscosity drift	Reduced stabilization efficiency
Thermal cycling	Transport temperature swings	Phase heterogeneity	Inconsistent rheology
Excessive shear	Recirculation, pumping	Delayed viscosity recovery	Segregation under stress
Underdosing	Meter calibration drift	Low admixture consumption	Bleeding and settlement
Overdosing	Pump pulsation	Elevated yield stress	Pumpability loss

Integrated Troubleshooting and Preventive Control

Effective troubleshooting at this level requires alignment between logistics management, plant maintenance, and admixture supply practices. Routine calibration of dosing systems, temperature-controlled storage, and minimization of unnecessary recirculation reduce variability at source. From an Industry 4.0 perspective, real-time monitoring of storage temperature, pump operating profiles, and dosing accuracy enables early detection of deviation from defined operating windows.

For technology transfer, explicit guidance on storage limits, acceptable shear exposure, and validated addition sequences must accompany product deployment to prevent misinterpretation of performance issues as formulation defects.

20.1.4 Bridging QA Metrics with Application Outcomes

Quality assurance metrics must be interpreted in the context of application performance rather than as standalone acceptance criteria. Correlating batch data, rheological measurements, and site observations enables identification of leading indicators for failure. Digital QA systems increasingly support this linkage by integrating production data with field feedback.

Limitations of Isolated QA Acceptance Criteria

Quality assurance programs for VMAs traditionally rely on compliance with predefined acceptance ranges for viscosity, density, solids content, and pH. While these parameters confirm manufacturing stability, they offer limited predictive value for in-concrete performance when interpreted in isolation. Polymer-based admixtures operate within complex, multivariate systems where functional response emerges from interactions with cement chemistry, fines distribution, mixing energy, and time. As a result, QA conformity does not necessarily equate to reliable field behavior, particularly in high-performance or tightly optimized concrete systems.

Correlation of Production Data with Rheological Behavior

Bridging the gap between QA metrics and application outcomes requires systematic correlation of batch-level production data with rheological response measured under representative conditions. Time-resolved viscosity development, yield stress evolution, and shear recovery behavior provide functional context to

conventional certificate values. When aligned with production parameters such as batch temperature, hydration time, and shear exposure, these rheological indicators reveal early deviations that precede visible performance failure.
Such correlations enable identification of leading indicators rather than post-failure diagnostics.

Integration of Field Observations into QA Frameworks

Field observations including segregation tendency, pump pressure stability, slump retention behavior, and finishing response represent the ultimate validation of VMA performance. Capturing these observations in structured, quantitative formats allows them to be linked back to specific production batches and QA data. Without this feedback loop, recurring issues are often treated symptomatically rather than addressed at root cause.
Standardization of field reporting protocols is therefore essential to transform experiential feedback into actionable quality intelligence.

Table 20.1.4: Linking QA Metrics to Application-Relevant Performance Indicators

QA Metric	Traditional Use	Application-Relevant Correlation	Failure Risk Signal
Viscosity	Batch conformity	Shear recovery and time dependency	Pumpability instability
Solids content	Dosage accuracy	Effective polymer availability	Segregation variability
pH	Chemical stability	Adsorption kinetics	Slump drift
Batch temperature	Process control	Hydration rate shift	Delayed thickening
Density	Material uniformity	Dosing precision	Over- or under-structuring

Digital QA and Industry 4.0 Enablement

Digital QA systems increasingly enable continuous integration of manufacturing, plant, and site data streams. Automated data capture from batching systems, mixer torque sensors, and site performance logs supports multivariate analysis of VMA behavior across its lifecycle. Predictive models can then identify combinations of QA deviations that correlate with application risk, even when individual parameters remain within specification limits.

For technology transfer, such digitally enabled QA frameworks provide a shared language between supplier and customer, improving transparency and accelerating corrective action. Aligning QA metrics with application outcomes ultimately transforms quality assurance from a gatekeeping function into a proactive performance management tool.

20.2 Application Sensitivity of Polymeric VMAs

Polymeric VMAs exhibit performance profiles that are highly dependent on their physicochemical environment. Adsorption equilibria, shear response, and hydration timing collectively define whether the polymer functions as a stabilizer or becomes a source of instability. Troubleshooting must therefore treat VMAs as responsive materials rather than inert additives.

20.2.1 Adsorption Competition with PCEs and Air Entrainers

VMAs often compete for adsorption sites with polycarboxylate ethers and air-entraining agents. Changes in addition sequence or dosage can shift surface coverage and disrupt intended rheological balance. Understanding competitive adsorption is central to diagnosing unexpected loss of stability or air control.

Competitive Adsorption at Cement and Filler Surfaces

Polymeric VMAs share adsorption sites with polycarboxylate ether superplasticizers and air-entraining agents on cement particles and fine mineral surfaces. These sites are finite and chemically heterogeneous, leading to preferential adsorption governed by charge density, molecular architecture, and functional group distribution. When PCEs with strong anchoring groups are introduced at high dosage, they can displace previously adsorbed VMA chains, reducing effective polymer availability for viscosity control.

Air-entraining agents further complicate this balance by adsorbing at solid–liquid and air–water interfaces, indirectly modifying surface energy and polymer distribution.

Influence of Addition Sequence and Dosage Strategy

The order and timing of admixture addition exert a decisive influence on adsorption outcomes. Early VMA addition may lead to premature surface occupation that is later disrupted by PCE dosing, whereas delayed addition can limit polymer access to active sites already saturated by dispersants. Minor dosage adjustments, often considered negligible, can trigger disproportionate shifts in surface coverage and destabilize the intended rheological balance.
Such sensitivities are magnified in low-water systems where surface area per unit water is high and competitive interactions intensify.

Manifestation in Rheology and Air Stability

Competitive adsorption effects typically manifest as loss of segregation resistance, unexpected yield stress reduction, or instability in air content. In some cases, partial VMA displacement leads to transient rheological behavior, where stability appears adequate immediately after mixing but degrades during transport or placement. Air systems may show increased variability, coalescence, or surface defects due to altered interfacial stabilization.
These manifestations are often misattributed to overdosing or raw material variability unless competitive adsorption is explicitly considered.

Table 20.2.1: Competitive Adsorption Effects between VMAs, PCEs, and Air Entrainers

Interaction Scenario	Dominant Adsorbate	Primary Shift	Concrete Performance Impact
High PCE dosage	PCE	VMA displacement	Reduced segregation control
Delayed VMA addition	PCE	Limited VMA adsorption	Rheology drift
High air entrainer level	AEA	Interfacial competition	Air instability
SCM-rich systems	Mixed	Increased surface demand	Dosage sensitivity

Low water content	PCE	Aggressive competition	Narrow operating window

Control and Mitigation Strategies

Mitigating competitive adsorption requires coordinated admixture design and application discipline. Optimized addition sequences, balanced dosage ratios, and selection of VMAs with tailored adsorption strength reduce displacement risk. Industry 4.0-enabled plant systems support this approach by tracking addition timing, dosage accuracy, and resulting rheological signatures in real time.
For technology transfer, explicit compatibility matrices and interaction guidelines should accompany VMA deployment to prevent reactive troubleshooting and support predictable field performance.

20.2.2 Shear History and Mixing Energy Effects

Shear exposure during mixing and pumping can partially degrade polymer structures or alter associative interactions. Excessive energy input may reduce viscosity contribution, while insufficient mixing can lead to delayed activation. Troubleshooting requires mapping shear history across the full production and placement chain.

Impact of Shear on Polymer Conformation and Network Formation

Polymeric VMAs are highly sensitive to mechanical energy input during mixing, pumping, and recirculation. Shear forces can realign polymer chains, disrupt weak associative networks, or, in extreme cases, induce partial chain scission in high-molecular-weight systems. These changes alter the polymer's ability to structure the aqueous phase, leading to a reduction in viscosity contribution or delayed thickening. Inadequate mixing, conversely, can leave polymer aggregates insufficiently hydrated, preventing full activation and compromising initial stabilization of the paste.

Cumulative Shear Across the Production Chain

Shear history is rarely confined to the mixer; it accumulates along the concrete production and placement chain. Truck mixer rotations, pipeline pumping, and recirculation loops impose variable and sometimes unpredictable energy input. The net effect on polymer performance depends on both intensity and duration of shear, as well as the timing relative to cement hydration and admixture addition. These interactions are nonlinear, meaning small variations in energy input can produce disproportionately large changes in fresh concrete rheology.

Manifestation in Field Performance

Improperly managed shear history often emerges as inconsistent slump, early or delayed yield stress development, reduced segregation resistance, or erratic pumpability. In high-performance mixes with low water content or high fines, sensitivity to shear is amplified, making field performance highly contingent on careful control of mixing and placement energy. Tracking shear exposure and understanding its cumulative impact are therefore critical for effective troubleshooting.

Table 20.2.2: Shear and Mixing Energy Effects on VMA Performance

Shear Scenario	Polymer Response	Rheological Consequence	Concrete Impact
Excessive shear	Network disruption, partial chain scission	Reduced viscosity, rapid structure breakdown	Loss of segregation resistance, pumpability issues
Insufficient shear	Incomplete hydration or dispersion	Delayed viscosity build-up	Poor initial stability, slump retention issues
Moderate, controlled shear	Optimal network formation	Predictable thickening	Stable yield stress, consistent placement

Mitigation Strategies

Mitigation requires a combination of formulation design and process control. Selecting VMAs with shear-tolerant structures, staging polymer addition, and controlling mixer and pump energy profiles help maintain functional performance. Industry 4.0-enabled monitoring, including torque sensors, inline rheology probes, and digital twin simulations, allows tracking of cumulative shear and early detection of deviations from optimal operating conditions.

20.2.3 Temperature, Time, and Hydration Kinetics

Temperature and elapsed time influence both cement hydration and polymer solvation dynamics. VMAs that perform well at moderate temperatures may underperform in hot or cold concreting due to altered adsorption rates and water structuring. Time-dependent rheological drift is often a combined chemical and thermal effect.

Thermal Effects on Polymer Functionality

Temperature strongly influences both cement hydration and polymer solvation behavior. Elevated temperatures accelerate hydration reactions, increasing early-age stiffening and potentially shortening the effective working window of VMAs. Conversely, low temperatures slow both hydration and polymer chain mobility, delaying network formation and reducing viscosity development. These thermal effects alter adsorption kinetics, water structuring, and associative interactions, shifting VMA performance away from the expected response observed under standard laboratory conditions.

Time-Dependent Evolution of Rheology

The interaction between temperature and elapsed time results in time-dependent rheological drift. Even VMAs that initially provide stable flow can lose effectiveness as hydration proceeds or as polymer chains reconfigure under evolving chemical conditions. This drift is particularly significant in low-water or high-SCM systems, where small changes in hydration rate or polymer mobility have amplified effects on yield stress and segregation resistance. Extended transport, delayed placement, or high ambient temperatures can therefore lead to premature stiffening or instability.

Practical Implications for Troubleshooting

Monitoring of temperature and timing is essential to predict and correct performance deviations. Time-resolved rheological measurements, combined with awareness of site thermal conditions, help distinguish polymer-related drift from cement or admixture variability. Corrective strategies may include adjustment of VMA dosage, sequencing, or use of temperature-tolerant formulations to maintain consistent performance across variable conditions.

Table 20.2.3: Temperature and Time Influence on VMA Performance

Condition	Polymer Response	Rheological Consequence	Field Impact
High temperature	Accelerated adsorption and network collapse	Early stiffening	Reduced workability, pumpability loss
Low temperature	Slower hydration and chain mobility	Delayed viscosity development	Slump retention instability
Extended holding time	Progressive adsorption changes	Gradual rheological drift	Segregation risk, finishing difficulty
Temperature fluctuations	Reversible chain conformational changes	Inconsistent thickening	Variability in placement quality

Mitigation Strategies

Formulation design must account for the expected thermal and temporal profiles of concrete placement. Use of VMAs with temperature-stable backbones, staged addition strategies, and adjustment of mixing or curing conditions can minimize undesired rheological drift. Industry 4.0 tools, such as inline temperature monitoring and predictive rheology models, enable proactive correction of deviations, ensuring stable performance under variable site conditions.

20.2.4 Water Chemistry and Recycled Water Interactions

Ionic strength, dissolved organics, and fines present in recycled water can significantly modify VMA effectiveness. These factors affect polymer charge screening and hydration behavior, leading to variability not observed with potable water. Robust troubleshooting must include routine characterization of mixing water.

Chemical Complexity of Mixing Water in Modern Concrete Production

Water used in concrete production increasingly contains recycled and wash water fractions that introduce chemical complexity beyond that of potable sources.

Elevated ionic strength, variable concentrations of calcium, sodium, sulfates, and chlorides, along with dissolved organic species and suspended fines, alter the physicochemical environment in which polymeric VMAs operate. These constituents directly influence polymer solvation, chain expansion, and associative interactions, shifting performance away from laboratory-qualified behavior.

Effects of Ionic Strength and Charge Screening

Increased ionic strength compresses the electrical double layer surrounding charged polymer segments, reducing effective chain extension and weakening viscosity contribution. For anionic VMAs, calcium-rich recycled water promotes ion bridging and premature adsorption, while monovalent salts accelerate charge screening without providing stabilizing interactions. These mechanisms reduce polymer availability in the pore solution and narrow the functional dosage window, particularly in low-water and high-fines systems.

Influence of Dissolved Organics and Suspended Fines

Organic residues originating from admixture carryover, hydration byproducts, or environmental contamination compete with VMAs for adsorption sites and modify surface energy at solid–liquid interfaces. Fine particles present in recycled water further increase specific surface area demand, consuming polymer through non-productive adsorption. Together, these effects introduce variability in rheological response that is absent when using clean, controlled water sources.

Table 20.2.4: Recycled Water Constituents and Their Impact on VMA Performance

Water Component	Primary Interaction Mechanism	Polymer Response	Concrete Performance Impact
High ionic strength	Charge screening	Reduced chain expansion	Lower viscosity contribution
Calcium-rich water	Ion bridging, premature adsorption	Reduced polymer availability	Slump instability

Dissolved organics	Competitive adsorption	Delayed or uneven activation	Rheological variability
Suspended fines	Non-productive adsorption	Increased dosage demand	Segregation sensitivity
Variable pH	Solvation and charge shift	Inconsistent hydration	Finishability issues

Implications for Troubleshooting and Control

Effective troubleshooting requires routine characterization of mixing water chemistry, including ionic composition, total dissolved solids, organic content, and fines concentration. Reliance on potable water assumptions leads to misdiagnosis of VMA performance issues and inconsistent corrective action. Incorporating water quality monitoring into Industry 4.0 plant data systems enables correlation between water variability and rheological response, supporting proactive adjustment of dosage and sequencing.

For technology transfer, clear guidance on acceptable water quality envelopes and validated adjustment strategies is essential to maintain reliable VMA performance under increasing reliance on recycled water.

20.3 Typical Failure Modes Observed in Concrete Systems

Failure modes associated with VMAs rarely manifest as complete loss of function. More commonly, they appear as marginal instabilities that accumulate into visible quality defects or operational difficulties. Recognizing these patterns early enables corrective action before significant production disruption occurs.

Progressive Nature of VMA-Related Failures

Failure modes associated with viscosity-modifying admixtures seldom present as abrupt or total loss of functionality. Instead, they typically emerge as incremental deviations in rheological behavior that accumulate over time and across process stages. Small shifts in yield stress, segregation resistance, or air stability may initially remain within acceptable limits, yet their combined effect results in visible quality defects, placement inefficiencies, or increased operational risk. Such progressive failure mechanisms are characteristic of polymer-sensitive systems operating close to their functional boundaries.

Interaction-Driven Manifestation of Instability

Most observed failure modes arise from interaction effects rather than intrinsic polymer deficiencies. Variations in SCM content, water chemistry, temperature, shear history, and admixture compatibility act synergistically to push the system beyond its stable operating window. VMAs that perform reliably in one mix design or production context may exhibit instability when even minor variables shift, underscoring the conditional nature of their performance. Recognizing interaction-driven patterns is therefore essential for accurate root-cause analysis.

Operational and Quality Implications

Marginal instabilities translate into practical challenges such as inconsistent pump pressure, fluctuating slump, surface defects, and reduced finishability. These issues increase production variability, raise labor and energy demands, and elevate the risk of rejection or rework. Early recognition of emerging failure signatures allows targeted corrective measures, including dosage adjustment, sequencing changes, or process control refinement, before disruptions propagate through the production chain.

Role of Early Detection and Digital Monitoring

Timely identification of failure trends depends on continuous observation rather than post-event inspection. Industry 4.0-enabled monitoring of rheological proxies, batching accuracy, and environmental conditions provides early warning signals that precede visible defects. Integrating these signals into troubleshooting workflows shifts failure management from reactive response to proactive control, preserving both product quality and operational efficiency.

20.3.1 Segregation and Bleeding under Variable SCM Content

Variations in supplementary cementitious material content alter particle packing and surface chemistry. VMAs optimized for one SCM blend may not adequately stabilize another, resulting in bleeding or aggregate settlement. Troubleshooting requires reassessing polymer demand across realistic SCM variability.

Influence of SCM Variability on Paste Structure and Stability

Supplementary cementitious materials introduce substantial variability in particle size distribution, surface chemistry, and water demand. Changes in fly ash class, slag fineness, limestone filler content, or calcined clay reactivity directly modify paste packing density and interparticle forces. VMAs optimized for a specific SCM blend may therefore lose effectiveness when SCM composition shifts, even if total binder content remains unchanged. Such sensitivity reflects altered surface area demand and modified adsorption environments rather than intrinsic polymer failure.

Polymer Demand and Adsorption Effects

Different SCMs exhibit distinct affinities for polymer adsorption due to variations in surface charge, mineralogy, and porosity. High-surface-area materials increase non-productive polymer adsorption, reducing the fraction available for aqueous phase structuring. Conversely, smoother or less reactive SCMs may reduce polymer demand, leading to under- or over-structuring when dosage is held constant. These adsorption imbalances manifest as insufficient yield stress development or excessive paste rigidity, both of which promote segregation or bleeding.

Field Manifestations of Instability

Segregation and bleeding under variable SCM content often appear intermittently, correlating with raw material supply changes rather than process deviations. Aggregate settlement, water accumulation at the surface, or paste-rich zones near formwork are typical indicators. These effects are frequently misattributed to batching errors or water addition unless SCM-driven polymer demand shifts are explicitly considered during troubleshooting.

Table 20.3.1: SCM-Related Drivers of Segregation and Bleeding

SCM Characteristic	Primary Effect	Polymer Response	Concrete Performance Impact

High surface area	Increased adsorption demand	Reduced effective VMA concentration	Bleeding, settlement
Variable fineness	Packing density change	Dosage sensitivity	Segregation risk
Reactive SCMs	Early hydration interaction	Time-dependent drift	Delayed instability
SCM blend changes	Surface chemistry shift	Altered adsorption balance	Inconsistent stability
Filler-rich systems	Lower water availability	Over-structuring	Pumpability loss

Corrective and Preventive Approaches

Effective troubleshooting requires reassessment of polymer demand across the full range of realistic SCM variability rather than reliance on a single reference blend. Screening VMAs against multiple SCM scenarios, adjusting dosage envelopes, and defining SCM-specific operating windows improve robustness. Industry 4.0-enabled tracking of SCM source, fineness, and replacement level allows correlation with rheological outcomes and supports predictive adjustment.
For technology transfer, explicit guidance on SCM sensitivity and validated adjustment protocols reduces the risk of segregation and bleeding as raw material supply evolves.

20.3.2 Pumpability Loss and Excessive Yield Stress

Over-structuring of the paste phase can increase yield stress beyond pumpable limits. This condition often arises from cumulative effects of VMAs, fines, and high-range water reducers. Diagnosis must balance stability requirements against energy demands of placement equipment.

Mechanistic Origin of Over-Structuring in the Paste Phase

Pumpability loss associated with VMAs is rarely the result of a single parameter but instead reflects cumulative over-structuring of the paste phase. Excessive yield stress develops when polymer-induced water structuring, fines content, and dispersant action interact synergistically to create a highly interconnected particle network. VMAs designed to enhance segregation resistance can, under certain conditions, amplify interparticle cohesion beyond the threshold compatible with pumping systems. This effect is particularly pronounced in mixes with high fines loading, low water-to-binder ratios, and aggressive polycarboxylate ether dosing.

Contribution of Admixture Interactions and Fines

High-range water reducers increase particle dispersion and expose additional surface area for polymer adsorption, effectively raising VMA demand. When dosage is not adjusted accordingly, VMAs may occupy a greater fraction of available surfaces or reinforce associative networks within the pore solution. Fines from SCMs, fillers, or recycled water further intensify this effect by increasing surface area and reducing free water. The resulting rheological profile is characterized by elevated static and dynamic yield stress, even when plastic viscosity appears acceptable.

Operational Consequences during Pumping and Placement

Excessive yield stress manifests operationally as increased pump pressure, unstable flow, line blockages, or intermittent surging. These symptoms increase energy consumption, accelerate equipment wear, and reduce placement reliability. In extreme cases, corrective water addition is applied in the field, compromising durability and consistency. Without a structured diagnostic approach, such issues are often misattributed to equipment limitations rather than rheological imbalance.

Table 20.3.2: Drivers of Pumpability Loss and Yield Stress Escalation

Contributing Factor	Primary Mechanism	Rheological Effect	Operational Impact
High VMA dosage	Excess network formation	Elevated yield stress	Pump pressure increase
High fines content	Reduced free water	Increased cohesion	Flow instability
Aggressive PCE use	Surface area exposure	Higher polymer demand	Dosage sensitivity
Low water ratio	Limited lubrication	Steep yield stress rise	Pump blockage risk

Recycled water fines	Additional adsorption sites	Over-structuring	Energy inefficiency

Balancing Stability and Pumpability

Effective diagnosis requires balancing the need for segregation resistance with the mechanical limits of placement equipment. Rheological characterization under representative shear rates, combined with pump pressure monitoring, provides actionable insight into whether yield stress exceeds operational thresholds. Industry 4.0-enabled systems support this balance by correlating mix design parameters, admixture dosing, and real-time pumping data.
For technology transfer, defining pumpability envelopes and clearly communicating the trade-off between stability and energy demand reduces reactive adjustments and supports reliable field execution.

20.3.3 Slump Retention Instability and Rheological Drift

Unstable slump retention reflects time-dependent changes in polymer adsorption and hydration kinetics. VMAs may initially stabilize the mix but later lose effectiveness as competing reactions progress. Monitoring rheological evolution rather than single time-point measurements is essential for accurate troubleshooting.

Time-Dependent Nature of Slump Loss Mechanisms

Slump retention instability associated with VMAs is fundamentally a time-dependent phenomenon driven by evolving polymer adsorption and cement hydration kinetics. Initial stabilization often reflects transient availability of polymer chains in the pore solution, while progressive hydration, ion release, and surface saturation gradually alter adsorption equilibria. As these competing reactions advance, the effective contribution of the VMA diminishes, leading to accelerated loss of workability or unpredictable rheological behavior despite unchanged dosage.

Polymer Reconfiguration and Competitive Interactions

Over time, polymer chains undergo conformational adjustment in response to increasing ionic strength and changing surface chemistry. Competitive adsorption with PCEs and interaction with hydration products further redistribute polymer away from the aqueous phase. These processes may not be apparent immediately after mixing but become dominant during transport, waiting periods, or extended placement operations. The resulting rheological drift is often nonlinear, with abrupt transitions from stable flow to rapid stiffening.

Limitations of Single Time-Point Assessment

Conventional assessment of VMA performance frequently relies on slump or flow measurements taken shortly after mixing. Such single time-point evaluations fail to capture delayed instability mechanisms and provide limited insight into long-term workability control. Mixes that satisfy initial acceptance criteria may therefore exhibit unacceptable performance under realistic holding times. Accurate troubleshooting requires time-resolved rheological monitoring that reflects the full duration of production and placement.

Table 20.3.3: Drivers of Slump Retention Instability and Rheological Drift

Driver	Primary Mechanism	Rheological Signature	Field Consequence
Progressive hydration	Surface saturation	Accelerated yield stress rise	Early slump loss
Ionic strength increase	Polymer contraction	Reduced viscosity contribution	Flow instability
Competitive adsorption	Polymer displacement	Sudden rheological shift	Placement disruption
Extended holding time	Time-dependent reconfiguration	Nonlinear drift	Unpredictable workability
Temperature elevation	Kinetics acceleration	Compressed working window	Finishing difficulty

Diagnostic and Preventive Strategies

Effective diagnosis of slump retention issues requires monitoring rheological evolution over time rather than relying on isolated measurements. Time-sweep rheology, staged slump testing, and correlation with temperature and shear history provide a more accurate representation of in-field behavior. Industry 4.0-enabled tracking of transport duration, ambient conditions, and mixer activity supports

early identification of drift patterns.
For technology transfer, clearly defining acceptable workability retention windows and training site personnel to recognize early drift indicators reduces reactive adjustments and improves consistency.

20.3.4 Air Entrainment Variability and Surface Defects

Interactions between VMAs and air-entraining agents can modify bubble size distribution and stability. Variability in air content often manifests as surface voids or durability concerns. Effective troubleshooting links chemical interactions with observed finishing and hardened concrete performance.

Chemical Interactions Governing Air Stability

Air entrainment variability in VMA-containing systems is primarily governed by interfacial interactions between polymeric VMAs and air-entraining agents. VMAs can modify surface tension, alter surfactant adsorption kinetics, or partially shield air-entrainer molecules from effective action. These effects influence bubble nucleation, coalescence resistance, and stability during mixing and transport, resulting in air systems that deviate from target specifications despite correct nominal dosing.

Impact on Bubble Size Distribution and Concrete Surface Quality

Even when total air content remains within specification limits, changes in bubble size distribution can significantly affect fresh and hardened concrete performance. Coarser or unstable air void systems increase the likelihood of surface defects such as bugholes, pinholing, and localized segregation during finishing. In hardened concrete, poorly stabilized air structures compromise freeze–thaw durability and surface scaling resistance, linking early rheological behavior directly to long-term performance outcomes.

Process Sensitivity and Field Variability

Air entrainment sensitivity is amplified under variable mixing energy, temperature fluctuations, and holding times typical of field operations. VMAs that enhance paste structuring may restrict bubble mobility, while others may promote excessive foam stabilization under high shear. These effects are rarely captured in laboratory air content tests conducted immediately after mixing, leading to discrepancies between certified admixture performance and observed site behavior.

Integration of Observations for Effective Troubleshooting

Effective troubleshooting requires correlating measured air content, bubble stability indicators, and observed surface finish quality rather than treating them as independent parameters. Time-resolved air measurements, combined with visual inspection of fresh concrete and hardened surfaces, provide a more reliable diagnostic framework. Industry 4.0-enabled data capture, including mixer energy input, transport duration, and ambient conditions, supports identification of patterns that predispose mixes to air instability and surface defects.

20.3.5 Finishability and Visual Quality Issues

Excessive surface stickiness or delayed bleed water release affects finishing operations and final appearance. These issues are frequently attributed to VMAs but often involve combined effects of fines, temperature, and finishing timing. Structured evaluation prevents misattribution and unnecessary formulation changes.

Mechanistic Drivers of Surface Stickiness and Bleed Delay

Finishability problems associated with polymeric VMAs originate from excessive paste structuring at the concrete surface and delayed water release from the capillary network. VMAs designed to enhance segregation resistance increase water retention and yield stress in the near-surface zone, which can suppress timely bleed water formation. When combined with high fines content or reactive SCMs, this effect is amplified, producing a sticky surface that resists troweling and responds poorly to conventional finishing practices.

From a physicochemical standpoint, delayed bleed is linked to enhanced polymer adsorption on fine particles and cement hydrates, increasing bound water fraction and reducing free water mobility. These interactions evolve with time and temperature, making finishability a dynamic property rather than a fixed formulation outcome. Standard laboratory bleed tests rarely capture this temporal complexity.

Influence of Temperature, Fines, and Finishing Window

Environmental and compositional factors strongly modulate visual quality outcomes. Elevated temperatures accelerate cement hydration and polymer solvation, narrowing the optimal finishing window and increasing risk of surface tearing or crust formation. Conversely, cooler conditions may prolong setting while maintaining high surface viscosity, leading to over-finishing and localized discoloration.

High limestone filler, recycled fines, or clay-contaminated sands further complicate surface behavior by increasing specific surface area and polymer demand. In such systems, VMAs are often blamed for finishability loss, although the root cause lies in cumulative water immobilization and mistimed finishing operations.

Operational Manifestations and Diagnostic Indicators

In practice, finishability defects appear as surface drag, trowel marks, uneven gloss, or delayed closure of pores. Visual defects such as mottling or shadowing often correlate with premature finishing before bleed completion or with repeated finishing passes on an over-structured surface. Diagnostic indicators include low early bleed rates, elevated surface yield stress measured by vane or penetrometer methods, and sensitivity of surface behavior to minor water adjustments.

Differentiating polymer-driven effects from aggregate or process-related issues requires synchronized observation of fresh concrete rheology, setting progression, and finishing timing rather than isolated acceptance tests.

Data-Driven Evaluation and Corrective Strategies

Structured evaluation integrates mix design data, ambient conditions, and real-time observations of surface response. Industry 4.0-enabled approaches increasingly deploy temperature sensors, maturity tracking, and digital logs of finishing time relative to batching and placement. These datasets allow identification of recurring patterns that lead to finishability complaints and support targeted corrective actions such as minor VMA dose adjustments, revised finishing schedules, or fines management strategies.

For technology transfer, clear guidance on finishing windows and environmental sensitivity is essential to prevent unnecessary reformulation and to align admixture performance with site practices.

20.4 Role of Industry 4.0 in Application Troubleshooting

Industry 4.0 frameworks fundamentally transform how viscosity-modifying admixture related deviations are detected, interpreted, and corrected at the application stage. By integrating cyber-physical systems, sensorized equipment, and real-time analytics, VMA troubleshooting evolves from post-failure diagnosis to anticipatory process control. For VMAs, whose performance is highly sensitive to shear history, temperature, water chemistry, and admixture interactions, digitalization provides the resolution and continuity of data that conventional QA practices cannot deliver. This shift has direct implications for production stability, liability reduction, and accelerated technology transfer across plants and projects.

20.4.1 Plant-Level Data Capture and Mix Tracking

Automated capture of mix design parameters, real-time dosing logs, mixer energy profiles, and ambient conditions establishes a digital fingerprint for every concrete batch. When VMAs are involved, such traceability is critical because small deviations in water content, cement fineness, or superplasticizer sequence can disproportionately affect rheological stability. Plant-level Manufacturing Execution Systems linked to batching software enable rapid back-tracing of field complaints to specific production events. This data backbone supports root-cause isolation and reduces reliance on operator memory or manual reporting.

Plant-level digitalization establishes a continuous and high-resolution record of concrete production variables, forming a unique operational fingerprint for each batch. Automated acquisition of mix design parameters, water addition, VMA and PCE dosing events, mixer energy consumption, and ambient temperature or humidity eliminates gaps inherent to manual reporting. For VMA-containing systems, this granularity is essential because rheological stability is highly sensitive to marginal fluctuations in shear input, water balance, and admixture sequencing. Digital fingerprints therefore convert batch production from a nominal recipe-based activity into a data-rich process state description.

Influence of Digital Traceability on VMA Performance Diagnosis

VMAs amplify the impact of small upstream deviations that may remain invisible in conventional QA frameworks. Variations in cement fineness, fines moisture correction, or delayed admixture injection can alter polymer hydration and network formation kinetics. When such events are digitally logged, deviations in fresh-state behavior observed at the site can be rapidly correlated with specific production signatures. This traceability allows differentiation between formulation-related issues and execution-related disturbances, which is critical for credible troubleshooting and supplier feedback.

Integration of MES and Batching Systems

Manufacturing Execution Systems integrated with batching control software enable synchronized tracking of raw material lots, dosing precision, and process timing. For VMA applications, MES platforms provide time-stamped alignment of polymer addition with mixing intensity and residence time. This alignment is particularly important for associative polymers and polysaccharide-based VMAs that exhibit shear history dependence. By enabling rapid back-tracing of field complaints to discrete production events, MES integration shortens diagnostic cycles and reduces dependency on operator recollection.

Operational Benefits for Root-Cause Isolation

Digital mix tracking supports statistical process control and anomaly detection, allowing identification of non-conforming batches before dispatch or placement. When VMA-related instability occurs, historical data enable comparison against stable reference fingerprints to isolate causal variables. This approach minimizes speculative troubleshooting and prevents unnecessary formulation changes. Over time, accumulated datasets support the definition of robust operating windows for VMA dosage and process conditions, strengthening both internal quality systems and external technology transfer.

Table 20.4.1: Digital Data Streams Supporting VMA Troubleshooting at Plant Level

Data Stream	Source System	VMA-Relevant Insight	Practical Impact

Mix design and water balance	Batching software	Detection of micro water deviations	Improved viscosity consistency
Admixture dosing logs	Automated dosing units	Verification of VMA and PCE sequence	Reduced rheology drift
Mixer energy and torque	Sensorized mixers	Shear history reconstruction	Early instability identification
Environmental conditions	Plant sensors	Temperature-driven viscosity shifts	Seasonal robustness
Raw material lot tracking	MES	Correlation with performance changes	Faster root-cause isolation

Corrective and Preventive Implementation Guidance

Effective use of plant-level data capture requires predefined VMA-relevant key performance indicators rather than indiscriminate data accumulation. Establishing alarm thresholds for water balance deviation, dosing delay, or abnormal mixer energy trends enables proactive intervention. For technology transfer, documenting typical digital fingerprints for stable VMA performance and training plant personnel to interpret deviations significantly reduces recurring application issues. Integration with Industry 4.0 analytics platforms further enables predictive flagging of high-risk batches before placement.

20.4.2 Correlation of Rheology Signals with Field Performance

Direct rheological measurements remain impractical in most ready-mix environments, but Industry 4.0 enables the use of robust indirect indicators. Mixer torque, mixing energy curves, pump pressure fluctuations, and discharge time distributions can be statistically correlated with slump retention, segregation resistance, and placement behavior in the field. For VMA-containing systems, these proxies provide early warning of over-structuring or viscosity collapse. When continuously trended, such signals enable objective performance assessment and minimize subjective judgments by site personnel.

Direct rheological measurements remain impractical in most ready-mix and site environments due to equipment complexity and time constraints. Industry 4.0 architectures compensate by leveraging indirect but continuously available process signals that reflect the evolving rheological state of fresh concrete. For VMA-containing systems, where viscosity development and yield stress are highly sensitive to shear history and water balance, these proxy signals provide actionable insight that bridges plant operation and field behavior. When systematically captured and analyzed, they transform subjective placement observations into quantifiable performance metrics.

Rheology Proxies Derived from Production and Transport Equipment

Mixer torque and cumulative mixing energy curves directly reflect resistance to flow development within the batch. In VMA-modified concretes, rising torque at constant water content often indicates progressive polymer structuring or excessive adsorption on fines. Pump pressure profiles and pressure fluctuation frequency during conveying further capture changes in apparent viscosity and structural rebuild under shear. Discharge time distributions from trucks or hoppers provide an integrated response signal, sensitive to both over-structuring and viscosity collapse. These proxies collectively approximate rheological evolution without intrusive measurement.

Statistical Correlation with Field Placement Behavior

When aligned with time-stamped field observations, indirect rheology signals can be statistically correlated with slump retention, segregation resistance, surface finish quality, and ease of placement. For example, increasing pump pressure variability often precedes observable blockage risk or poor mold filling. Conversely, declining mixer energy combined with rapid discharge may indicate insufficient yield stress development, leading to bleeding or settlement. Such correlations enable objective interpretation of field complaints and reduce dependence on operator perception or anecdotal feedback.

Early Warning Indicators for VMA Instability

VMAs exhibit non-linear responses to combined variations in shear, temperature, and admixture interaction. Continuous trending of rheology proxies allows identification of characteristic signatures associated with over-structuring, delayed viscosity build-up, or structural breakdown during transport. Early warning thresholds can be defined based on historical stable performance envelopes. This proactive detection is particularly valuable for applications sensitive to placement windows, such as SCC, pumped concrete, and mixes with high SCM variability.

Table 20.4.2: Indirect Rheology Signals and Corresponding Field Performance Indicators

Proxy Signal	Measurement Source	Indicative Rheological Condition	Field Performance Impact
Mixer torque trend	Sensorized mixer	Progressive viscosity increase	Reduced workability, pump resistance
Mixing energy curve	Batching control system	Polymer structuring kinetics	Slump retention behavior
Pump pressure level	Pump sensors	High apparent viscosity	Blockage risk, slow placement
Pressure fluctuation frequency	Pump sensors	Structural instability	Inconsistent flow, segregation
Discharge time distribution	Truck or hopper logs	Low yield stress or collapse	Bleeding, settlement

Corrective and Preventive Application Strategies

Effective implementation requires calibration of proxy signals against validated field performance benchmarks during commissioning. Once established, digital dashboards can flag deviations in real time, enabling adjustment of VMA dosage, water balance, or mixing protocol before placement. For technology transfer, sharing correlation models between plant operators and admixture suppliers accelerates refinement of operating windows and polymer selection. Integration with predictive analytics further enhances robustness by forecasting instability trends based on evolving signal patterns.

20.4.3 Predictive Analytics for Application Risk Identification

Machine learning models trained on historical plant and site data can identify multivariate patterns preceding VMA-related failures such as bleeding, blockage, or poor finishability. By incorporating variables including cement lot variability, temperature profiles, admixture age, and shear intensity, predictive analytics enable risk-based decision making before concrete is dispatched. VMAs benefit disproportionately from these tools due to their non-linear response to interacting parameters. Predictive dashboards therefore act as digital safeguards, supporting proactive adjustments to dosing or process windows.

Predictive analytics extends Industry 4.0 frameworks beyond descriptive monitoring toward anticipatory control of VMA performance at the application stage. Machine learning models trained on historical plant and site datasets can identify complex, multivariate patterns that precede rheology-related failures. For VMA-modified concretes, where performance emerges from non-linear interactions between polymer chemistry, fines characteristics, shear history, and temperature, these tools provide a level of foresight unattainable through rule-based quality control. Risk identification therefore shifts from reactive troubleshooting to probabilistic decision support prior to dispatch.

Multivariate Inputs Relevant to VMA Sensitivity

Effective predictive models integrate variables that directly influence polymer structuring and stability. Cement lot-to-lot variability in fineness and alkali content alters adsorption kinetics, while SCM variability modifies surface area demand. Temperature profiles during mixing and transport affect polymer hydration and viscosity build-up rates. Admixture age and storage history influence active concentration, and shear intensity derived from mixer energy or pump profiles captures structural breakdown and rebuild dynamics. The combined influence of these parameters explains why univariate thresholds are insufficient for VMA risk management.

Pattern Recognition and Failure Precursor Identification

Supervised and semi-supervised learning approaches can be trained to recognize precursor signatures associated with bleeding, segregation, pump blockage, or poor finishability. Subtle combinations of rising mixer energy, delayed discharge, and elevated temperature may indicate impending over-structuring, even when individual parameters remain within nominal limits. Conversely, declining energy coupled with high SCM surface area may precede viscosity collapse. By quantifying these patterns as risk scores, predictive systems provide actionable guidance rather than binary pass–fail judgments.

Predictive Dashboards as Operational Safeguards

Predictive outputs are most effective when translated into intuitive dashboards linked to plant operations. Risk indices can be visualized alongside recommended corrective actions such as VMA dosage adjustment, water fine-tuning, or modified mixing protocols. For VMA applications, these dashboards function as digital safeguards, preventing high-risk batches from being dispatched without intervention. Over time, continuous model retraining using new production and field feedback improves robustness and adapts to evolving raw material supply conditions.

Table 20.4.3 Predictive Analytics Elements for VMA Application Risk Control

Predictive Element	Data Inputs	Identified Risk	Preventive Action
Cement and SCM variability model	Fineness, replacement level, lot ID	Segregation, bleeding	Dosage window adjustment
Thermal response model	Mix and ambient temperature	Over-structuring	Mixing energy or water correction
Shear history model	Mixer energy, pump pressure	Blockage risk	Protocol modification
Admixture condition model	Age, storage temperature	Viscosity drift	Concentration verification
Composite risk index	Integrated multivariate data	Application failure	Dispatch hold or corrective dosing

Corrective and Preventive Deployment Strategy

Successful deployment requires collaboration between plant operators, data scientists, and admixture suppliers to ensure chemical relevance of model inputs. Predictive thresholds should be validated against real field outcomes rather than laboratory surrogates alone. For technology transfer, documenting model logic, assumptions, and recommended interventions builds trust and facilitates cross-plant standardization. Integration with MES and cloud-based analytics platforms ensures scalability and continuous improvement across production networks.

20.4.4 Digital Feedback Loops between Site and Supplier

Closed-loop digital communication between construction sites, batching plants, and admixture suppliers accelerates learning cycles and formulation optimization. Structured field data, including placement feedback, sensor trends, and curing observations, can be securely shared to support rapid root-cause analysis. For VMA suppliers, this real-world data stream enables faster refinement of polymer architecture, molecular weight distribution, or solution concentration. Such feedback loops institutionalize technology transfer and embed continuous improvement into the commercial relationship.

Digital feedback loops create a continuous and structured exchange of performance data between construction sites, batching plants, and admixture suppliers. Within an Industry 4.0 framework, these loops shorten learning cycles by converting field experience into actionable formulation and application knowledge. For VMA systems, where laboratory optimization often fails to capture full-scale variability, real-time feedback enables rapid reconciliation between designed polymer behavior and actual site performance. This connectivity strengthens accountability and reduces the lag between problem detection and corrective action.

Structured Field Data as a Foundation for Root-Cause Analysis

Effective feedback loops rely on standardized and structured field data rather than informal verbal reporting. Placement behavior, finishing response, pumpability observations, and curing surface quality can be digitally logged and time-stamped. When combined with sensor-derived trends such as pump pressure or discharge time, these datasets provide a coherent narrative of performance evolution. For VMA troubleshooting, such structure allows suppliers to distinguish between formulation limitations and application-driven disturbances with significantly higher confidence.

Supplier-Level Interpretation and Polymer Optimization

Access to real-world operational data enables VMA suppliers to refine polymer architecture based on actual stress conditions. Insights into shear exposure, temperature gradients, and adsorption environments inform adjustments to molecular weight distribution, side-chain density, or solution concentration. These refinements are particularly relevant for associative polymers and polysaccharide-based VMAs, whose performance envelopes are narrow and context-dependent. Digital feedback therefore transforms formulation development from episodic troubleshooting into continuous, evidence-based optimization.

Institutionalizing Technology Transfer and Continuous Improvement

When digital feedback loops are formalized through secure platforms and agreed data protocols, they become an integral part of the commercial relationship. Lessons learned from one project can be codified and transferred across plants, regions, or mix families. For end users, this institutional memory reduces repeated errors and accelerates commissioning of new materials. For suppliers, it enables scalable improvement without reliance on frequent site visits, aligning operational efficiency with consistent VMA performance.

Table 20.4.4 Digital Feedback Loop Elements Supporting VMA Optimization

Feedback Element	Data Contributor	Technical Insight Generated	Long-Term Benefit
Digital placement reports	Construction site	Workability and finishability trends	Faster issue diagnosis
Sensor data sharing	Plant and site equipment	Shear and viscosity evolution	Targeted formulation tuning
Centralized data platforms	Producer and supplier	Cross-project pattern recognition	Robust operating windows

Secure collaboration portals	Supplier–customer interface	Accelerated root-cause analysis	Reduced downtime
Knowledge codification	Supplier technical teams	Standardized adjustment protocols	Scalable technology transfer

Corrective and Preventive Collaboration Framework

For effective implementation, clear data ownership, access rights, and confidentiality agreements must be established. Feedback loops should focus on predefined VMA-relevant indicators rather than exhaustive data transfer. Corrective actions derived from shared insights should be documented and validated at both plant and site levels. Over time, these practices embed continuous improvement into daily operations and elevate VMA deployment from product supply to integrated technical partnership.

20.5 Technology Transfer and Customer Interface Implications

Effective troubleshooting extends beyond technical diagnosis to how knowledge is transferred to customers. Clear definition of operating limits, supported by data and digital tools, reduces misuse and misinterpretation of VMAs. The customer interface becomes a critical control point for performance reliability.

20.5.1 Defining Application Operating Windows for VMAs

Application operating windows convert intrinsic polymer performance into actionable limits for dosage, temperature, mixing intensity, and transport duration. For viscosity-modifying admixtures, these windows define the conditions under which rheological control remains predictable and stable. Clear boundary definition reduces the likelihood of off-spec application caused by environmental fluctuation, raw material variability, or process deviations. Operating windows therefore function as practical risk-management tools rather than theoretical formulation descriptors.

Translation of Polymer Rheology into Operational Limits

VMAs exhibit non-linear rheological responses governed by molecular weight distribution, functional group density, and adsorption behavior on cementitious surfaces. Translating these characteristics into operational limits requires linking laboratory rheology outcomes with plant and site process parameters. Dosage ranges must account for temperature-dependent viscosity development and shear-induced structural breakdown or rebuild. Mixing limits define the minimum energy required for polymer activation and the maximum shear beyond which performance degradation occurs.

Influence of Environmental and Material Variability

Temperature, cement fineness, SCM replacement level, and fines moisture content shift the effective operating window of VMAs. Elevated temperatures accelerate polymer hydration and structuring, narrowing upper dosage limits. High-surface-area SCMs increase adsorption demand, shifting minimum effective dosage upward. Without explicit recognition of these influences, fixed recommendations become unreliable. Defining operating windows as multidimensional envelopes rather than single-point targets improves robustness under variable field conditions.

Field Data Integration and Periodic Window Recalibration

Operating windows should be treated as living frameworks that evolve with accumulated production and placement data. Industry 4.0-enabled data capture allows correlation of stable and unstable performance with specific operating conditions. Periodic recalibration using validated field outcomes ensures continued relevance as raw material sources, process equipment, or climatic conditions change. This data-driven updating process reinforces credibility and supports long-term technology transfer.

Table 20.5.1: Key Parameters Defining VMA Application Operating Windows

Parameter	Lower Boundary Driver	Upper Boundary Driver	Performance Risk Outside Window

VMA dosage	Insufficient yield stress	Over-structuring	Bleeding or pump blockage
Temperature	Delayed viscosity build-up	Accelerated structuring	Slump loss or instability
Mixing energy	Incomplete polymer activation	Structural degradation	Inconsistent rheology
Transport time	Insufficient structural rebuild	Excessive thixotropic recovery	Placement difficulty
SCM content	High adsorption demand	Reduced free water	Segregation or finishability loss

Corrective and Preventive Guidance for Technology Transfer

For effective technology transfer, operating windows must be communicated using clear visual formats and practical decision logic rather than abstract laboratory metrics. Training materials should emphasize how and why boundaries shift under changing conditions. Embedding operating windows into digital tools and plant dashboards enables real-time compliance monitoring and proactive adjustment. Such integration aligns operating guidance with Industry 4.0 practices and significantly reduces misuse-related VMA failures.

20.5.2 Standardized Troubleshooting Protocols for Technical Service

Standardized troubleshooting protocols provide a common technical language for diagnosing VMA-related issues across regions, plants, and project types. Given the sensitivity of viscosity-modifying admixtures to interacting variables, inconsistent diagnostic approaches often lead to conflicting conclusions and ineffective corrective actions. Structured protocols align technical service responses with polymer science and field realities, ensuring repeatable and defensible outcomes. They also protect organizational knowledge from fragmentation and personnel dependency.

Structured Checklists for Systematic Data Collection

Effective protocols begin with structured checklists that enforce disciplined data capture at both plant and site levels. Critical variables include mix design parameters, water corrections, admixture sequence, temperature history, SCM characteristics, and observed placement behavior. For VMA systems, omission of even a single parameter can distort root-cause analysis due to non-linear response behavior. Checklists therefore act as safeguards against incomplete diagnosis and premature formulation changes.

Decision Trees Linking Symptoms to Root Causes

Decision trees translate collected data into logical diagnostic pathways. Observable symptoms such as bleeding, pump resistance, or poor finishability are linked to probable causes including adsorption imbalance, shear overexposure, or thermal acceleration. By guiding technical personnel through predefined branches, decision trees reduce subjective interpretation and improve response speed. For technology transfer, these tools also serve as training instruments, reinforcing consistent analytical thinking across service teams.

Scalability and Consistency in Technical Service Operations

Standardized protocols enable scalable technical service by ensuring uniform diagnostic depth regardless of location or experience level. Digitalization of protocols through mobile forms or MES-linked platforms further enhances consistency and data quality. Aggregated troubleshooting cases can be analyzed for recurring patterns, informing formulation refinement and application guidance. This feedback reinforces continuous improvement while maintaining consistent customer experience.

Table 20.5.2: Elements of Standardized VMA Troubleshooting Protocols

Protocol Element	Purpose	VMA-Specific Value	Operational Benefit
Structured checklists	Complete data capture	Avoids missed sensitivity factors	Faster diagnosis
Defined data requirements	Comparable cases	Enables pattern recognition	Improved solution quality

Decision trees	Guided root-cause analysis	Reduces subjective bias	Consistent outcomes
Digital workflow tools	Standard execution	Higher data integrity	Scalable service
Case databases	Knowledge retention	Evidence-based guidance	Continuous improvement

Corrective and Preventive Implementation Considerations

For effective deployment, protocols must be validated against real case histories and periodically reviewed. Overly complex frameworks reduce adoption, while oversimplified ones fail to capture VMA sensitivity. Training programs should focus on disciplined protocol use rather than individual intuition. Integration with Industry 4.0 data platforms ensures that troubleshooting outputs feed directly into product development and operating window refinement.

20.5.3 Digital Tools for Customer Guidance and Training

Digital tools transform customer guidance from static instruction into interactive, context-aware decision support. For viscosity-modifying admixtures, whose performance depends on tightly coupled chemical and process variables, digital platforms provide a practical mechanism to translate polymer science into operational understanding. Interactive tools reduce misuse driven by oversimplified rules and enable customers to respond correctly to changing conditions. Their adoption aligns naturally with Industry 4.0 strategies already present in modern concrete production environments.

Interactive Training Modules and Knowledge Transfer

Digital training modules allow structured learning based on real application scenarios rather than abstract theory. Visualizations of viscosity development, yield stress evolution, and shear sensitivity improve comprehension of VMA behavior. Modular content enables differentiation between operator-level guidance and advanced technical training. For technology transfer, such platforms ensure consistent messaging across regions and reduce dependency on in-person training.

Dosage Calculators and Context-Aware Decision Support

Dosage calculators embedded within digital platforms translate operating windows into actionable recommendations. By incorporating variables such as temperature, SCM content, target slump retention, and mixing energy, these tools provide context-aware dosage guidance. For VMAs, this approach minimizes under- or over-dosing associated with static dosage tables. Embedded safeguards can prevent recommendations outside validated operating envelopes, reinforcing correct application discipline.

Scenario Simulation and Risk Awareness

Scenario simulation tools allow users to explore the impact of parameter changes on fresh concrete behavior before implementation. Simulations illustrating the effects of temperature rise, delayed placement, or SCM substitution increase user awareness of non-linear VMA responses. This proactive learning reduces trial-and-error adjustments in the field. For suppliers, scenario data also reveal recurring user challenges, guiding refinement of guidance and training content.

Table 20.5.3: Digital Guidance Tools and Their Contribution to VMA Reliability

Digital Tool	Primary Function	VMA-Specific Benefit	Customer Impact
Interactive training modules	Conceptual learning	Improved understanding of rheology	Reduced misuse
Dosage calculators	Operational guidance	Context-aware dosing	Consistent performance
Scenario simulations	Risk visualization	Awareness of non-linear behavior	Fewer field failures
Digital dashboards	Real-time feedback	Compliance with operating windows	Faster corrective action
Knowledge portals	Centralized access	Standardized guidance	Scalable training

Implementation and Adoption Considerations

Successful deployment requires intuitive user interfaces and alignment with existing plant workflows. Tools must be validated against field data to maintain credibility and avoid overreliance on modeled outputs. Regular updates reflecting new formulations, operating windows, and field learnings are essential. When integrated with Industry 4.0 data systems, digital guidance platforms become living components of the customer interface, supporting long-term VMA performance reliability.

20.5.4 Capturing Field Learnings for Continuous Improvement

Systematic capture of field learnings creates a feedback loop that connects real-world performance to product development, formulation optimization, and customer guidance. For viscosity-modifying admixtures, whose behavior is highly context-dependent, undocumented field observations limit the ability to refine operating windows, troubleshoot recurrent issues, or anticipate performance under variable conditions. Discipline in documentation and integration of data into centralized platforms ensures that empirical insights inform continuous improvement rather than remaining anecdotal.

Structured Documentation of Field Observations

Capturing field experiences requires standardized formats for logging placement behavior, workability trends, slump retention, bleeding, segregation, and finish quality. Sensor-derived process data, such as mixer torque, pump pressure, discharge timing, and temperature, should be correlated with these observations. For VMA applications, structured capture allows differentiation between formulation limitations, process deviations, and environmental effects. Consistency in documentation ensures comparability across projects and locations.

Integration into Formulation and Qualification Processes

Field learnings directly inform refinement of polymer chemistry, dosage ranges, and qualification protocols. Observed deviations under specific SCM compositions, mixing regimes, or climatic conditions can guide adjustments to molecular weight distribution, side-chain architecture, or solution concentration. Incorporating these insights into laboratory testing protocols and qualification methods ensures that formulations remain robust under evolving operational **scenarios, reducing repeat failures and enhancing product reliability.**

Feedback into Guidance and Technology Transfer

Lessons learned from the field should be systematically fed back into customer guidance, digital tools, and training platforms. Updated operating windows, dosage calculators, and scenario simulations reflect the latest empirical evidence, enabling more accurate decision-making at the plant and site levels. For technology transfer, this approach institutionalizes continuous improvement, creating a living knowledge base that strengthens customer competence and fosters trust in VMA performance.

Table 20.5.4: Field Learning Capture and Continuous Improvement Workflow

Step	**Data Source**	**Integration Target**	**VMA Performance Benefit**
Field observation logging	Placement notes, slump tests	Centralized database	Identify recurring issues
Sensor data correlation	Mixer torque, pump pressure	Analytical dashboards	Quantify rheology deviations
Case analysis	Aggregated project data	Formulation refinement	Improve polymer robustness
Update guidance	Operating windows, calculators	Customer-facing tools	Reduce misuse and variability
Knowledge retention	Technical manuals, digital portals	Training and onboarding	Scalable technology transfer

Implementation Considerations

Successful capture of field learnings requires a culture of disciplined documentation and digital integration. Clear responsibility for data collection, secure data sharing, and defined update protocols are critical. Integration with MES systems, cloud-based analytics, and digital dashboards ensures that insights flow efficiently from field to R&D and back to the customer interface. Over time, this structured feedback loop establishes evidence-based continuous improvement, strengthening both VMA performance reliability and customer confidence.

21. Industry 4.0 Roadmap for VM Admixtures

The adoption of Industry 4.0 principles in the context of viscosity-modifying admixtures (VMAs) establishes a structured framework for proactive process control, data-driven formulation optimization, and enhanced technology transfer. By integrating digitalization, connectivity, and advanced analytics across the concrete value chain, VMA performance can be continuously monitored, predicted, and adjusted. This roadmap defines a phased approach encompassing plant-level instrumentation, predictive modeling, customer interface digitalization, and closed-loop continuous improvement.

21.1 Phase 1: Digital Instrumentation and Data Capture

The first phase of Industry 4.0 implementation focuses on capturing high-resolution process and environmental data to create a digital foundation for VMA monitoring and control. Viscosity-modifying admixtures are highly sensitive to subtle variations in water content, SCM composition, temperature, and shear exposure. Deploying sensors, automated dosing systems, and MES-linked batching software enables continuous recording of these critical variables, providing a reliable operational fingerprint for each batch.

Critical Data Streams for VMA Monitoring

Key parameters for digital capture include: mixer torque and cumulative mixing energy, real-time admixture dosing logs, ambient temperature and humidity, SCM lot and fineness information, and pump or transport pressure profiles. For VMAs, which respond non-linearly to shear history and adsorption dynamics, this detailed monitoring allows detection of deviations that might not manifest immediately in field performance but could compromise stability during placement.

Implementation Considerations

Sensor Integration: Ensure that torque, flow, and pressure sensors are calibrated for the rheological range of VMA-modified concretes.

MES Alignment: Batch and dosing data should be synchronized within the MES for traceability and historical comparison.

Data Resolution: Sampling frequency must be sufficient to capture short-term fluctuations that influence polymer network development.

Benefits for Root-Cause Analysis and Technology Transfer

Digital instrumentation enables rapid back-tracing of field complaints to specific production events, reducing reliance on operator memory or anecdotal reporting. By establishing a baseline of stable performance, deviations can be quantified and correlated with upstream variables, allowing suppliers to provide targeted recommendations. This approach strengthens technology transfer by converting empirical observation into structured, actionable knowledge.

Table 21.1: Key Digital Instrumentation Components for VMA Monitoring

Component	Measurement Target	VMA-Relevant Insight	Operational Benefit
Mixer torque & energy sensors	Shear exposure, viscosity development	Detect over-structuring or collapse	Early intervention and adjustment
Automated dosing units	VMA and PCE sequence	Verify polymer delivery and timing	Reduce dosage errors
Environmental sensors	Temperature, humidity	Assess thermal influence on viscosity	Adjust operating windows proactively
MES-linked batch tracking	Mix design, raw material lot	Trace anomalies to specific events	Faster root-cause isolation
Pump/transport pressure sensors	Flow resistance	Identify potential blockage risk	Optimize placement strategy

Corrective and Preventive Guidance

To maximize benefits, plant personnel must be trained in interpreting sensor outputs and integrating them with field observations. Alarm thresholds for critical deviations should be defined based on historical stable performance. Periodic review of instrumentation accuracy and data integrity ensures reliability of the captured information. Embedding this digital foundation sets the stage for advanced analytics, predictive modeling, and closed-loop feedback in subsequent Industry 4.0 phases.

21.2 Phase 2: Data Integration and Visualization

The second phase consolidates heterogeneous data streams into a unified digital architecture, transforming raw measurements into interpretable and actionable information for VMA formulation, production, and application control. Data integration bridges plant instrumentation, laboratory characterization, and field performance, enabling a coherent view of polymer–cement interactions across time and location. For VMAs, whose efficiency depends on subtle balances between molecular architecture, adsorption kinetics, and shear history, integrated datasets are essential to move from descriptive monitoring to diagnostic understanding.

A centralized data lake or industrial cloud platform should aggregate MES outputs, sensor feeds, laboratory rheology results, and quality control indicators. Standardized data models and time stamping allow correlation of mixer torque evolution with dosage sequence, SCM variability, and environmental conditions. Applying Industry 4.0 connectivity principles ensures interoperability between batching plants, admixture production sites, and technical service laboratories, facilitating technology transfer across organizational boundaries.

Visualization layers translate complex datasets into intuitive dashboards tailored to different stakeholders. Process engineers benefit from trend maps linking cumulative mixing energy to apparent viscosity development, while formulation chemists can visualize correlations between SCM fineness distributions and VMA demand. For customers, simplified interfaces focusing on robustness indicators and placement risk indices enhance transparency without exposing proprietary formulation details.

Key Visualization and Integration Elements for VMAs

Effective industrial deployment of viscosity-modifying admixtures increasingly depends on structured visualization and integration of heterogeneous data streams. When polymer–particle interactions are translated into actionable digital indicators, operational decisions can shift from corrective intervention to predictive control. The combination of process sensors, laboratory characterization, and field feedback enables VMAs to function as adaptive rheology regulators within Industry 4.0 concrete production environments.

Table 21.2: Key Integration Elements for VMAs

Element	Integrated Data Sources	VMA-Specific Insight	Value for Operations
Rheology dashboards	Torque sensors, lab rheometer data	Consistency window and structural build-up	Rapid validation of mix stability
Dosage-sequence maps	Automated dosing logs, MES	Interaction between VMA and PCE timing	Reduction of incompatibility risks
Environmental overlays	Temperature, humidity, curing profiles	Thermal sensitivity of polymer networks	Proactive adjustment of dosage
SCM variability panels	Supplier data, fineness, LOI	Adsorption demand shifts	Early reformulation guidance
Field feedback loops	Pump pressure, placement logs	Correlation with flow resistance	Improved placement reliability

Data Processing and Analytics Considerations

Raw data must be filtered, normalized, and statistically processed to remove noise and enable meaningful comparison across plants and projects. Moving averages, control charts, and multivariate correlation matrices are recommended to distinguish true formulation effects from operational variability. Where possible, automated data validation routines should flag outliers linked to sensor drift or manual intervention, preserving data integrity for downstream analytics.

Impact on Technology Transfer and Decision Support

Integrated visualization accelerates knowledge transfer by converting complex polymer behavior into shared digital representations. Best-practice envelopes derived from historical datasets can be exported across regions, supporting consistent VMA performance despite differences in raw materials or climate. This digital transparency reduces dependency on tacit knowledge and supports scalable deployment of advanced VMA technologies within an Industry 4.0 framework.

21.3 Phase 3: Predictive Analytics and Risk Management

The third phase advances Industry 4.0 adoption from descriptive and diagnostic insight to predictive control of VMA performance and associated operational risks. By applying advanced analytics and machine learning models to integrated historical and real-time datasets, potential instability, segregation risk, or pumpability loss can be anticipated before manifesting in production or placement. For viscosity-modifying admixtures, predictive capability is particularly valuable due to their nonlinear rheological response to composition, shear history, and environmental variability.

Predictive models should be trained on processed datasets derived from Phase 1 and Phase 2 infrastructures, incorporating mixer torque profiles, cumulative mixing energy, SCM mineralogy and fineness, temperature trajectories, and admixture sequencing data. Multivariate regression, random forest, and gradient boosting algorithms have demonstrated robustness in capturing nonlinear interactions relevant to polymer-cement systems. Model outputs should be expressed as probability-based risk indices rather than binary pass or fail criteria, aligning with industrial decision-making practices.

Risk Taxonomy for VMA-Modified Concretes

A structured risk framework supports systematic interpretation of predictive outputs. Typical risk categories include rheological instability, delayed structural build-up, excessive thixotropy, pump line pressure escalation, and segregation during transport or placement. Each risk category should be associated with quantified leading indicators derived from model sensitivity analysis, enabling targeted mitigation actions rather than generic dosage adjustments.

Table 21.3: Key Risk Categories for VMA-Modified Concretes

Risk Category	Primary Predictive Inputs	Processed Indicator	Preventive Action

Loss of viscosity stability	Torque decay rate, water variability	Stability risk index (%)	Adjust VMA molecular weight or dosage
Over-structuring	Cumulative mixing energy, temperature	Thixotropy growth coefficient	Modify mixing protocol or sequence
Pumpability failure	Pressure gradient trends	Pump risk probability	Rebalance VMA/PCE ratio
Segregation tendency	SCM fineness variance, density contrast	Segregation score	Reformulate with hybrid VMA systems
Placement variability	Environmental fluctuations	Robustness envelope deviation	Narrow operating window digitally

Data Processing and Model Governance

All predictive analytics must be supported by rigorous data preprocessing, including normalization across plants, dimensionality reduction, and cross-validation to prevent overfitting. Model performance should be continuously monitored using rolling prediction error metrics, with retraining triggered when raw material sources or climatic regimes shift. From an Industry 4.0 perspective, model governance is critical to ensure traceability, reproducibility, and regulatory confidence in digitally assisted formulation decisions.

Integration into Operational Risk Management

Predictive outputs should be embedded directly into MES or decision-support dashboards, generating early-warning alerts when risk thresholds are exceeded. These alerts enable preventive intervention at the batching or formulation stage rather than reactive correction on site. Over time, accumulated prediction–outcome pairs refine the risk models, creating a closed-loop learning system that progressively reduces uncertainty in VMA performance across diverse applications.

Value for Technology Transfer and Industrial Deployment

Predictive analytics convert empirical formulation know-how into transferable digital assets. By codifying the relationships between polymer architecture, processing conditions, and field behavior, suppliers can support customers with proactive guidance tailored to local materials and equipment. This approach accelerates adoption of advanced VMA technologies, reduces trial-and-error cycles, and aligns admixture development with the principles of resilient, data-driven construction manufacturing.

21.4 Phase 4: Digital Feedback Loops and Supplier Collaboration

The fourth phase completes the Industry 4.0 roadmap by establishing bidirectional digital feedback loops between concrete producers, admixture suppliers, and upstream raw material providers. For viscosity-modifying admixtures, continuous feedback is essential to manage variability in cement chemistry, SCM availability, and site-specific placement conditions. Digital collaboration transforms the supplier–customer relationship from reactive troubleshooting to proactive co-optimization of formulations and processes.

Closed-loop feedback systems should integrate field performance data, including pump pressure evolution, placement observations, surface finish quality, and early-age stability indicators, with plant-level predictive outputs. These datasets must be processed into normalized performance descriptors that can be securely shared with suppliers through cloud-based platforms or digital twin environments. Applying Industry 4.0 interoperability standards ensures that data exchange remains scalable while protecting proprietary formulation details.

Collaborative Digital Interfaces and Data Exchange

Supplier collaboration platforms should provide tiered access to processed indicators rather than raw datasets, supporting trust and data governance. For VMAs, key shared metrics include robustness indices, deviation from validated operating envelopes, and predicted risk trends. Integration with supplier formulation databases allows rapid assessment of whether observed deviations are best addressed through dosage adjustment, molecular architecture modification, or complementary additive selection.

Table 21.4: Key Collaboration Layers and Data Exchange

Collaboration Layer	Shared Digital Output	VMA-Relevant Insight	Joint Value Creation
Producer to supplier	Performance deviation index	Sensitivity to local materials	Faster root-cause alignment
Supplier to producer	Formulation response scenarios	Polymer adjustment pathways	Reduced trial cycles
SCM supplier interface	Variability alerts	Adsorption demand shifts	Anticipatory reformulation
Technical service loop	Field outcome correlation	Validation of predictions	Continuous model refinement
Management dashboards	KPI trends	Stability and risk reduction	Strategic partnership alignment

Data Processing and Security Considerations

Before external sharing, all data must be anonymized, aggregated, and statistically processed to remove plant-specific identifiers while preserving functional relevance. Control charts and confidence intervals should accompany shared indicators to communicate uncertainty transparently. Cybersecurity and access control are integral to this phase, aligning with Industry 4.0 reference architectures and ensuring compliance with data protection regulations in multinational supply chains.

Impact on Continuous Improvement and Risk Reduction

Digital feedback loops enable suppliers to detect emerging incompatibilities early, such as shifts in cement alkali content or SCM fineness that alter VMA adsorption behavior. By correlating these signals with predictive models, corrective actions can be deployed upstream, often before the producer experiences a field issue. Over time, this collaborative learning reduces variability, enhances robustness, and supports standardized deployment of advanced VMA technologies across regions.

Role in Technology Transfer and Industrial Scaling

Digitally enabled collaboration converts localized experience into scalable knowledge assets. Best-performing formulation–process combinations can be codified and transferred across plants and markets with minimal loss of fidelity. This approach strengthens long-term partnerships, shortens innovation cycles, and aligns admixture development with the principles of resilient, data-driven construction manufacturing under Industry 4.0.

21.5 Phase 5: Customer Interface Digitalization and Continuous Improvement

The fifth phase extends Industry 4.0 implementation to the customer interface, embedding continuous improvement mechanisms directly into daily operational decision-making. For viscosity-modifying admixtures, digitalized customer interaction ensures that formulation robustness, placement reliability, and performance consistency are maintained across diverse projects and geographical regions. By transforming customer-facing tools into data-enabled platforms, feedback from application conditions becomes a structured input for ongoing optimization.

Digital customer interfaces should consolidate project-specific mix data, predicted performance envelopes, and validated operating windows into accessible dashboards. These platforms translate complex rheological and polymer-interaction models into practical guidance, such as recommended dosage ranges, sequencing constraints, and sensitivity alerts. Processing customer-generated data through standardized normalization routines allows aggregation across projects, supporting statistically meaningful continuous improvement cycles.

Digital Touchpoints for VMA Users

Well-designed interfaces provide differentiated access levels for plant operators, site engineers, and technical managers. For VMAs, critical touchpoints include mobile-enabled placement risk indicators, pumpability trend visualization, and alerts linked to environmental deviations. Integration with predictive analytics engines enables real-time adjustment recommendations while preserving supplier intellectual property.

Table 21.5: Key Interface Elements for Real-time Adjustment Recommendations

Interface Element	Processed Customer Input	Delivered Insight	Improvement Mechanism
Operator dashboard	Batch and placement logs	Real-time robustness status	Immediate corrective action
Site engineer app	Pump pressure and flow data	Placement risk prediction	Reduced on-site variability
Technical service portal	Aggregated project data	Cross-project trend analysis	Targeted reformulation
Training modules	Performance deviations	Root-cause scenarios	Knowledge standardization
Management KPI view	Stability and complaint rates	ROI of digital adoption	Strategic refinement

Continuous Improvement Through Closed Learning Loops

All customer-facing data streams must be statistically processed using control charts, trend analysis, and variance decomposition to distinguish systemic issues from isolated events. Feedback loops linking customer observations to supplier predictive models create a learning system in which each project refines future recommendations. Over time, this reduces uncertainty in VMA dosage selection and narrows operating windows without sacrificing robustness.

Alignment with Industry 4.0 and Technology Transfer Objectives

Customer interface digitalization completes the transition from product-centric supply to solution-oriented collaboration. By embedding data-driven guidance into customer workflows, advanced VMA technologies become easier to deploy consistently, even in regions with limited technical support. This approach accelerates technology transfer, supports industrial scaling, and aligns admixture application with the principles of smart, connected construction manufacturing.

21.6 Key Takeaways

The structured Industry 4.0 roadmap for viscosity-modifying admixtures establishes a progressive transition from reactive quality control to predictive, collaborative, and continuously improving deployment models. Each phase builds on a digitally secured foundation, ensuring that polymer-sensitive phenomena such as adsorption kinetics, shear-dependent structuring, and temperature-driven viscosity shifts are no longer managed empirically but through quantified, traceable signals. The cumulative effect is a measurable reduction in variability and an increase in formulation robustness under industrial conditions.

A critical insight across all phases is that data alone does not generate value unless it is processed, contextualized, and translated into actionable decision support. High-frequency sensor signals, predictive risk indices, and customer feedback must be statistically filtered, normalized, and correlated to isolate true VMA performance drivers from background operational noise. This disciplined data processing approach enables consistent replication of best-performing solutions across plants, projects, and regions.

From a technology transfer perspective, the roadmap converts tacit know-how into codified digital assets. Predictive models, validated operating envelopes, and customer-facing guidance tools reduce dependency on individual expertise and shorten learning curves for new materials or applications. This aligns VMA deployment with Industry 4.0 manufacturing principles, supporting scalable innovation while maintaining formulation integrity and intellectual property protection.

Strategically, the integration of suppliers, producers, and customers through digital feedback loops reshapes the admixture value chain into a collaborative ecosystem. Early detection of incompatibilities, proactive reformulation, and evidence-based customer guidance collectively lower technical risk and improve lifecycle performance of VMA-modified concretes. The roadmap therefore functions not only as a digitalization pathway but as a risk management and competitiveness framework for advanced admixture technologies.

Table 21.6: Industry 4.0 Roadmap Phases for VMA Deployment

Phase	Core Focus	VMA-Relevant Actions	Key Benefit
Phase 1	Digital instrumentation	Sensor installation, MES integration, batch fingerprinting	High-resolution monitoring and traceability

Phase 2	Data integration and visualization	Dashboards, KPI tracking, trend analysis	Objective interpretation of field and plant behavior
Phase 3	Predictive analytics	Machine learning risk models, multivariate correlation	Early detection and risk-based intervention
Phase 4	Digital feedback loops	Supplier–site collaboration, structured field data	Accelerated troubleshooting and formulation refinement
Phase 5	Customer interface and continuous improvement	Digital training, scenario simulation, feedback integration	Evidence-based guidance, scalable technology transfer

22. Technology Transfer and Industrial Implementation

Technology transfer for viscosity-modifying admixtures represents the controlled migration of polymer science from laboratory-defined performance into reproducible, economically viable industrial reality. Successful implementation requires alignment between formulation chemistry, manufacturing capability, digital process control, and application practice. Within an Industry 4.0 framework, technology transfer is no longer a linear handover but a continuously validated system supported by data integrity, predictive analytics, and structured feedback across the value chain.

Industrial implementation must therefore be designed as a risk-managed process, where formulation intent, process capability, and field variability are explicitly connected through digital workflows. For VMAs, whose performance is governed by adsorption kinetics, polymer conformation, and shear history, even minor deviations can translate into disproportionate rheological effects. A robust transfer strategy minimizes these deviations through standardized data-driven controls rather than post hoc corrective action.

22.1 From Formulation Science to Industrial Practice

Translating VMA formulation science into industrial practice begins with deconstructing laboratory performance into controllable industrial variables. Molecular weight distribution, backbone chemistry, and functional group density define water retention efficiency and structural build-up, yet industrial realization depends on how these parameters interact with mixing energy, dosing accuracy, temperature, and raw material variability. The first step is therefore defining formulation operating envelopes rather than fixed targets.

Laboratory and pilot-scale datasets must be processed to quantify sensitivity coefficients linking polymer structure to rheological response under varying shear and environmental conditions. Statistical normalization of viscosity evolution, torque response, and stability indices allows definition of acceptable industrial tolerances. These processed envelopes form the basis for manufacturing specifications, dosing logic, and application guidance, ensuring that formulation intent survives scale-up and routine production.

Title 22.1: Translation of VMA Formulation Science into Industrial Control Parameters

Translation Element	Laboratory Descriptor	Processed Industrial Parameter	Implementation Purpose
Polymer architecture	Mw, charge density	Dosage sensitivity curve	Robust dosage control
Solution rheology	Intrinsic viscosity	Pumpability margin	Placement reliability
Adsorption behavior	Isotherm shape	SCM compatibility index	Raw material adaptability
Shear response	Thixotropy loop	Mixing energy window	Process reproducibility
Thermal sensitivity	Viscosity–temperature slope	Climate robustness factor	Seasonal stability

Industrial practice further requires embedding these parameters into digital control systems. MES-linked dosing, automated sequence verification, and real-time torque monitoring ensure that formulation science is enforced consistently, independent of operator variability. Applying Industry 4.0 principles at this stage converts polymer chemistry from expert-dependent knowledge into executable manufacturing logic, significantly reducing scale-up risk.

The outcome of this structured translation is a formulation that behaves predictably across plants, projects, and markets. By anchoring industrial practice in processed scientific descriptors rather than empirical adjustment, technology transfer becomes repeatable, auditable, and scalable. This foundation is essential for subsequent steps involving standardization, training, and collaborative optimization across the admixture value chain.

22.2 Standardization, Documentation, and Digital Workflows

Standardization translates validated formulation knowledge into repeatable industrial execution. For viscosity-modifying admixtures, standardization does not imply rigid prescriptions but controlled flexibility defined by digitally managed operating envelopes. Documentation and workflows must therefore capture not only nominal formulation parameters but also allowable variation ranges, sensitivity limits, and corrective pathways derived from processed performance data.

Digital documentation should be structured as formulation passports that link polymer identity, dosage logic, sequencing constraints, and compatibility boundaries with cement and SCM classes. These passports must be version-controlled within quality management systems and synchronized with MES platforms to ensure that only approved parameter sets are executable at plant level. Statistical processing of historical performance data allows continuous refinement of these documents as raw material sources or operating conditions evolve.

Table 22.2: Digital Standardization Artifacts and Workflow Integration for Viscosity-Modifying Admixtures

Standardization Artifact	Digital Content	Processed Data Basis	Industrial Benefit
Formulation passport	Polymer type, Mw range, charge density	Stability and viscosity indices	Consistent scale-up behavior
Application guideline	Dosage window, sequencing rules	Sensitivity coefficients	Reduced field variability
Raw material matrix	Cement and SCM compatibility	Adsorption demand metrics	Faster adaptation to change
Change control log	Version history, approvals	Deviation frequency analysis	Traceability and compliance

Audit dashboard	KPI trends, alarm events	Control chart evaluation	Continuous improvement

Digital workflows enable automated propagation of approved standards across plants and regions. When predictive analytics indicate a shift in robustness margins, updated operating envelopes can be released through controlled workflows without disrupting ongoing production. Applying Industry 4.0 connectivity ensures alignment between formulation updates, operator interfaces, and customer-facing guidance, reducing lag between learning and implementation.

From a technology transfer perspective, standardized digital workflows decouple knowledge from individuals and embed it into systems. Training materials, corrective actions, and best practices can be directly linked to documented standards, creating a closed knowledge loop. This approach accelerates deployment of advanced VMA technologies while maintaining formulation intent, data integrity, and regulatory readiness across industrial operations.

22.3 Training, Competence Development, and Change Management

Competence Architecture for VMA Industrialization

Effective industrial deployment of viscosity-modifying admixtures depends on structured competence development that aligns polymer science, process engineering, and application practice. Training programs must translate molecular-level understanding of adsorption kinetics, hydrodynamic volume, and shear-dependent conformation into operational decision-making at plant and site levels. In industrial environments, this requires tiered competence models where formulation scientists define robustness logic, production teams execute parameterized workflows, and field engineers interpret rheological behavior within predefined operating envelopes. The objective is not encyclopedic knowledge transfer but functional literacy that enables consistent execution under variable raw material and climatic conditions.

Digital Training Platforms and Industry 4.0 Enablement

Modern training strategies increasingly rely on digital platforms integrated with manufacturing execution systems and quality databases. Interactive modules linked to real production data allow operators and engineers to visualize how deviations in mixing energy, dosing sequence, or temperature influence viscosity evolution and stability margins. Simulation-based learning, supported by historical torque and rheology datasets, accelerates competence development while reducing reliance on trial-and-error during live production. Such digitally enabled training frameworks align with Industry 4.0 principles by embedding learning directly into operational contexts rather than isolating it as a standalone activity.

Structured Change Management for Formulation and Process Updates

Change management for VMAs must address the intrinsic sensitivity of rheology-modifying polymers to formulation and process adjustments. Any modification to polymer grade, dosage logic, or sequencing protocol requires coordinated updates across documentation, control systems, and training content. Structured change management workflows should therefore include impact assessment based on sensitivity coefficients, automated update of formulation passports, and targeted retraining triggered by material changes. By formalizing these steps, organizations minimize the risk of performance drift during transitions and preserve the integrity of validated formulation envelopes.

Table 22.3: Competence and Change Management Elements for VMA Implementation

Element	Digital Enablement	Data Foundation	Industrial Outcome
Role-based training matrix	LMS linked to MES	Process variability analysis	Reduced operator-induced deviation
Simulation learning modules	Rheology and torque datasets	Sensitivity modeling	Faster troubleshooting
Change impact assessment	Predictive analytics tools	Historical deviation trends	Controlled formulation updates
Automated retraining triggers	QMS–LMS integration	Change frequency metrics	Sustained process robustness

Field feedback capture	Mobile data entry platforms	Placement performance indices	Continuous improvement loop

Organizational Adoption and Cultural Alignment

Beyond technical training, successful technology transfer requires cultural alignment toward data-driven decision-making. Operators and engineers must trust digitally defined operating envelopes and resist empirical adjustments that undermine formulation intent. Leadership plays a critical role by reinforcing adherence to standardized workflows and by using performance dashboards as primary management tools. Over time, this cultural shift transforms VMAs from experience-dependent additives into predictable, system-integrated technologies that support consistent concrete performance across regions and applications.

22.4 Pilot Projects, Scaling Strategy, and Risk Mitigation

Role of Pilot Projects in VMA Industrialization

Pilot projects represent the critical interface between controlled formulation knowledge and full-scale industrial variability. For viscosity-modifying admixtures, pilot validation must reproduce realistic shear histories, dosing dynamics, and raw material heterogeneity that are absent in laboratory testing. The objective is to stress-test formulation operating envelopes under semi-industrial conditions where adsorption kinetics, polymer hydration rate, and mixer-specific energy input interact simultaneously. Well-designed pilots therefore function as quantitative learning systems rather than demonstration trials, generating statistically relevant datasets for subsequent scale decisions.

Pilot execution should prioritize representativeness over throughput. Mixing geometries, residence times, and dosing architectures must mirror target plants as closely as possible, while allowing intensified sampling of rheological evolution, torque response, and stability under transient conditions. Short pilot campaigns that intentionally explore boundary conditions yield more transferable knowledge than extended runs at nominal settings, particularly for VMAs with steep dosage-response curves.

Scaling Strategy from Pilot to Multi-Plant Deployment

Scaling VMA technology requires formal translation of pilot-derived robustness margins into executable plant-level logic. Rather than scaling by linear factor, successful strategies rely on dimensionless parameters such as specific mixing energy, polymer-to-fines surface area ratio, and normalized adsorption demand. These parameters allow pilot outcomes to be mapped onto diverse plant configurations without re-optimizing formulation chemistry at each site. Scaling thus becomes a controlled parameter mapping exercise supported by processed data rather than empirical adjustment.

Digital twins of admixture production and concrete mixing processes increasingly support this phase. By calibrating models with pilot data, organizations can simulate plant-specific scenarios and pre-define safe operating windows before physical deployment. Integration of these simulations with MES platforms enables automatic parameter instantiation during commissioning, reducing start-up volatility and minimizing off-spec production during ramp-up.

Risk Identification and Quantification in VMA Scale-Up

Risk mitigation for VMAs must address both technical and operational failure modes. Technical risks include unexpected shifts in adsorption efficiency due to cement variability, temperature-dependent viscosity drift, and shear-induced polymer degradation. Operational risks often arise from dosing inaccuracies, sequencing deviations, or insufficient operator competence during transition phases. Effective mitigation therefore depends on early identification, quantitative ranking, and assignment of digital controls to each risk category.

Failure mode and effects analysis tailored to polymer-based admixtures provides a structured framework. By linking each identified risk to measurable indicators such as viscosity deviation rate, torque instability, or pump pressure fluctuation, organizations can convert abstract concerns into monitorable parameters. These indicators form the basis for alarm thresholds, interlocks, and corrective workflows embedded in Industry 4.0 control architectures.

Table 22.4: Pilot, Scaling, and Risk Mitigation Framework for VMAs

Phase	Key Activity	Digital Tool	Risk Mitigated	Industrial Value

Pilot validation	Boundary condition testing	High-frequency rheology logging	Hidden sensitivity zones	Reliable scale decisions
Parameter mapping	Dimensionless scaling	Process simulation models	Non-linear scale effects	Cross-plant consistency
Commissioning	Digital twin deployment	MES parameter instantiation	Start-up instability	Faster ramp-up
Risk analysis	VMA-focused FMEA	Deviation analytics	Undetected failure modes	Proactive control
Monitoring	KPI and alarm integration	Control charts and dashboards	Performance drift	Sustained robustness

Continuous Learning and Adaptive Risk Control

Risk mitigation does not terminate after successful scale-up. Continuous capture of production and field performance data allows recalibration of robustness margins as raw materials, climate, or application practices evolve. Advanced analytics can detect weak signals of emerging instability, such as gradual increases in viscosity variance or longer stabilization times after mixing. These insights enable preemptive adjustment of operating envelopes before customer-visible issues occur.

Embedding pilot learnings into adaptive control systems closes the technology transfer loop. Over time, the organization accumulates a statistically grounded understanding of VMA behavior across scales and contexts, transforming scale-up risk from an episodic challenge into a managed, continuously optimized process.

22.5 Industrialization Under Industry 4.0 Principles

Cyber-Physical Integration of VMA Manufacturing

Industrialization of viscosity-modifying admixtures within an Industry 4.0 framework is defined by tight coupling between physical formulation processes and digital control layers. Polymer solution preparation, dosing, and blending operations are instrumented with sensors capturing torque, flow rate, temperature, and viscosity proxies at high frequency. These data streams feed cyber-physical systems that continuously compare real-time performance against validated operating envelopes derived from formulation science. For VMAs, where adsorption kinetics and shear history exert nonlinear influence on rheology, such closed-loop integration converts process variability into actionable control signals rather than post-production deviations.

Data-Driven Formulation Execution and Autonomous Control

Industry 4.0 implementation moves VMA production from recipe execution toward data-driven formulation logic. Dosage and sequencing decisions are no longer static setpoints but adaptive parameters adjusted in response to raw material variability and environmental conditions. Machine learning models trained on historical rheology and stability datasets can predict viscosity evolution trajectories and recommend micro-adjustments within approved limits. Autonomous control remains bounded by formulation passports and compliance rules, ensuring that adaptive behavior enhances robustness without violating validated chemical intent or regulatory constraints.

Digital Traceability and Compliance-by-Design

Full digital traceability is a core pillar of Industry 4.0 industrialization. Each VMA batch is digitally linked to polymer lot characteristics, process conditions, and downstream application outcomes through integrated ERP, MES, and quality systems. Such end-to-end traceability enables compliance-by-design, where audit readiness and regulatory alignment are intrinsic properties of the production system rather than after-the-fact documentation exercises. For global admixture operations, this approach simplifies cross-border technology transfer by ensuring that identical digital logic governs execution regardless of plant location.

Table 22.5: Industry 4.0 Enablers for VMA Industrialization

Industry 4.0 Element	Digital Implementation	Data Source	Operational Impact

Cyber-physical systems	Inline sensors and control loops	Torque, flow, temperature	Real-time rheology control
Adaptive dosing	ML-assisted parameter tuning	Historical batch data	Reduced variability
Digital twins	Virtual plant models	Pilot and production datasets	Faster commissioning
Traceability backbone	ERP–MES–QMS integration	Batch and lot records	Audit readiness
Predictive analytics	Trend and anomaly detection	KPI time series	Early risk mitigation

Interoperability Across the Value Chain

Industry 4.0 principles extend beyond the admixture plant to encompass the full concrete value chain. Digital exchange of formulation passports, compatibility matrices, and application guidance allows ready-mix producers to align mixing energy, sequencing, and placement practices with VMA design intent. Feedback from field sensors and placement performance data can be reintegrated into central analytics platforms, enabling continuous refinement of robustness margins. Such interoperability transforms technology transfer into a collaborative, data-supported ecosystem rather than a one-directional handover.

Strategic Impact on VMA Technology Transfer

By embedding formulation science into digital infrastructure, Industry 4.0 industrialization fundamentally changes the economics and risk profile of VMA deployment. Scale-up becomes faster, reproducibility improves across regions, and reliance on expert intervention decreases. More importantly, the continuous learning loops inherent to Industry 4.0 ensure that VMA technologies evolve alongside changing cement chemistries, sustainability drivers, and application demands, securing long-term industrial relevance.

22.6 Supply Chain Integration and Raw Material Variability Management

Raw Material Variability as a System-Level Constraint

For viscosity-modifying admixtures, raw material variability is not a peripheral disturbance but a primary determinant of performance robustness. Variations in cement mineralogy, alkali content, sulfate form, and SCM surface chemistry directly alter polymer adsorption demand and hydration-mediated viscosity development. On the polymer side, batch-to-batch fluctuations in molecular weight distribution, residual monomer content, and ionic strength of solutions influence hydration kinetics and shear response. Effective industrialization therefore requires treating raw material variability as a system-level constraint that must be digitally integrated into formulation execution rather than corrected downstream.

Digital Integration Across the Supply Chain

Industry 4.0 enables vertical and horizontal integration of supply chain data relevant to VMA performance. Supplier certificates, inline raw material analytics, and historical performance records can be aggregated into centralized platforms that contextualize formulation behavior in real time. Linking cement supplier data, SCM origin, and polymer lot characteristics to rheological KPIs allows early recognition of incompatibility trends before they manifest as field failures. Such integration shifts supply chain management from reactive qualification to predictive compatibility management.

Quantification and Classification of Variability

Not all variability carries equal risk. Statistical processing of historical data allows classification of raw material changes according to their impact on viscosity stability, water retention, and segregation resistance. Cement and SCM sources can be clustered based on adsorption demand metrics and rheology sensitivity coefficients, while polymer lots can be ranked by hydration rate and shear stability indices. These classifications support differentiated control strategies, where high-risk combinations trigger tighter operating envelopes or enhanced monitoring, and low-risk combinations allow broader flexibility without compromising performance.

Table 22.6: Supply Chain Integration Framework for VMAs

Supply Chain Element	Digital Input	Processed Metric	Control Strategy	Industrial Benefit
Cement sources	Mineralogy, alkali, sulfate data	Adsorption demand index	Adaptive dosage window	Stable rheology
SCM variability	Fineness, surface chemistry	Compatibility coefficient	Sequencing adjustment	Reduced segregation
Polymer lots	Mw distribution, residuals	Hydration rate index	Lot-specific calibration	Consistent viscosity
Logistics conditions	Storage time, temperature	Aging sensitivity factor	Shelf-life management	Reduced drift
Field feedback	Placement and finishability data	Performance deviation score	Continuous refinement	Closed-loop learning

Adaptive Control and Supplier Collaboration

Adaptive control strategies rely on real-time recognition of variability signals and predefined response logic. When supply chain data indicate a shift in raw material characteristics, dosing algorithms and mixing energy targets can be adjusted within validated limits. Equally important is structured collaboration with suppliers, where performance data are shared to refine material specifications and reduce upstream variability. For VMAs, such partnerships often yield greater robustness gains than formulation changes alone.

Strategic Implications for Technology Transfer

Integrated supply chain management strengthens technology transfer by stabilizing the context in which formulation science is executed. By embedding raw material intelligence into digital workflows, organizations reduce the dependency on local empirical adjustments and preserve formulation intent across regions and projects. Over time, this approach builds a resilient, data-informed ecosystem capable of accommodating evolving cement chemistries and sustainability-driven material substitutions without compromising VMA performance.

22.7 Performance Benchmarking, Lifecycle Value, and Sustainability Alignment

Benchmarking Frameworks for VMA Performance

Performance benchmarking for viscosity-modifying admixtures must extend beyond initial rheological targets to encompass robustness, reproducibility, and contribution to system efficiency. Industrial benchmarking frameworks therefore combine fresh-state rheology metrics with indicators of process stability such as viscosity variance, dosage drift frequency, and recovery after shear disruption. Comparative benchmarks against internal references and market standards enable objective positioning of VMA technologies under realistic production and placement conditions. Such multidimensional benchmarking is essential for identifying formulations that deliver not only peak performance but sustained value across diverse operating environments.

Lifecycle Value Assessment Across the Concrete Chain

Lifecycle value of VMAs is realized through cumulative benefits accrued from raw material optimization, reduced waste, and improved placement reliability. Quantifying this value requires integrating admixture performance data with upstream and downstream cost drivers, including cement reduction potential, lower rework rates, and extended workability windows. Digital lifecycle assessment models can incorporate these parameters to translate rheological control into economic and environmental indicators. For industrial stakeholders, this approach reframes VMA selection from unit cost comparison to total value contribution over the concrete lifecycle.

Sustainability Alignment and Low-Carbon Concrete Systems

Sustainability alignment of VMA technologies is increasingly linked to their ability to enable low-clinker, high-SCM concrete systems without compromising performance. VMAs that stabilize rheology in blends with calcined clays, limestone fillers, or recycled fines directly support decarbonization strategies. Benchmarking must therefore include compatibility with emerging cement chemistries, tolerance to variability, and contribution to reduced water and cement demand. Polymer design choices, such as reduced residual monomers or bio-based backbones, further influence sustainability profiles and regulatory acceptance.

Table 22.7: Benchmarking and Sustainability Metrics for VMAs

Dimension	Key Metric	Data Source	Evaluation Purpose	Strategic Outcome
Rheological performance	Viscosity stability index	Inline and lab rheology	Fresh-state robustness	Reliable placement
Process efficiency	Dosage variance	MES and batch records	Manufacturing consistency	Reduced waste
Lifecycle value	Cement reduction factor	LCA and cost models	Total cost assessment	Economic optimization
Sustainability	CO_2 reduction potential	Material balance data	Decarbonization impact	Regulatory alignment
Durability support	Segregation resistance	Field performance logs	Long-term quality	Asset longevity

Digital Benchmarking and Continuous Improvement

Industry 4.0 platforms enable continuous benchmarking by aggregating performance data across plants, projects, and regions. Automated dashboards compare current performance against historical baselines and sustainability targets, highlighting deviations and improvement opportunities. Machine learning tools can identify formulation or process patterns associated with superior lifecycle outcomes, guiding iterative optimization. This continuous benchmarking transforms sustainability from a static claim into a measurable, evolving attribute of VMA technology.

Strategic Positioning Through Aligned Value Propositions

Aligning performance benchmarking with lifecycle and sustainability metrics strengthens the strategic positioning of VMA technologies in increasingly regulated and sustainability-driven markets. By demonstrating quantifiable contributions to low-carbon concrete, resource efficiency, and operational reliability, suppliers can engage customers on value creation rather than price competition. Such alignment also facilitates transparent communication with regulators and certifying bodies, supporting broader adoption of advanced VMA solutions.

23. Future Trends and Technology Transfer Pathways

The next development phase of viscosity-modifying admixtures will be driven by convergence between advanced polymer design, pervasive digitalization, and structured technology transfer mechanisms that shorten the distance between laboratory discovery and industrial adoption. Emerging directions include stimuli-responsive VMA architectures capable of modulating rheology in response to shear, temperature, or ionic environment, as well as closed-loop admixture concepts embedded within semi-autonomous and autonomous concrete production systems. Parallel advances in bio-based, low-carbon, and self-regulating polymer chemistries further expand the functional envelope of rheology control under increasingly constrained sustainability targets.

Equally critical to these innovations are the industrial frameworks that enable their deployment at scale. Pilot-scale validation, multi-stakeholder testing platforms, and digitally enabled formulation passports provide the backbone for transferring adaptive VMA technologies into routine production. When integrated with predictive analytics, digital twins, and plant-level execution systems, these pathways ensure that innovation remains aligned with operational reliability, regulatory compliance, and construction performance. Collectively, these trends define a roadmap toward VMAs that are intrinsically adaptive, data-connected, and positioned as enabling technologies for the future of high-performance and low-carbon concrete systems.

23.1 Smart and Responsive VMAs

Strategic Context for Next-Generation VMAs

Future viscosity-modifying admixtures are evolving from passive rheology modifiers into active, responsive components of the concrete system. This transition is driven by increasing raw material variability, low-clinker cement chemistries, and the rise of digitally controlled production environments. Smart VMAs are designed to sense and respond to changes in shear, ionic strength, temperature, or hydration state, thereby maintaining rheological stability without continuous external correction. From a technology transfer perspective, their industrial value lies not only in performance enhancement but in reducing formulation sensitivity and operational intervention.

23.1.1 Stimuli-Responsive Polymers

Molecular Design Principles and Trigger Mechanisms

Stimuli-responsive viscosity-modifying admixture polymers are engineered to undergo reversible physicochemical changes when exposed to specific environmental triggers inherent to cementitious systems. Relevant stimuli include ionic strength, multivalent cation concentration, pH, temperature, and applied shear. Polymers incorporating weak polyelectrolyte backbones, associative hydrophobes, or switchable side chains adjust their hydrodynamic volume and intermolecular association state in response to these triggers. In highly alkaline pore solutions, calcium-mediated charge screening and complexation alter chain conformation, enabling controlled transition between low-viscosity transport states and high-viscosity structural build-up after placement.

Rheological Functionality in Cementitious Environments

The functional advantage of stimuli-responsive VMAs lies in decoupling workability from stability. During mixing and pumping, elevated shear rates and temperature suppress intermolecular association, minimizing viscosity and energy demand. Upon shear reduction or changes in ionic equilibrium, polymer chains reassociate or expand, increasing yield stress and enhancing segregation resistance. This behavior directly addresses the non-linear rheological requirements of modern concrete placement, particularly in self-consolidating and automated casting systems where fixed-viscosity additives are suboptimal.

Industrial Formulation and Manufacturing Constraints

Translating responsive polymer concepts into industrial products requires strict control over molecular weight distribution, trigger sensitivity, and response kinetics. Manufacturing processes must ensure narrow tolerance in functional group density to avoid premature activation during storage or inconsistent response across polymer lots. Solution stability, shelf-life under variable temperature, and compatibility with superplasticizers are critical formulation constraints. Digital characterization of response curves and embedding these descriptors into formulation passports allow predictable industrial execution within Industry 4.0 control frameworks.

Process Control and Industry 4.0 Integration

Stimuli-responsive VMAs are inherently suited to data-driven process control. Inline monitoring of torque, power consumption, and temperature provides indirect measurement of polymer state during mixing and placement. These signals can be interpreted by predictive models to anticipate rheological transitions and adjust dosing or sequencing in real time. Within cyber-physical systems, responsive polymers act as material-level actuators, complementing algorithmic control and reducing the need for manual intervention. Successful integration depends on bounding adaptive behavior within validated operating envelopes to maintain compliance and reproducibility.

Technology Transfer and Risk Management

Technology transfer of responsive VMAs must address perceived risk associated with dynamic behavior. Structured pilot trials are required to validate response repeatability under realistic variability in cement chemistry and climate. Risk mitigation strategies include defining fail-safe viscosity floors, conservative trigger thresholds, and fallback operating modes compatible with conventional VMAs. Clear communication of functional logic to production and field personnel is essential to prevent misinterpretation of adaptive behavior as instability.

Table 23.1.1: Stimuli-Responsive VMA Polymer Concepts and Industrial Implications

Stimulus	Polymer Design Feature	Rheological Response	Control Parameter	Industrial Benefit
Ionic strength	Weak polyelectrolyte backbone	Chain contraction or expansion	Pore solution conductivity	SCM adaptability
Calcium ions	Chelating or complexing groups	Controlled adsorption	Ca^{2+} activity proxy	Stable build-up

Temperature	Thermo-responsive side chains	Viscosity modulation	Mix temperature	Climate robustness
Shear rate	Associative hydrophobes	Reversible network formation	Mixer power draw	Pumpability
pH	Ionizable functional groups	Conformation switching	Alkalinity window	Placement reliability

Strategic Outlook

Stimuli-responsive VMA polymers represent a transition toward intelligent materials capable of self-adjusting within complex cementitious systems. Their successful industrial adoption will depend less on novelty of polymer chemistry and more on disciplined integration into digital control architectures, standardized documentation, and practitioner training. When properly transferred, these materials offer a pathway to higher robustness, lower energy demand, and improved compatibility with future low-carbon concrete technologies.

23.1.2 Adaptive Rheology Concepts

Conceptual Shift from Fixed Targets to Dynamic Rheology

Adaptive rheology represents a fundamental shift from static viscosity targets toward time- and context-dependent control of flow behavior. In cementitious systems, rheological requirements vary continuously during mixing, transport, placement, and early structuration. Adaptive concepts recognize that a single viscosity or yield stress value cannot satisfy all stages of the process. Instead, VMAs are designed and deployed as dynamic regulators that adjust the balance between yield stress, plastic viscosity, and thixotropic recovery in response to process conditions and material state.

Multiscale Mechanisms Enabling Adaptation

Adaptive rheology emerges from the combined action of polymer conformation dynamics, adsorption equilibria, and microstructural evolution of the cementitious matrix. Associative polymer networks reversibly form and dissociate under shear, while competitive adsorption with superplasticizers and fine particles modulates effective polymer availability. At the mesoscale, flocculation and deflocculation cycles alter particle packing density, feeding back into macroscopic flow behavior. Effective adaptive systems are engineered so that these mechanisms operate synergistically rather than competitively, ensuring smooth transitions between flow and stability regimes.

Industrial Implementation Through System Integration

Industrial realization of adaptive rheology depends on integrating material design with process control. Inline proxies such as mixer torque, power draw, and discharge pressure provide continuous insight into rheological state without requiring direct rheometry. These signals are processed through predictive models that correlate observed trends with expected performance envelopes. Control actions may include micro-adjustments in VMA dosage, sequencing relative to superplasticizers, or controlled water correction within predefined limits. Such integration transforms adaptive rheology from an intrinsic material property into a controllable industrial function.

Data-Driven Adaptation and Industry 4.0 Control

Within an Industry 4.0 framework, adaptive rheology is governed by closed-loop control architectures. Historical datasets linking raw material characteristics, process conditions, and rheological outcomes are used to train machine learning models capable of forecasting viscosity evolution and structural build-up. These models enable anticipatory control rather than reactive correction, reducing variability and improving robustness. Importantly, adaptive behavior remains bounded by formulation passports and compliance constraints, ensuring reproducibility across plants and regulatory environments.

Risk Management and Technology Transfer

Adaptive rheology introduces perceived complexity that must be managed during technology transfer. Clear definition of adaptation limits, fail-safe modes, and diagnostic indicators is essential to maintain operator confidence. Pilot trials should focus on validating response stability under extreme but realistic conditions, such as high SCM variability or temperature fluctuations. Documentation must translate adaptive logic into actionable rules, preventing misinterpretation of dynamic behavior as formulation instability.

Table 23.1.2: Adaptive Rheology Elements and Industrial Control

Adaptive Element	Material or System Lever	Monitoring Signal	Control Action	Industrial Benefit
Yield stress tuning	Associative VMA networks	Mixer torque	Dosage micro-adjustment	Segregation resistance
Viscosity modulation	Polymer conformation	Power consumption	Sequencing optimization	Pumpability
Thixotropic recovery	Reversible adsorption	Rest-time response	Placement timing	Shape stability
Raw material compensation	Competitive adsorption	Trend deviation	Envelope shift	Robustness
Climate adaptation	Temperature-sensitive response	Mix temperature	Control window update	Seasonal consistency

Strategic Implications

Adaptive rheology concepts align closely with the requirements of automated and low-carbon concrete production. By allowing materials to self-adjust within controlled limits, these systems reduce dependence on narrow raw material specifications and manual intervention. For admixture suppliers, adaptive rheology offers a pathway to differentiated value propositions centered on reliability, digital compatibility, and lifecycle performance rather than isolated rheological metrics.

23.2 Integration with Autonomous Concrete Production

Viscosity-modifying admixtures are progressively transitioning from formulation-dependent stabilizers into active control elements within autonomous concrete production environments. As batching, mixing, transport, and placement operations adopt sensor-rich and algorithm-driven execution, VMA performance must be characterized by reproducible rheological fingerprints that can be interpreted and adjusted through closed-loop control systems. This evolution places new demands on polymer consistency, adsorption predictability, and responsiveness to dynamic inputs such as temperature gradients, hydration kinetics, shear history, and real-time workability indicators.

Within automated production architectures, advanced VMAs function as rheological actuators that translate digital control signals into physical flow and stability responses. When coupled with inline monitoring, digital twins, and predictive analytics, such admixtures enable continuously adaptive mix designs with reduced operator dependence. This integration supports high-precision manufacturing routes, including robotic placement and additive construction, while improving efficiency, repeatability, and process resilience across next-generation concrete production platforms.

23.2.1 Closed-loop mix optimization

Control Logic for Continuous Mix Adjustment

Closed-loop mix optimization replaces predefined mix recipes with continuously adjusted formulations governed by real-time performance feedback. In concrete production, rheology, workability retention, and stability evolve dynamically as functions of raw material variability, temperature, and shear history. Closed-loop systems continuously compare measured proxies such as mixer torque, power draw, discharge flow, and temperature against target trajectories derived from validated rheological models. Deviations trigger bounded corrective actions, allowing the mix to converge toward optimal performance without manual intervention.

Role of VMAs in Feedback-Controlled Optimization

Viscosity-modifying admixtures are uniquely suited for closed-loop optimization because small dosage or sequencing adjustments generate measurable and rapid rheological responses. VMAs act as fine-tuning actuators that complement superplasticizers, enabling independent control of yield stress and plastic viscosity. In closed-loop systems, VMA response curves are embedded into control algorithms, allowing predictive estimation of adjustment magnitude and direction. This approach minimizes oscillation and avoids overcorrection, which is critical for maintaining production stability at high throughput.

Sensing, Modeling, and Decision Layers

Effective closed-loop optimization relies on a layered architecture. The sensing layer captures high-frequency process data, while the modeling layer interprets these signals through digital twins and data-driven rheology models. The decision layer applies optimization logic, often supported by machine learning, to select corrective actions that maximize robustness and efficiency. Importantly, all decisions are constrained by formulation passports and compliance limits, ensuring that adaptive behavior remains auditable and reproducible across plants.

Industrial Deployment and Robustness Management

Industrial deployment of closed-loop optimization must account for sensor noise, time delays, and variability in material response. Robust systems incorporate signal filtering, confidence scoring, and fallback modes that revert to conservative settings when uncertainty exceeds defined thresholds. Pilot trials are essential to tune control gains and validate system stability under extreme but realistic conditions. Once commissioned, continuous learning from production and field data incrementally improves optimization accuracy and reduces variability.

Table 23.2.1: Closed-Loop Mix Optimization Architecture

Layer	Function	Key Input	Control Output	Industrial Benefit
Sensing	State observation	Torque, power, temperature	Rheology estimate	Real-time insight
Modeling	Prediction	Digital twin, ML models	Expected trajectory	Anticipatory control
Decision	Optimization	Deviation analysis	Dosage or sequencing	Stable correction
Execution	Actuation	Automated dosing	VMA adjustment	Consistent quality
Learning	Refinement	Batch and field data	Model update	Continuous improvement

Sustainability and Lifecycle Impact

Closed-loop optimization directly supports sustainability objectives by reducing overdesign and waste. Precise control of rheology enables lower cement content, reduced water demand, and fewer rejected batches. Over the lifecycle of concrete production, these incremental efficiencies translate into measurable reductions in CO_2 emissions and energy consumption. VMAs designed for predictable, reversible response amplify these benefits by allowing fine control without compromising durability or placement reliability.

23.3 Bridging Research, Production, and Field Adoption

Bridging laboratory-scale innovation with industrial manufacturing and consistent field performance remains one of the most decisive challenges in the advancement of viscosity-modifying admixtures. Translation of novel polymer concepts into reliable construction practice depends not only on scientific merit but on disciplined pilot validation, manufacturability assessment, and repeatable scale-up protocols that preserve rheological intent under variable production conditions. Without such structure, even technically superior VMA chemistries fail to achieve sustained market adoption.

Effective technology transfer is enabled through collaborative ecosystems that align researchers, admixture producers, concrete manufacturers, and contractors around shared performance metrics and decision frameworks. Staged validation programs, digital twin-supported scale-up, and standardized rheological benchmarks provide objective pathways for de-risking implementation. When coupled with transparent documentation and digitally supported feedback loops, these tools accelerate movement from concept to routine use, ensuring that next-generation VMAs achieve predictable, confident deployment across diverse concrete construction environments.

23.3.1 Pilot Trials and Scale-Up Frameworks

Control Logic for Continuous Mix Adjustment

Closed-loop mix optimization represents a fundamental departure from static, predefined concrete recipes toward adaptive formulations governed by continuous performance feedback. In industrial concrete production, key fresh-state properties such as rheology, workability retention, and segregation resistance evolve in response to raw material variability, ambient and mix temperature, moisture fluctuations, and cumulative shear exposure during mixing and transport.

Autonomous control systems continuously compare measured process proxies, including mixer torque, electrical power draw, discharge flow rate, and internal temperature, against target trajectories generated from validated rheological and process models. When deviations are detected, the system executes bounded corrective actions that progressively steer the mix toward the desired performance envelope, eliminating the need for operator-driven interventions while maintaining process stability at scale.

Role of VMAs in Feedback-Controlled Optimization

Viscosity-modifying admixtures occupy a central functional position within closed-loop optimization frameworks due to their rapid, sensitive, and largely reversible influence on plastic viscosity and yield stress. Minor adjustments in VMA dosage, addition timing, or dilution state produce measurable rheological responses within practical control-cycle times. In advanced control architectures, VMA response surfaces are embedded directly into control algorithms, enabling predictive estimation of both magnitude and direction of corrective actions. This separation of roles, where superplasticizers primarily govern dispersion and flow initiation while VMAs provide fine-scale viscosity modulation, allows multivariable control without destabilizing the mix. Such predictability is critical for avoiding oscillatory behavior, overdosing, or cumulative drift during high-throughput production.

Sensing, Modeling, and Decision Layers

Effective closed-loop mix optimization relies on a layered digital architecture aligned with Industry 4.0 principles. The sensing layer acquires high-frequency, plant-level data from mechanical, thermal, and flow-related sensors integrated into mixers and conveying systems. These signals are interpreted by the modeling layer using hybrid digital twins that combine first-principles rheology with data-driven machine learning models trained on historical production data. The decision layer applies optimization logic to determine corrective actions that balance robustness, efficiency, and compliance. All decisions are constrained by formulation passports, regulatory limits, and durability requirements, ensuring that adaptive behavior remains traceable, auditable, and reproducible across different plants and production lines.

Industrial Deployment and Robustness Management

Deployment of closed-loop optimization in industrial environments must explicitly address sensor noise, signal latency, actuator resolution, and variability in material response. Robust systems integrate signal filtering, confidence metrics, and hierarchical fallback modes that revert the process to conservative operating windows when uncertainty exceeds predefined thresholds. Structured pilot trials under controlled but realistic conditions are essential to tune control gains, validate model assumptions, and assess stability under extreme raw material or environmental variability. Following commissioning, continuous learning from batch records and field performance progressively refines model accuracy, reduces variability, and enhances resilience across the production network.

Table 23.2.1: Closed-Loop Mix Optimization Architecture

Layer	Function	Key Input	Control Output	Industrial Benefit
Sensing	State observation	Torque, power, temperature	Rheology estimate	Real-time insight
Modeling	Prediction	Digital twin, ML models	Expected performance path	Anticipatory control
Decision	Optimization	Deviation analysis	Dosage or sequencing change	Stable correction
Execution	Actuation	Automated dosing systems	VMA adjustment	Consistent quality
Learning	Refinement	Batch and field performance data	Model update	Continuous improvement

Sustainability and Lifecycle Impact

Closed-loop mix optimization delivers direct and quantifiable sustainability benefits by minimizing overdesign, reducing raw material waste, and lowering energy consumption. Precise rheology control enables reductions in cement content and water demand while maintaining placement reliability and surface quality. Across the lifecycle of concrete production, these incremental efficiencies accumulate into meaningful reductions in CO_2 emissions and operational costs. VMAs engineered for predictable, reproducible, and reversible response enhance these outcomes by enabling fine control without compromising durability, robustness, or long-term performance, aligning autonomous production with both economic and environmental objectives.

23.3.2 Knowledge transfer to practitioners

Bridging Advanced Control Systems and Operational Practice

Effective knowledge transfer is a decisive success factor for the industrial adoption of autonomous and semi-autonomous concrete production systems. Advanced control logic, digital twins, and adaptive admixture strategies provide limited value if plant personnel, quality managers, and field engineers cannot interpret system outputs or trust automated decisions. Knowledge transfer must therefore translate abstract concepts such as rheological state estimation, feedback control, and model confidence into operationally meaningful guidance aligned with daily production workflows. Structured communication ensures that practitioners understand not only how systems behave, but why specific corrective actions are executed under given conditions.

Operational Training and Competency Development

Training programs for practitioners must move beyond conventional admixture handling or mix design instruction and incorporate core principles of data-driven process control. Operators should be familiar with key sensor signals, acceptable variability bands, and the functional roles of VMAs within adaptive control loops. Scenario-based training using historical production data and simulated disturbances enables personnel to recognize normal adaptive behavior versus true system anomalies. Such competency development increases acceptance of autonomous adjustments, reduces unnecessary manual overrides, and supports consistent performance across shifts and production sites.

Standardized Knowledge Artifacts and Digital Documentation

Codification of expert knowledge into standardized digital artifacts is essential for scalable technology transfer. Formulation passports, control logic summaries, and response maps for VMAs provide practitioners with concise, auditable references that link admixture behavior to control outcomes. These artifacts should be embedded directly into plant information systems and human–machine interfaces, enabling contextual access during production. Version control and centralized updates ensure that improvements derived from learning algorithms are propagated uniformly, avoiding knowledge fragmentation across decentralized operations.

Human–Machine Collaboration and Decision Transparency

Autonomous systems achieve higher adoption when practitioners remain engaged as informed supervisors rather than passive observers. Transparent decision logic, including confidence indicators and rationale summaries for control actions, allows operators to validate system behavior against their experiential knowledge. In the context of VMAs, visualizing expected rheological impact before and after dosage adjustments strengthens trust and accelerates learning. Such human–machine collaboration aligns with Industry 4.0 principles by combining algorithmic consistency with practical expertise accumulated in real production environments.

Cross-Organizational Learning and Continuous Improvement

Knowledge transfer extends beyond individual plants to organizational and inter-organizational levels. Aggregated production data, anonymized where necessary, enable benchmarking and dissemination of best practices across regions and customers. Feedback from practitioners, including placement observations and finishing behavior, enriches model training datasets and closes the loop between production and application. This bidirectional flow of knowledge transforms autonomous concrete production into a continuously improving system rather than a static technological deployment.

Table 23.3.2: Knowledge Transfer Mechanisms for Autonomous Concrete Production

Mechanism	Primary Audience	Key Content	Industrial Outcome
Structured training modules	Operators, QC staff	Control logic, sensor interpretation	Reduced manual intervention
Digital formulation passports	Plant engineers	Admixture limits, response maps	Compliance and consistency

HMI-integrated guidance	Operators	Real-time explanations, confidence scores	Improved trust and adoption
Scenario-based simulations	Supervisors, trainers	Disturbance response, failure modes	Faster troubleshooting
Field feedback integration	R&D and operations	Placement and finishing observations	Continuous system refinement

24. Glossary of Terms and Acronyms

The following glossary consolidates technical terminology and abbreviations used throughout the volume, with emphasis on polymer chemistry, rheology control, admixture formulation, and digitally enabled concrete production. Definitions are framed to support industrial formulation, manufacturing practice, and technology transfer.

24.1 Admixture and Polymer Science Terms

Associative Polymer

Water-soluble polymer containing hydrophobic groups that form transient physical junctions in aqueous media, contributing to viscosity enhancement and thixotropic rebuild through reversible network formation.

Charge Density

Number of ionic functional groups per polymer chain or per unit molecular weight, governing adsorption strength on cementitious surfaces and interaction with dissolved ions.

Degree of Substitution (DS)

Average number of substituent groups per anhydroglucose unit in modified polysaccharides, influencing solubility, adsorption behavior, and rheology control efficiency.

Hydrodynamic Volume

Effective volume occupied by a polymer coil in solution, determined by molecular weight, conformation, and ionic environment, directly linked to viscosity contribution.

Molecular Weight Distribution (MWD)

Statistical distribution of polymer chain lengths within a product, affecting shear stability, robustness, and batch-to-batch consistency.

Water Immobilization

Physical and physicochemical binding of free water by polymers and fines, reducing bleeding and segregation without excessive yield stress increase.

24.2 Rheology and Fresh Concrete Behavior

Plastic Viscosity

Slope of the shear stress–shear rate relationship beyond yield stress, reflecting resistance to flow once motion is initiated.

Yield Stress

Minimum stress required to initiate flow in a cementitious suspension, closely related to shape stability and segregation resistance.

Structural Build-Up

Time-dependent increase in static yield stress at rest due to flocculation, hydration, and polymer network development.

Thixotropy

Reversible reduction in viscosity under shear followed by recovery upon rest, critical for pumping, placement, and extrusion-based processes.

Rheological Fingerprint

Characteristic, reproducible response pattern of a concrete mix under defined shear, time, and temperature conditions, used for digital identification and control.

24.3 Cementitious Materials and Binder Systems

Supplementary Cementitious Materials (SCMs)

Materials such as fly ash, slag, limestone filler, calcined clays, or silica fume used to partially replace clinker and modify performance and sustainability.

Low-Clinker Binder

Binder system with reduced Portland clinker content, often exhibiting higher fines content and altered adsorption demand.

Mineralized Binder

Binder incorporating carbonated or mineralized phases formed through CO_2 utilization or curing, characterized by modified surface chemistry.

Residual Adhered Mortar (RAM)

Old cement paste remaining on recycled aggregates, contributing to increased water demand and rheological variability.

24.4 Processing, Manufacturing, and Field Operations

Dosage Window

Range of admixture concentration over which stable and acceptable performance is achieved under variable production conditions.

Sequencing

Order and timing of admixture addition during batching and mixing, strongly influencing competitive adsorption and effectiveness.

Extended Dormant Period

Time interval between mixing and placement during which concrete remains at rest, common in transport-intensive or hot-climate operations.

Scale-Up Robustness

Ability of a formulation to maintain performance when transferred from laboratory to industrial production volumes.

24.5 Digitalization, Industry 4.0, and Control Systems

Digital Twin

Dynamic virtual representation of a physical process or material system, continuously updated with real-time data to predict behavior and guide control actions.

Closed-Loop Control

Automated control strategy in which sensor feedback is used to adjust process variables, such as VMA dosage, to maintain target rheological states.

Formulation Passport

Digitally managed documentation linking polymer identity, dosage logic, compatibility limits, and validated performance envelopes.

Manufacturing Execution System (MES)

Plant-level digital system coordinating production scheduling, batching, quality control, and data capture.

Predictive Analytics

Data-driven modeling techniques used to anticipate deviations in performance before they manifest in production or field non-compliance.

24.6 Sustainability and Performance Metrics

Clinker Reduction Factor

Fraction of Portland clinker replaced by SCMs or alternative binders, often correlated with increased rheological sensitivity.

Stability Index

Composite metric describing resistance to segregation, bleeding, and phase separation under defined test conditions.

Carbon Footprint (kg CO_2-eq)

Life-cycle greenhouse gas emissions associated with a material or formulation, increasingly used as a design constraint.

24.7 Acronyms

Acronym	Meaning
AI	Artificial Intelligence
ASTM	American Society for Testing and Materials
EN	European Norm
ESG	Environmental, Social, and Governance
LOI	Loss on Ignition
MES	Manufacturing Execution System
Mw	Weight-Average Molecular Weight
OB	Optical Brightener
PCE	Polycarboxylate Ether
RAM	Residual Adhered Mortar
SCM	Supplementary Cementitious Material
SCC	Self-Compacting Concrete
UHPC	Ultra-High-Performance Concrete
VMA	Viscosity-Modifying Admixture

25. Reference Library

25.1 ACI Committee Guides and Reports

- ACI Committee 207. Guide to Mass Concrete. ACI Materials Journal, 2021.
- ACI Committee 212. (2022). "Chemical Admixtures for Concrete." ACI Report.
- ACI Committee 213. Guide for structural lightweight aggregate concrete. Latest editions.
- ACI Committee 304. Guide for measuring, mixing, transporting, and placing concrete.
- ACI Committee 304. Guide for Measuring, Mixing, Transporting, and Placing Concrete Underwater. ACI Materials Journal, 2021.
- ACI Committee 305. Guide to hot weather concreting. Latest editions.
- ACI Committee 305. Hot Weather Concreting. ACI Materials Journal, 2021.
- ACI Committee 306. Cold Weather Concreting. ACI Materials Journal, 2021.
- ACI Committee 349. Code requirements for nuclear safety-related concrete structures.

25.2 ASTM and ISO Standards

- ASTM C1602 revisions, 2023.
- ASTM C1610, Standard Test Method for Static Segregation of Self-Consolidating Concrete Using a Column Technique, updated technical notes 2024.
- ASTM C1712, Standard Test Method for Static Segregation of Self-Consolidating Concrete, latest revision 2023.
- ASTM C231/C231M-24. Air content of freshly mixed concrete by the pressure method.

- ASTM C494/C494M-24. Chemical admixtures for concrete.
- ASTM D4332-21, Standard practice for conditioning containers, packages, or packaging components for testing.
- ISO 11357-2:2023. Plastics – Differential scanning calorimetry – Determination of glass transition temperature.
- ISO 11930:2019, Evaluation of the antimicrobial protection of a cosmetic product, applied by analogy to aqueous polymer systems.
- ISO 14040 and ISO 14044, Life Cycle Assessment Principles and Framework/Requirements, latest consolidated editions, 2023.
- ISO 14067, Carbon Footprint of Products, application to construction materials, 2023.
- ISO 15747:2022, Plastic containers for liquids, performance requirements and testing.
- ISO 15883-5:2022, Process Sensor Validation and Performance Monitoring.
- ISO 17359:2022, Condition Monitoring and Diagnostics of Machines.
- ISO 20816-1:2023. Mechanical vibration – Measurement and evaluation of machine vibration.
- ISO 22400-2:2023, Manufacturing Operations Management and Key Performance Indicators.
- ISO 22403:2023. Mixing equipment – Performance characterization and scale-up principles.
- ISO 22904-2, Concrete Rheology and Process Monitoring, technical update 2024.
- ISO 23247-1:2023, Digital Twin Framework for Manufacturing.
- ISO 23270, Digital Data Structures for Concrete Production, technical report, 2024.
- ISO 30401:2022, Knowledge Management Systems Requirements.
- ISO 3219:2023/2024. Plastics – Determination of viscosity of polymers under defined shear conditions/using rotational viscometers.

- ISO 45001, Occupational Health and Safety Management Systems, guidance revision 2023.
- ISO 9001:2015/14001:2015 Digital integration and management system guidance, various updates 2024.
- ISO 9001:2015 and ISO 9004:2018. Guidance documents on digital quality management and process monitoring.

25.3 Journal Articles and Academic Publications

- Ahmed, S., et al. (2024). Transient effects in continuous polymer processing and rheological stability. Journal of Applied Polymer Science, 141(18), 55231.
- Andrade, C.; et al. Durability and quality control of post-tensioning grouts. Construction and Building Materials, 2022, 344, 128215.
- Angst, U., et al. Durability of marine concrete structures. Materials and Corrosion, 2020.
- Assaad, J. (2020-2023). Various papers on Admixture Interaction, Rheology Control, UHPC, and VMAs. Construction and Building Materials and Materials.
- Austin, S. A., et al. (2015-2021). Papers on Shotcrete rheology and rebound control. Construction and Building Materials.
- Avet, F., Snellings, R., et al. (2022). Papers on low-clinker concrete, SCM variability, and sustainability. Construction and Building Materials.
- Baldyga, J., et al. (2022). Residence time distribution effects in non-Newtonian reactor systems. Chemical Engineering Science, 247, 117048.
- Banfill, P. (2021). Thixotropy and Rheology of Fresh Cement Paste: Industrial Implications. Cement and Concrete Research, 146, 106471.
- Banthia, N., et al. (2006-2021). Papers on Fiber dispersion, workability, and shotcrete materials. Cement and Concrete Composites and Materials and Structures.

- Barnes, H. A. (2021). Reviews of filled viscoelastic systems and in-line rheometry. Journal of Non-Newtonian Fluid Mechanics.
- Bishnoi, S., et al. (2020-2023). Various papers on Rheology control, low-clinker systems, and SCM influence. Cement and Concrete Research.
- Brumaud, C., et al. (2014-2017). Papers on Cellulose ethers and cement hydration rheology. Cement and Concrete Research.
- Buswell, R. A., et al. (2018-2022). Papers on 3D printing and digital control for concrete extrusion. Cement and Concrete Research and Construction and Building Materials.
- Chen, B., et al. (2021). Shelf-life prediction of liquid construction chemical formulations. Cement and Concrete Composites, 124, 104257.
- Chen, J., et al. (2022). Rheological behavior of concrete with recycled aggregates. Construction and Building Materials.
- Chen, L., et al. (2024). Effects of residual monomers on rheology and stability. Polymer Degradation and Stability, 225, 110511.
- Cepuritis, R., et al. Manufactured sand in concrete. Construction and Building Materials, 2016.
- De Schutter, G., et al. (2023). Series of papers on Data-driven admixture optimization, Digital Concrete, and Industry 4.0. Cement and Concrete Composites.
- Favier, A., et al., Life Cycle Assessment of Chemical Admixtures, Construction and Building Materials, 2024.
- Ferrara, L., et al. (2021-2022). Papers on fresh-state behavior of HPFRC and fiber orientation. Materials and Structures.
- Ferraris, C. F. (2023). Concrete rheology: Measurement and practical relevance. Cement and Concrete Research, 166, 107049.
- Fehling, E.; et al. Production and quality control of UHPC. Materials and Structures, 2022, 55, 189.

- Flatt, R. J., et al. (2001-2023). Extensive papers on chemical admixtures, superplasticizers, rheology, and thermal degradation. Cement and Concrete Research and Construction and Building Materials.
- Gartner, E., et al. (2022-2023). Reviews of low-carbon pathways, sustainable technologies, and Industry 4.0. Cement and Concrete Research and related journals.
- Graybeal, B. A. (2022). Rheology and processing of ultra-high-performance concrete. Cement and Concrete Research, 156, 106785.
- Gram, A., et al. (2024). Digital monitoring of storage stability, viscosity drift, and quality assurance. Cement and Concrete Composites and Materials and Structures.
- Habert, G., et al., Environmental Impacts and Benefits of Admixtures, Construction and Building Materials, 2024.
- Heinze, T., et al. (2022). Polysaccharide modification and stability. Progress in Polymer Science.
- Houst, Y. F., and Bowen, P. (2022). Air-entraining agents and polymer interactions. Advances in Cement Research, 34, 187–198.
- Holm, T. A., & Bremner, T. W. (2000-2021). Lightweight aggregate concrete performance and stability. ACI Materials Journal and Construction and Building Materials.
- Kourti, T. (2022-2023). Statistical Process Control and Maintenance in Industry. Journal of Process Control and Technometrics.
- Kawashima, S., et al. (2022-2024). Extensive papers on admixture interactions, VMA systems, rheology optimization, and build-up mechanisms. Construction and Building Materials and Materials and Structures.
- Khayat, K. H., et al. (1997-2022). Research on anti-washout admixtures, pumping rheology, SCC stability, and VMAs.
- Khinast, J., et al. (2023-2024). Digital twins, recipe management, and inline sensing in chemical manufacturing. Computers and Chemical Engineering.

- Koehler, E. P., & Fowler, D. W. Summary of concrete workability test methods. ICAR Report, 2019.
- Kong, X., Zhang, Y., et al. (2020-2022). Structure–property relationships of polyacrylamide-based and advanced VMAs. Construction and Building Materials.
- Lee, J., et al. (2022-2024). Series on Data Analytics, Digital Twins, and Knowledge Management in Smart Manufacturing. Journal of Manufacturing Systems.
- Lei, L., Plank, J. (2023). Polymer structure–processing relationships and synergistic effects in cement. Construction and Building Materials, 365, 130048.
- Li, L., et al. (2022-2024). Rheology control in recycled aggregate and precast concrete manufacturing. Construction and Building Materials and Materials.
- Li, X., et al. (2021-2024). Extensive studies on polymer solution processing, stability, hydration kinetics, and electrolyte effects. Cement and Concrete Research and Construction and Building Materials.
- Li, Y., et al. (2023). Supply chain and degradation mechanisms of biopolymer admixtures. Construction and Building Materials.
- Li, Z., et al. (2023-2024). Admixture strategies for long-haul/hot-weather concrete and digital rheology control in UHPC. Materials and Automation in Construction.
- Liu, J., et al. (2023). Data-driven flocculation assessment and strategies for fines-rich concrete. Construction and Building Materials and Materials.
- Liu, Y., et al. (2024). Storage-induced aging effects in aqueous polymer feedstocks. Polymer Degradation and Stability, 223, 110472.
- Ma, B., et al. (2024). Practical implications and rheological control of modern concrete using VMAs. Cement and Concrete Composites.
- Mantellato, S., et al. (2019-2021). Comb polymer design and associative polymers in cement. Cement and Concrete Research.

- Marchon, D., Flatt, R., et al. (2018-2024). Deep investigation into polymer adsorption, hydration, thixotropy, buildability, and industrial processing effects. Cement and Concrete Research and Advances in Cement Research.

25.4 Regulatory and Industry Specifications

- EFNARC. European specification for sprayed concrete. Latest revisions.
- European Chemicals Agency (ECHA). Guidance on the compilation of safety data sheets, updated 2024.
- European Chemicals Agency (ECHA). Guidance on Safe Storage of Temperature-Sensitive Chemicals, 2024.
- European Chemicals Agency (ECHA). Various reports on compliance, enforcement, and inspection under REACH and CLP, 2023–2024.
- IAEA. Concrete materials for nuclear applications. Technical reports series.
- IEC 61511. Functional safety for the process industry sector, latest edition 2023.
- IEC 62264 and ISA-95 standards for ERP–MES integration and data structures, latest revisions 2023.

25.5 Books and Handbooks

- Baldyga, J., Bourne, J.R. Turbulent Mixing and Chemical Reactions. Wiley, 2023 edition.
- Biermann, C. J., Handbook of Pulping and Papermaking, 3rd ed., Academic Press, 2021.
- Ebnesajjad, S., Plastic Films in Packaging, William Andrew, 2022.
- Heinze, T., et al. (2022-2023). Cellulose Derivatives and Functional Polysaccharides/Polymers. Springer.
- Israelachvili, J. N., Intermolecular and Surface Forces, 4th ed., Academic Press, 2024.
- Lapasin, R., Pricl, S., Rheology of Industrial Polysaccharides, Springer, 2021.

- Larson, R.G. The Structure and Rheology of Complex Fluids, 2nd ed., Oxford University Press, 2023.

www.ingramcontent.com/pod-product-compliance
Ingram Content Group UK Ltd.
Pitfield, Milton Keynes, MK11 3LW, UK
UKHW051416290926
15471UKWH00009B/1628

9 798233 423192